I0815023

Analecta
Gregoriana
326

MIGUEL RAMÓN FUENTES

THE CONCEPT OF FUNCTION IN MOLECULAR BIOLOGY

A theoretical framework and a case study

Premio Bellarmino 2013

Vidimus et approbamus ad normam Statutorum Universitatis
Roma, Pontificia Università Gregoriana
18/06/2012
Prof. Gennaro AULETTA
Dott. Paolo VEZZONI

Progetto grafico di copertina: Serena Aureli

Impaginazione: Scuola Tipografica San Pio X

Gregorian & Biblical Press
Piazza della Pilotta 35, 00187 - Roma
books@biblicum.com

ISBN 978-88-7839-**343**-1

PREFACE

When I entered the Seminary of Barcelona in 2001 I started a very different life from the one I was living until then. I was working as a biochemist in a laboratory, and it seemed as if my scientific interest should leave some room for another kind of knowledge. Then I studied for a bachelor's degree in Philosophy and Theology. All the courses were very different from the ones I had followed in the Faculty of Biology. The course of Theological Anthropology slightly touched the frontier with science, but it was just to note that their fields of interest do not overlap with each other and do not have contradictions, and that is all.

In 2007 my bishop sent me to study Philosophy at the Pontifical Gregorian University in Rome. It was there I in came into contact with professor Gennaro Auletta. He was one of the professors of the Specialization of "Science and Philosophy". Speaking with him and following his lessons I realized that there are philosophical instruments to better understand the sciences. Auletta conceived, inspirited by the work of Charles Sanders Peirce, a philosophical approach to quantum mechanics. At that time he was also working on his greatest work to date, "Cognitive Biology", where he explains from the simplest cell to the complex universe and the human mind with the paradigm of information. He proposed I find an example of molecular biology feasible to be explained with that new paradigm. I have tried to do my best in these pages. I want to thank professor Auletta and the Pontifical Gregorian University warmly, because they have taught me that I do not have to choose between biochemistry and philosophy, but I can integrate both of them in a coherent framework that helps me to understand the world better.

Miguel Ramón

Introduction

1. Why a philosophical thesis about molecular biology?

The question that probably emerges when the title of this work is read is if it is possible to establish any relation between philosophy and a scientific discipline such as molecular biology. At first glance it seems as if this and other specialized biological disciplines have nothing in common and each one approaches the living world according to their own methods, without taking the other disciplines into account. The common opinion, shared even by many renowned biologists and philosophers, is that each of these disciplines has to go its own way without taking into account the results of the other ones.

This reduced approach to reality can be traced back to the 17th century, when scientific revolution introduced a very powerful methodology to investigate nature. The scientific method divided nature in different objects in order to study their parts. Following that method, science committed itself to a detailed study of reality, but lost the overall framework. It divided reality into pieces, and was then unable to reconstruct the object under study again. Scientific method led to a world where all its pieces are scattered over the laboratory table, but there was no further effort to reconstruct it. It was thought that knowing how the parts worked individually would be enough to infer how the whole functioned.

On the contrary, philosophy, and particularly Aristotelian philosophy, has a non-reduced approach to nature, investigating its objects, their dynamism and their properties. Philosophy has studied the features of life, but without penetrating the ultimate details of its processes. Philosophy has been characterized by a phenomenological and even metaphysic approach to life, identifying its most proper features in order to demarcate it from the non-living world; but it is difficult to find philosophical works where detailed biological data are taken into account. There has been a tacit agreement through which philosophy would think about life as the peculiarity of certain objects of nature while, on the other hand, molecular biology would investigate the detailed mechanisms of the processes that take place in living organisms.

It is not my intention to change the scientific method into a philosophical one or vice versa; that would be a misleading claim. On the contrary, I want to understand biological processes in detail, studying their biochemical characteristics, according to the method and object of science. From that point of view, the scientific method is a very powerful tool that helps us to delve into the details of molecular processes.

However, I cannot deny that, from my point of view, the knowledge of reality cannot be accomplished with just a detailed analysis of the elements we are interested in. I think that a proper philosophical framework, and more precisely a paradigm based on the notion of information, can be of great help not only to better understand what life is but also to provide new tools to continue biological research with a wider perspective.

I do not want to omit the scientific method, but it is also true that science has used a reductionist approach to nature that does not take into account all the factors involved in biological processes. It is assumed that the method science uses is to divide its object of study in its more basilar items. Arriving at the deepest structure of reality, individuating the ultimate elements from which everything is composed, seems as the cornerstone of all the knowledge about nature. As mentioned, science has cut nature in its most fundamental items, but has not reconstructed the entire picture once the features of those items were discovered. Philosophy can be of great help to make this reconstruction; it can provide a wider point of view to science in order to organize empirical data in a framework that takes into account all the factors at stake[1].

The dialogue between philosophy and science is not optional, but something necessary to enhance in the investigation of the unique truth of reality. This dialogue is not a discussion to decide to which part of the truth each discipline is devoted. This would lead to a sort of theory of the double truth, with which I disagree completely. On the contrary, I think that the dialogue is an occasion to share and illuminate each other the different approximations they have to the same phenomena.

[1] Cf. G. Auletta, *Integrated Cognitive Strategies*, 66.

2. The reductionistic approach of molecular biology

The discovery of DNA's structure in 1953 by James Watson and Francis Crick, was one of the paramount achievements in biology. The detailed structure of the molecule of life was discovered. The result of years of investigation it lead to the identification of every single atom that forms part of DNA.

Some years ago, it was thought that the proteins present in the nucleus of the cell were the ones that convey the information from generation to generation. It seemed that DNA just played a structural role, supporting the proteins that would be responsible of carrying the cellular information. But it was the DNA, a molecule simpler than proteins, the one that carries the information to synthesize the proteins of the cell. It is in the long strands of DNA where genes are aligned and controlled to express their information and yield the proteins they code for.

The discovery of the DNA structure has lead to the common opinion that everything about the mystery of genes would be completely revealed. It seemed as if knowing the ultimate structure of the molecule that conveys genetic information was the last difficult to overcome in order to have a fully account of what life is. The characterization of all the genes of a given organism would provide the necessary information to understand the basilar elements that define a living being. The importance of the discovery was too impressive to exclude thinking that the key to understand life was finally uncovered.

The characterization of the structure of the DNA molecule is probably what has put biology in the centre of the scientific debate. During the first half of the 20th century, physics was the most important science. This science studied the fundamental composition of matter, the structure and dynamics of the universe, and had very important technological applications. However, since the discovery of the DNA molecule, the second half of the 20th century witnessed an explosion of biological applications derived from the study of DNA. Pharmacological investigation have improved the drugs used to treat genetic problems, genetic analysis has helped to evaluate the propensity to develop certain diseases, the unique genetic imprinting of every human being has led to the development of identification techniques in criminal investigation, and many other applications that have also derived from the discovery of the molecular structure of DNA.

The powerfulness of genetic investigation has been feeding the opinion that humankind has already achieved a point where the secrets of life have been totally revealed. It seems as if with the discovery of the chemical nature of genes, investigators can fully characterize the structure and the behaviour of living beings. It is enough to see how in the press some articles appear with certain frequency about the discovery of a gene that explains a particular disease, a particular behaviour or even a particular way of thinking.

Behind this conception of molecular biology there is a reductionistic paradigm that takes for granted that knowing the most basilar items that characterize biological systems would lead to a fully account of the structure and behaviour of all living beings. In the end, biology would be just a very complex example of a physicochemical system. It has been thought that the complexity of life can be reduced to the physic and chemical features of its most basilar items.

Moreover, the discovery of DNA was not like the discovery of other biological molecules such as lipids or carbohydrates. DNA is the biological molecule that carries genetic information and, consequently, it was rapidly tagged as the most important molecule of the cell. It cannot be denied that DNA has a paramount role in the cell, but sometimes it is seen as the molecule that organizes and dictates the instructions for the functioning of the whole cell.

3 The aim of this work

The relation between science and philosophy is nowadays once of the most challenging topics. There are many fields in which this dialogue is at work. This dialogue is always characterized by a certain tension: science would be ready to reduce any explanation to the most fundamental items that can be experimentally identified according to its method. On the other side, philosophy would try to widen the scientific framework to set the question in a different perspective, but the measurable dimension of the items is often left out. To give some examples of this new dialogue, it is worth noting the relation between neurobiology and the question of free will; psychology and moral thinking; the difference between non-living and living matter; the appearance of man in the evolutionary story, and so on.

My proposal is to study an example taken from molecular biology. I want to study in detail a very precise example of protein synthesis; the incorporation of an amino acid that is not present in the genetic code and, consequently, is not expected to be found in proteins. I have chosen this kind of work because I think that it would be very helpful to see how the philosophical discourse about the non-reducibility of reality to its most fundamental parts finds a correspondence when a biological example is studied in detail. The aim of this work is to see if, in the dynamics of the biological example under study, there are elements that fit with the informational paradigm that is proposed as a philosophical framework.

It has to be said that the informational paradigm is not only appropriate for molecular biology, but is a philosophical framework that can be applied to any dimension of reality, from the physical world to the appearance of symbolic thinking. Information pervades any level of organization of the universe. However, I limit the object of study to molecular biology and, more precisely, to the process of insertion of Selenocysteine during protein synthesis in the cell.

4.Structure of the work

4.1. *The insufficiency of the reductionist approach*

The first part of my thesis is devoted to the introduction of two different approaches to reality. On the one hand, reductionism is presented as the method most frequently used to study the world from the scientific point of view. The most important tenet of the reductionist approach is that something is totally characterized when the pieces that compose the whole are identified and studied singularly. This reductionist point of view reveals that, behind it there is the conception that all sciences can ultimately be reduced to physics and chemistry, and any object of the world can be fully comprehended when its more fundamental items are fully characterized. Chapter I introduces different kinds of reductionism (ontological, methodological and epistemological) and considers if a true reduction between sciences has ever been achieved.

A reductionist approach would say that, if we really want to know what happens in a complex structure like a cell, it is enough to describe its parts: nucleus, endoplasmic reticulum, etc. In turn, to fully characterize those structures, it would be enough to individuate the molecules that are involved in their formation. And the reasoning can be followed until the most fundamental items of reality were reached. Then, it could be inferred that, knowing which basilar elements are involved in a given structure it is enough to fully characterize such structure, no matter how complex it is.

4.2 *Emergence and top-down causation*

Chapter II, on the other hand, introduces emergence as a wider framework to study biological phenomena. Emergence is a philosophical approach that maintains that the study of an object cannot be reduced only to the characterization of its most basilar items. To have a fully characterization of a complex object, it is also necessary to see *how* those basilar items are interconnected in space and time. A brief historic consideration is made to identify the philosophical background from which the concept of emergence has been coined.

Emergence shows that all throughout the history of the universe, matter has been organized in more and more complex structures. The novelty of those structures is not due to the addition of a different material element that at first was not there, but to a new pattern of organization among the already existing matter. Novelty in nature arises because new relations are introduced, and more and more complex structures and processes appear.

The notion of emergence, then, is linked to that of a layered structure of the world and to the fact that reality is organized in a modular way. Indeed, it can be seen that basilar elements are used to give rise to more complex items; which in turn can be related to form even more complex structures, and so on (this idea can therefore be connected with Aristotle's concept of form and formal cause). Atoms link each other to yield different molecules; molecules can form more complex structures as proteins, lipids, nucleic acids and carbohydrates. Those complex molecules can interact among them, forming cellular structures such as the endoplasmic reticulum, the Golgi apparatus, the ribosome or the nucleus. All these structures are spatially related thanks to the cellular membrane, that encloses all these partners in a functional unit (the cell). In turn, a given cell can be related to other cells in order to form a multicellular organism or a tissue. In the case of superior organisms, different tissues are used to form organs and systems and, eventually, the coordinated assembly of all of them in an individual.

The concept of emergence supports the idea that the chemical composition is not enough to fully characterize the cellular structures. There is something more when a novel structure or process arises in nature that cannot be traced back to the material characters of its most basilar items. The novelty that arises, that emerges, is something that has to do with the way those elements are structured, how they are ordered in space and time.

Emergence is a phylogenetic concept; that is, it is a concept that has to do with the way matter has been organized all throughout the history of the universe. If we have a look at the universe as it appears to us nowadays, we can see the result of an emergent process that has been at work since the very beginning of the universe. Emergence cannot be attested when we study a biological system. The biological system under study is already an emerged system that has arisen through a history of a complex assembly and interaction among its items.

However, the system can show that the novelty emerged has some influence over the very components whose combination has lead to that complex structure. This way of influence is known as top-down causation, because is a kind of cause exerted by the higher levels of organization of the system over the lower ones (Chapter II, section 4).

The most organized level of a biological system is the highest instance that can exert top-down causation. There is no needed to postulate that a complex system such as the cell has to have a mind or a sort of decisional instance to perform such causality. When I talk about top-down causation, I am rather referring to a formal constraint that the higher structures exert over the lower level items that are the basilar units of those very structures.

There is no violation of the closure of the physical world. Efficient causality only rules elements at the same level of organization. Higher levels of organization cannot have efficient power on lower ones. For example, a cell as a whole cannot have a direct causal-efficient influence on a given molecular reaction of lipid metabolism. However, the way into which the cell is organized, how the different structures are coordinated and how these structures canalize the distribution and availability of the most basilar molecules, is a formal constraint that concurs to facilitate certain causal effects.

If emergence is a phylogenetic concept that allows us to understand how the complexity and novelty of biological systems have arisen all throughout the history of evolution, top-down causation is concerned with how this novelty concurs to achieve certain causal incidence over the lower level items. Top-down causation is something that is seen at work, is something that happens, is an ontogenetic concept. It is the result of the coordinated activity of a higher level of organization in order to have a causal incidence over the lower levels. The notion of top-down causation invites us to make a reformulation of the classification of causes, introducing further distinctions than those found in the Aristotelian classification.

4.3 *Informational paradigm*

In the second part of the work, the notion of information used in biology is studied. It can be seen that when biologists talk about information, they are frequently using the Shannon paradigm of information. Shannon's model of information was formulated in the context of human communication, and was thought of as a mathematical model to reduce the loss of information when a message is send by a sender to a receiver through a certain channel (Chapter III, section 2). This framework leads to a vision of biology as if the notion of information only made reference to the activation of the information that is in the genes and its translation into protein. Genes (the genotype) would be the senders of information and the traits that they express (the phenotype), would be the receivers of such information.

I think that the picture offered by this informational approach is not complete. Indeed, life is not just the activation of genes, as if they were the only informational items that conduct how the rest of the cell has to behave. There have been many attempts to deal with the concept of information in biology: "The coding-for paradigm", "The developmental systems theory" and others (Chapter III, section 3). All of them try to show how the concept of information is paramount to understanding biology, but all of them make this attempt from Shannon's informational model.

The concept of information *per se*, is not enough to describe biological phenomena. Information, as Shannon understands it, is a concept that has to do with the reliability of information transmission despite the content of such information. It is not important what the content of the message is; its poroper transmissions is only what matters.

However, in the cell there is lots and lots of information of varying importance, and which is required in different contexts and moments. The cell cannot deal with all the potential information that it has at the same time; it has to deal with it in an orderly fashion, choosing which is the actual information that is needed in a certain moment. Information has to do with the characterization of the different elements that conform the cell; the amount of information of a molecule depends on the number of "yes/no" choices that have to be made to fully characterize the composition and structure of that molecule. From that point of view, the information needed to characterize a protein (a normal protein can have 300 amino acids) is greater than the number of particles that are in the universe.

The point is that the cell has to somehow structure all this information in order to coordinate all its molecules to perform the reactions that are needed for its own survival. Functions need to properly deploy certain reactions in response to environmental cues in order to maintain certain parameters within the narrow window that is needed for life.

Semiotics establishes the link between the huge amount of information in a cell and the functions that have to be deployed to keep the cell alive. Semiotics is a category that allows to deal with items not only according to their material composition but also regarding its linkage to the whole system they are part of. In fact, the cell deals semiotically with its own molecules. That means that the molecular structure of biological partners is important, but it is also important that this structure can be properly linked to the whole net of metabolic reactions that form the cell.

This leads us to consider the triadic semiotic relations proposed by Charles Sanders Peirce instead of the dyadic relations offered by Shannon's informational model (Chapter IV). A given molecule in a cell is no more just a piece of information that swims inside the cell with other molecules and structures. A molecule is also a sign that conveys information about a referent, which is important for the cell to perform the functions needed to keep it alive. Then, that molecule is a sign that stands for another thing (its referent) that is semiotically linked to other molecules of the cell forming a pattern (icon) of reactions and processes that are important to perform certain functions.

4.4 *Biological functions*

The semiotic approach shows that biological systems are more than the mere sum of their parts. Their coordinated assembly in space and time makes those systems related in such a way that allows them to perform certain functions (Chapter V). There have been two main philosophical approaches to the notion of function. The causal-role analysis establishes that the function of an item is determined by the role it plays in a given system. Its function depends on how it causally contributes to the global capacity of the system. The problem of this approach is that, from that point of view, an item should always play the same role; and this is something that does not fit in at all with what is found in the biological world, where some items play different roles depending on the context in which they are placed, as happens in the case study.

The second approach to function is the etiological analysis, where the main functional contribution of an item is linked to its etiological history; that is, to understand the role a biological trait plays in a cell, it is paramount to trace back its evolutionary history to see which role it had been performing in the past. The function of a given molecule depends on the history that molecule had all along its evolutionary path. However, there can be many traits that have persisted in an organism despite not being positively maintained by natural selection. Those traits can be without any role in the cell's functions for many generations, and at a given moment, they can be recruited and used to concur with some of them.

This shows that neither the role of a singular item nor its role in the past is enough to characterize the way it concurs with the cellular functioning. The cell has to perform certain functions, and to do that it uses everything that is at hand. For the constant performance of the necessary functions, the cell deals semiotically with its own molecules, not looking for singular structures and behaviours, but overall patterns of interactions that allow to link environmental and internal cues to proper cellular responses to those cues.

A certain item can be used differently depending on the surrounding conditions the structures of the cell provide to that item. Then, the new role it can play does not depend on the changing of its chemical nature, but on the surrounding conditions a higher level of organization provides, allowing the item to be treated differently from a semiotic point of view.

Dealing semiotically with its own molecules allows the cell to control information (Chapter V, section 5). The layered structure of cell's organization and the semiotic relation established through the whole metabolic network, allows the cell to exert an informational control over its own processes and structures. The cell as a whole cannot intervene in any single metabolic reaction, but it is organized in such a way (it is a pattern, or an icon) that the perturbation that enters at any point of that metabolic network, is semiotically treated, and the whole can respond in the proper way to the stimulus received.

This semiotic approach to the cell has also consequences for conceiving evolution. From that point of view, it can be said that natural selection does not select single molecules, but semiotic linkages among items that better respond to cellular needs. Nature, then, is not a designer that looks for the best version of a singular item of the mechanism. Nature is rather a tinkerer that tries to establish a semiotic linkage between the information that can be conveyed by certain molecules and the function they can be related to. Nature selects relations rather than singular partners.

4.5 *Case study*

The philosophical concepts regarding information, causality, semiotics and function are concepts that can be seen as very adequate for a philosophical approach to biological topics. From that point of view, they have had and they still have a very important role in the philosophical discourse about the nature and organization of life. However, I think that it would be useful to show the importance of those concepts with an example. The conceptual building could be solid, but only when it is applied to a case study is there the opportunity to see if it is explanatory powerful or not. That is why I think that it is extremely important to show with a biological example whether the philosophical categories introduced can explain biological phenomena and, furthermore, if they bring a wider conceptual framework to understanding biology from a non-reductive point of view.

The example selected is a very particular case of protein synthesis. As it is known, protein synthesis is an important process of cell metabolism. Proteins are necessary in many metabolic reactions and can be involved in many cellular functions. Then, it is very important that their synthesis and their activity be properly regulated. Protein synthesis is a very well known cellular pathway that translates the information contained in the DNA into protein. Hence, it is a key step in cell metabolism.

However, what really interests me is a particular feature of the synthesis of a reduced group of proteins called selenoproteins (Chapter VI). These proteins have a very special trait: the amino acid Selenocysteine has been found in their sequence. However, the important point is that Selenocysteine is not among the twenty canonical amino acids that can be found in the genetic code. That means that the cell has to somehow bypass the general system of protein synthesis in order to insert Selenocysteine in the sequence of selenoproteins.

The first step to have a protein is the transcription of a gene (DNA) into a messenger RNA (mRNA). Then, this mRNA is translated into protein following the rules of the genetic code: every three nucleotides of mRNA (that form a triplet or codon) are paired to a specific tRNA, to which a single and specific amino acid is bound. This matching occurs in the ribosome, which provides the proper conditions for the synthesis. The problem is that Selenocysteine does not have a specific RNA triplet that codes for it, so it seems difficult to explain how it can enter in the protein synthesis process.

The cell does achieve the insertion of Selenocysteine through an astonishing and complex mechanism where many extra partners are involved. Biologists talk abut this mechanism as a natural expansion of the genetic code; as if life, after having closed the needed "vocabulary" used by the genetic code, is constrained to insert a new "meaning" without being allowed to introduce a new "word". The mechanism is complex and involves many partners: secondary structures of RNA and special proteins. The scope of all this machinery is to provide a new chemical environment that allows the cell to use a codon that normally stands for the end of protein synthesis differently. Since there are no triplets left in the genetic code to assign one to Selenocysteine, one of the existing triplets is surrounded and constrained to be linked to a different event, which is the insertion of Selenocysteine instead of finishing translation. In fact, one of the three triplets that codify for the "stop" signal that indicates the end of translation is read as "Selenocysteine" by the ribosome, and this amino acid can then be inserted in the polypeptide chain.

The insertion of Selenocysteine can be done because the cell provides, with those additional partners, the necessary surrounding conditions needed for concurring to a change in the semiotic linkage. The UGA codon, in the new context, is not a signal to "stop" translation, but to codify for Selenocysteine. The chemical nature of the codon has not changed; only the surrounding conditions are modified in order to facilitate the insertion of Selenocysteine. Therefore, what has been changed is its functional significance.

Besides the molecular mechanism that performs this insertion, it is shown that this process is regulated depending on the availability of selenium (an essential element for the synthesis of Selenocysteine). There are different partners that exert a very precise control in the insertion of Selenocysteine (Chapter VII). The redefinition of "stop" into "Selenocysteine" is not a straightforward mechanism that always takes place whatever the conditions may be. Depending on the "boundary conditions", there will be a "stop" event or a "Selnocysteine insertion" event. So we can talk about the cell as a system that has a semiotic exchange with the environment in order to "decide" which option is better: to stop protein synthesis or to insert Selenocysteine.

Chapter VIII is devoted to seeing how the philosophical concepts introduced in the two previous parts are a great help to better understand the case study. This philosophical framework allows a non-reductionist approach to biology, where all the factors are taken into account, not only the chemical nature of the items involved but also their organization and their role in the whole. Concepts as "emergence", "top-down causation", "sign", "semiotics" and "function", among others, can be identified in the case study, showing that the informational paradigm is a strong epistemological framework to have a more complete comprehension of biological phenomena. The notions of teleonomy and teleology are also considered, seeing that the informational paradigm also allows the introduction of final causality in the study of biological phenomena.

FIRST PART

PHILOSOPHICAL BACKGROUND

Chapter I

Reductionism

1. Reductionism *versus* emergence

Reductionism, in general, can be defined as the epistemological framework that approaches to reality assuming that to fully explain something, you have to do nothing more than characterize the parts that compound it. A given system is nothing but the sum of its parts. The reductionist approach holds that to study an object or a process, what is required is to divide it into its most fundamental parts in order to completely analyze its traits. Once those traits are identified, also the whole that is compound by those parts is fully explained.

Emergence, on the contrary, states that the whole is more than the sum of its parts. It is not enough to fully characterize the properties of the parts to give a complete explanation of the whole. The parts of the whole, of course, can be isolated and its properties and traits studied individually. But what emergence says is that there are new properties of the whole that come from the fact that the parts are integrated and interconnected in a certain way. The new item is composed by a certain number of parts, but the whole is no more the sum of the properties of its parts. There is a novelty that arises, that emerges from the parts to constitute the new properties of the whole. Emergence is a genetic process that has been at work since the beginning of the universe. If we see how the universe has evolved since its initial stages, we can see that there has been an increase of complexity in its structure and an emergence of novelty that is not due to the appearance of new matter, but to the organization of the already existing one. The emergence of novelty is a process that has needed a lot of time; nothing less than the age of the universe. The whole history of the universe can be understood as the history of the emergence of novelty[1].

---

[1] Cf. H.J. Morowitz, *The Emergence of Everything*.

We can trace historically how scientists and philosophers of different periods have taken the concepts of reductionism and emergence into account. In fact, it is difficult to understand reduction without the concept of emergence. Every time that a fully reductive explanation of reality has been proposed, an anti-reductionist reaction has also appeared pointing out the difficulties of that reductive approach and stressing the irreducibility of certain issues. It is also difficult to understand emergence without reductionism, since emergence intends to be a wider framework to comprehend the very things reductionism studies. The concept of emergence appears as an answer to the intuition that the universe could not be fully explained appealing only to the properties of its fundamental stuff and the laws that govern it. That entails that nature cannot be ultimately explained only by physics.

What I will try to show in this chapter is that reductionism has focused on the question about what the world is made of and how we can epistemologically study it. This has led to investigate reality under the optics of what Aristotle defined as the material and the efficient causes. Reductionism wants to know the material nature of things and the laws they follow when they interact each other.

Emergence, for its part, does not reject the reductionist analysis, but claims that it is not the whole story. In the background of the emergent thesis we can glimpse Aristotle's formal and final causes. However, the relationship between science and philosophy is not so well established at present as to force a straightforward correspondence between Aristotelian philosophy and contemporary science. Nevertheless, it shall be shown that emergence shares many commonalities with formal and final causes, and I will try to elucidate some key concepts that would help to re-establish the intuitions that philosophy can offer science to better understand the phenomena it studies.

In this chapter I shall expose the different kinds of reductive approaches that are found in the philosophy of science. After that, a particular case of reductionism will be studied, that of the reduction of thermodynamics to statistical mechanics. Finally, some issues about reductionism in biology are considered.

2. **Some historical landmarks**

Reductionism is not something new that has appeared out of nothing in our epistemological horizon and that has recently shaken our worldview. As it happens with many philosophical tenets, reductionism has come a long way from its first appearing until it reached our time. Reductionism can be considered as a theoretical category that has been always at hand, a point of view about nature that can be adopted to study the universe.

From an historical point of view, we can see that it was during the ancient Greek philosophy that this question appeared for the first time. Democritus and Leucippus thought that atoms were the ultimate level of reality. If we look closely at things, it can be seen that they can be divided into parts. When there cannot be further division, the simplest elements of nature are reached, the indivisible units with which all the things are made. For Democritus and Leucippus these simplest elements of which everything is made are atoms. Nature is composed of atoms that, in different proportions and combinations, can constitute the variety of things we find in the world. Then, what really distinguishes one object from another is not the quality of its stuff, but the arrangement of the atoms that form them. This is a formal aspect of reality that cannot be cached with a reductionistic approach. The same atoms can be involved in the constitution of two different objects. The difference between them, then, is not the kind or number of atoms, but the way they are arranged and related to each other. The formal structure of the object is an essential characteristic that cannot be totally reduced to the material stuff it is made of. Of course, here it would be useful to clarify, as Paul Davies does, that

> what we today call atoms are clearly no the atoms of ancient Greece, for they are composite bodies that may be broken apart. But most physicists believe that on a much smaller scale of size there does exist a set of entities which play the same role conceptually as the atoms of ancient Greece, that is, they constitute a collection of fundamental, primitive objects from which all else is put together[2].

The reductionist approach became stronger in the 17th century, when the scientific method appeared as a very powerful tool to understand and manipulate nature. The conception of the universe as a machine, as a fully deterministic system that is inexorably governed by the laws of physics began to be popular. Laplace's deterministic model of the universe is famous, as is his affirmation that in such universe, God is a not necessary hypothesis. This conception had very important consequences not only for natural sciences, but also for philosophy and theology. Indeed, if everything is *absolutely* determined by physics, could

---

[2] P. Clayton – P. Davies, ed., *The Re-emergence of Emergence*, ix.

we continue to defend free will? Do we have to accept, for example, that mind is *just* the product of the electrical activity of the brain? And in the theology field, how can we understand the notion of providence or supernatural in a *fully* deterministic world that is only characterized by physical laws? Álvaro Moreno and Jon Umerez summarize as follows how reductionism has changed our worldview:

> This narrow but powerful conception paved the way for an undeniable scientific advancement which implied neglecting other more encompassing views, but less able to yield applied results. This process might be described as a gradual replacement of a previous Aristotelian and, hence, plural approach by a monist one. There was therefore an impulse to build an unified view of causality, aimed at fulfilling the whole range of possible relations meant by the cause-effect link, following the successful model of physical causality. The question now is whether the current state of scientific knowledge may still develop further keeping that model[3].

In this scenario, there were some voices that arose against this reductionist approach and tried to give alternative explanations to physicalism. Vitalism was one of the tentative answers. Vitalism holds that there is an essential difference between the non-living and living world, and this difference is due to a "biological principle", an "*élan vital*" or "entelechy" that gives to matter the properties we see in living systems[4]. However, vitalism was an uncomfortable tenet for scientists, since it is difficult to admit in science something whose existence could not be tested in a laboratory. From that point of view, emergence, as it shall be shown, is seen as a third way that, without adding nothing more than the material stuff of which the world is made, tries to account of the fact that this stuff could give raise to certain novelties. My guess is that emergence in general, and the informational paradigm in particular, could be a powerful tool to help philosophy and science have a richer approach to nature.

---

[3] A. Moreno – J. Umerez, «Downward Causation at the Core», 106.

[4] Vitalism is a philosophical tenet opposed to that of mechanicism. Vitalism defends that life is not just the result of physical and chemical interactions, but there is also a non-physical principle that drives the dynamics of life. A short description of this philosophical perspective can be found in M.O. Beckner, «Vitalism».

3. **Reduction and reductionism**

Before beginning to deal with what reductionism means, which kind of reductions we can individuate and whether a real reduction has ever been achieved or not, we have to introduce a distinction to explain the difference between reduction and reductionism.

As Evandro Agazzi states, reduction is a scientific procedure, an instrument to investigate the universe. As such, it has to be described, well interpreted and understood within the limits of its method. This means that reduction is a methodological procedure in science that has to be submitted to critical consideration. On the other hand, reductionism has a higher claim; it does not seem a mere instrument for scientific investigation but a philosophical doctrine with ontological implications for our worldview. Reductionism is a metaphysical tenet of the method of reduction, and it presupposes a monistic conception of reality[5]. Reductionism is the view that claims that to fully understand a complex system, you only need to account for the structure, the behaviour and the laws of each of its parts. A whole system can be exhaustively explained appealing exclusively to the features of the items that compose it[6].

Therefore, there is an ontological assumption in reductionism's position; it is implicitly accepted that the most fundamental physical level is the real ontological basis of the world. The comprehension of the features of those elemental items allow scientists to understand any complex combination of them. Any complex system is nothing but the more basilar elements that form it. Those basilar elements are not only the particles that have the ontological primacy, but also the elements to which all the features of complex systems can be reduced. The structure, behaviour and laws that characterize and rule complex systems can be reduced to the structure, behaviour and laws of its fundamental items.

Hence, reductionism is a relation established between any given complex level of reality[7] and the items that are grounding this very level from the level

---

[5] Cf. E. Agazzi, «Introduction».

[6] Cf. M. Silberstein, «Reduction, Emergence and Explanation», 81.

[7] The concept "level of reality" will be further developed below. Roughly speaking, we can say here that nature is structured in different levels, according to the items that constitute those levels. Then we can talk about the atomic level, the molecular level, the cellular level, tissues, organs, organisms, communities, ecosystems and, finally, the universe as a whole.

below. Reductionism has to explain this relation and whether the lower level properties can fully explain the properties of higher ones. If this is true, the whole life on earth could be reduced to its most basilar elements. Every complex level can be reduced to the following lower level and, at least, to the most fundamental one.

This ontological consideration about reductionism has also an epistemological consequence. What is said about the ontological properties of the items of each level could be also predicated from the different sciences that study those very levels. Then, not only could a cell be reduced to the atoms that compound it, but also biology could be reduced to physics. Reductionism is a philosophical tenet that claims that every item and law of our universe can be explained by the items and laws that rule the most fundamental ontological level of reality. This monistic worldview is formulated and articulated through the reduction of theories and sciences into the most fundamental epistemological issues to which every theory or branch of science could be referred. This relation is characterized by the fact that one theory is reduced by another theory; that is, the latter explains the former. Reductionism establishes a relation between theories in order to determine which one is more fundamental and, then, which has to be considered as the most explanatory power. This work can be done not only with theories but also with whole sciences; for example, the attempt to reduce biology to physics.

The question about reductionism was one of the most interesting debates in the scientific and philosophical community during the 20th century. As often happens in such debates, a concise definition of the terms used is required. Most of the misunderstandings in the reduction and reductionism debate have been because of the use of vague terms. Indeed, it has been seen that there were different conceptions about what reduction is. That is the reason why we need to clarify the terminology and introduce a classification of the types of reductionisms[8].

Emergence would be a third way between materialist monism and the dualistic vitalism. Materialist monism is a reduced philosophical tenet because it focuses just on the material aspect of nature. On the other hand, vitalism

[8] Francisco Ayala was among the firsts that propose such a classification of reductionisms. We partially follow here the characterization he does in F.J. Ayala, «Introduction».

introduces a non-material principle in order to explain the non-reducible aspects of nature. Emergent monism, for its part, tries to explain the unity and diversity of the universe, showing that there is a process of emerging novelty in nature (then, it is not ultimately reduced to the material stuff of its elements), but this novelty is not achieved through the introduction of different principles to explain it (as does dualism) but through the introduction of new arrangements among the already existing material. Those formal aspects that are important for the emergence of novelty are considered in the informational paradigm followed in this work.

3.1 *Ontological reductionism*

The fundamental claim of ontological reductionism is that higher-level entities are nothing but the sum of their parts. Francisco Ayala states that this kind of reductionism only admits physicochemical entities and processes, and this «implies that the laws of physics and chemistry fully apply to all biological processes at the level of atoms and molecules»[9]. Ayala says that this kind of reductionism is applied both to the stuff from which every level of reality is made and to the laws that govern them. Agazzi defines ontological reductionism as

> the conception according to which there exists one single fundamental stuff of the world and that, therefore, the science which concerns itself with this fundamental stuff will arrive sooner or later to encompass the other disciplines concerned only with certain "manifestations" or more "superficial" properties of this stuff[10].

Nancey Murphy faces also the question of ontological reductionism making a further distinction. She distinguishes between ontological reductionism in the strict sense and atomist reductionism. The former states that the same unities or ultimate particles are found at every level of reality. From that point of view, no "vital force", "entelechy", "*élan vital*", or any other immaterial principle would be found. All levels of reality are made ultimately of the same units. However, the existence of a hierarchy of levels in nature it is recognized and that, although

[9] F.J. Ayala, «Introduction», viii.

[10] E. Agazzi, «Reductionism as Negation», 15.

being made of the same stuff, they have different degrees of organization and complexity.

On the other hand, atomist reductionism not only defends the existence of the same fundamental entities in all nature, but also says that those entities are the only ones that could be considered *really* real. For this position, any other level of reality that would be above physics is just an epiphenomenon[11].

Murphy considers another kind of reductionism; the one she names causal reductionism. Furthermore, she considers that it is the gate that leads to the acceptance of the other types of reductionism. «Causal reductionism is the thesis that, in the hierarchy of sciences, all causal influences are "bottom-up" –from part to whole. This thesis has been so pervasive for so long that it should be counted as one of the central metaphysical assumptions of the modern era»[12]. Causal reductionism entails ontological reductionism, since it is assumed that everything is determined in a bottom-up way. That means that what really counts is the nature and behaviour of the lower-level items, because it is their causal properties that really establishes what would happen at higher levels. Furthermore, those higher levels would only be interesting as a description. «If causal reductionism is true then it is indeed physics that is doing all the work, and human thought and behavior are all epiphenomenal»[13].

I think that causal reductionism does not add anything to ontological reductionism. If only the most fundamental items had ontological consistency, then it follows that only those items would exert causal influence. It is not necessary to distinguish the reduction of all the stuff to its most fundamental parts and the reduction of the behaviour of the system to the causes of those very elements.

3.2 *Methodological reductionism*

This type of reductionism is a research strategy for the acquisition of knowledge. Most scientists agree that reduction is a powerful tool to investigate nature. That means that, whatever level of organization is being studied, its division and the study of its fundamental parts would be of great help. Methodological reductionism does not state that the information achieved by this method gives

[11] Cf. N. MURPHY, «Reductionism», 23.

[12] N. MURPHY, «Reductionism», 19.

[13] N. MURPHY, «Reductionism», 25.

a complete explanation of what is under study, but its knowledge is important to get the full picture. In any case, the study of higher-level features is important so as not to bias scientific investigation against ontological reductionism.

Then, as I have shown in Agazzi's distinction between reduction and reductionism, it would seem that methodological reductionism is what he understands as reduction. Reduction is a very useful procedure to get information about phenomena. Under this perspective, reduction is an indispensable way to investigate the nature of the world. However, it could fall into reductionism if this methodological procedure leads to an ontological tenet.

Methodological reductionism, however, does not give a full account of phenomena. That is the reason why even if it is a very powerful method, it also needs a wider philosophical framework to have a more complete interpretation of its results. The reductionist approach gives precious information about the chemical and physical characters of reality but, «when problems of high complexity like those occurring in life and cognition are studied, the specific physical and chemical interactions that could be involved are also understood in the context of more complex aspects of reality, such as structures and networks, information and correlations, functions and biological goals, cognitive processes and intentional purposes»[14].

3.3 *Epistemological reductionism*

Murphy defines epistemological reductionism as «the view that laws or theories pertaining to higher levels of the hierarchy of the sciences can (and should) be shown to follow from lower-level laws, and ultimately from the laws of physics»[15]. We are faced with a type of reductionism that tries to unify laws in a way that every phenomenon could be reduced to the laws of physics. It is an effort to discover how the fundamental laws of physics can explain everything that happens at any level of reality. Michael Silberstein defines this type of reduction in a more general way saying that «in epistemological reduction one set of representational items is reduced to another. These representational items

[14] G. AULETTA, *Integrated Cognitive Strategies*, 62.
[15] N. MURPHY, «Reductionism», 23.

are all human constructions and often taken to be linguistic or linguistic surrogates»[16].

Epistemological reduction is also known as theory reduction or inter-theoretic reduction, and tries to find the necessary conditions to reduce one branch of science to another. The aim of this epistemological approach is to reduce the different theories to a single one; and, consequently this would also imply the reduction of all the branches of sciences into the science to which the reducing theory belongs. Reductionism is always linked to the attempt to achieve the unity of science. This means that, despite all the branches that we can find in science, in the end, all of them could be reduced to a fundamental one, which is physics. Paul Oppenheim and Hilary Putnam say in their famous paper *Unity of science as a working hypothesis*:

> It would indeed be fantastic to suppose that the simplest regularity in the field of psychology could be explained directly –i.e., "skipping" intervening branches of science– by employing subatomic theories. But one may believe in the attainability of unitary of science without thereby committing oneself to this absurdity. It is not absurd to suppose that psychological laws may eventually be explained in terms of the behavior of individual neurons in the brain; that the behavior of individual cells –including neurons– may eventually be explained in terms of their biochemical constitution; and that the behavior of molecules –including the macro-molecules that make up living cells– may eventually be explained in terms of atomic physics[17].

John Kemeny and Paul Oppenheim reflect on the features that the process of reducing one theory to another has to have. The new theory has to explain (or predict, they use both concepts as equivalent) all the facts that the older theory has dealt with. One of the characteristics of the new theory is that it has to be simpler than the older. However, it could also be possible that the strength of the reducing theory would also imply certain degree of complexity. In theses cases, at least the increase in complexity should be counterbalanced by a greater strength of the reducing theory. This combination of complexity and strength is what the authors call a systemized theory. So, simplicity is not a value *per se*, because also an increase of complexity of the reducing theory could be seen

[16] M. Silberstein, «Reduction, Emergence and Explanation», 84.

[17] P. Oppenheim – H. Putnam, «Unity of Science», 7.

as an economical effect on the vocabulary of the theory if it also yields a well systematized theory[18]. Finally, it has to have also an economical aspect in the use of the vocabulary; that is, once the old theory is reduced to the new one, there has to be a simplification in the vocabulary, otherwise we are just superposing two different sciences but without establishing a relation of reduction between them[19].

The background of this tentative to reduce every explanation to a physical explanation is the old debate between mechanicism and vitalism; and in particular, the possibility of reducing biology to physics. Then, we find the relation between reductionism and the nature of science and the relation among its different disciplines again. The different branches of science correspond to different levels of reality that are organized from the lowest to the highest: elementary particles, atoms, molecules, cells, (multicellular) living beings and social groups. The relation between levels is as follows: «any thing of any level except the lowest must possess a decomposition into things belonging to the next lower level. In this sense each level, will be as it were a "common denominator" for the level immediately above it»[20]. This relation implies also that the laws that rule a determinate level, also rule combinations of the things belonging to that very level and also to those that pertain to the higher level.

The reduction of all levels to the most fundamental one could be done because it is reasonable to think that all the levels arose from the most fundamental one. There is a temporality and evolutionary appearance of these levels we are talking about; and this is achieved only through causal determinism[21].

---

[18] In fact, this is what any given theory looks for. «The role of a theory is not to give us more facts but to organize facts into a practically manageable form. In place of an infinite set of observation statements we are given a reasonably simple theory. Such a theory has the same explanatory ability as the long (or infinite) list of statements, but no one will deny that it is vastly simpler and hence preferable to such a list», J.G. Kemeny – P. Oppenheim, «On Reduction», 11-12.

[19] Cf. J.G. Kemeny – P. Oppenheim, «On Reduction», 7.

[20] P. Oppenheim – H. Putnam, «Unity of Science», 9.

[21] «Let us, as is customary in science, assume causal determination as a guiding principle; i.e., let us assume that things that appear later in time can be accounted for in terms of things and processes at earlier times. Then, if we find that there was a time when a certain whole did not exist, and that things on a lower level came together to form that whole, it is very natural to suppose that the characteristics of the whole can be causally explained by reference to these

3.3.1 Nagelian model of reduction

Ernest Nagel was the first to establish the conditions that have to be achieved to reduce a theory into another. Nagel acknowledges that there had been problems in the past when an attempt was made to reduce a certain field of sciences to physics. The reason why these attempts were unfruitful is that «expressions associated with certain established habits or rules of usage in one context of inquiry are frequently adopted in the exploration of fresh fields of study because of assumed analogies between the several domains»[22]. Then, the application of concepts from one field to another is far from being unproblematic. In some cases, the adoption of the terminology of one field of inquiry by another could be done straightforwardly, as happened in the reduction of Galileo's laws to Newton mechanics; but not always could this reduction be done without trouble.

Nagel defines reduction as «the explanation of a theory or a set of experimental laws established in one area of inquiry, by a theory usually though not invariably formulated for some other domain»[23]. Reduction is a relation between the reduced or "secondary theory" and the reducing or "primary theory" to which the former is reduced. This relation can be more or less problematic. In cases where there is a qualitative similarity between the phenomena of the secondary and the primary theory, no special problems are found. Is what happens, for example, with the mechanical model of point-masses, where dimensions are negligibly compared with the distances between bodies. This model is applicable to rigid bodies, although in this case also a rotation movement can also be made. But the fact that in both cases the same qualitatively phenomena are studied, makes reduction unproblematic, what Nagel calls a *homogeneous* reduction. The qualitatively similarity of the phenomena studied allows both sciences to employ the same vocabulary, making reduction an easy issue. In fact, «reductions of this

---

early events and parts; and that the theory of these characteristics can be micro-reduced by a theory involving only characteristics of the parts». P. Oppenheim – H. Putnam, «Unity of Science», 15.

[22] E. Nagel, *The Structure of Science*, 337.

[23] E. Nagel, *The Structure of Science*, 338.

type can therefore be regarded as establishing deductive relations between two sets of statements that employ a homogeneous vocabulary»[24].

It is difficult to find an example of such a clear and straightforward reduction in literature. Usually the vocabulary of the secondary theory does not exactly overlap that of the primary one. The assimilation of the traits of the secondary theory to the primary theory is not trivial. There are traits in both sciences that could be qualitatively different, and that makes their logical dependence problematic. In these cases of heterogeneous reduction, it seems as if the primary science left some terms of the secondary science out.

Lawrence Sklar establishes also what he calls a taxonomy of inter-theoretical reductions, seeking to put the different types of reductions that can be found in order[25]. There are cases in which there is no reduction at all, but just the replacement of one theory with another. However, what is a more challenging issue is the *inhomogeneous* reduction; that is, the reduction of one theory T_2 to a theory T_1 that does not contain some of the concepts of the former. Sklar focuses the problem with a sharp question:

> How can it be possible for one theory to be reducible to another when it is patently clear that the sentences of the former could not be deduced, or even approximately deduced, from those of the latter, even allowing for linguistic transformations such as substitutions of synonymous terms and grammatical transformations?[26]

The question could not be answered only from a linguistic point of view. If it were so, reduction would be just a matter of clarifying vocabulary. Reduction would be then the search of an appropriate linguistic apparatus that would lead us to state that, in fact, we are saying the same thing with different words that could better explain phenomena. Reduction would be just the replacement of one theory with other, as we have seen above.

Leaving aside the question of vocabulary, there must be empirical data that

---

[24] E. NAGEL, *The Structure of Science*, 339. This nagelian distinction is also expressed by Sklar as follows: «Two theories will be said to be *homogeneous* if they share the same conceptual apparatus, and *inhomogeneous* if one contains a concept not found in the other», L. SKLAR, «Types of Inter-theoretic Reduction», 110.

[25] Cf. L. SKLAR, «Types of Inter-theoretic Reduction».

[26] L. SKLAR, «Types of Inter-theoretic Reduction», 113.

can bridge the two theories. Empirical research is needed in order to discover some hypothesis or laws that can establish a bridge between the concepts of the reducing and the reduced theory. The differences in vocabulary must be a reflection of the differences of the items each theory or science is dealing with. And then, once the differences have been established, it has to be seen if a truly reduction can be achieved.

Nagel states that to reduce a theory T_2 to a theory T_1, the laws of T_2 must be deduced from T_1. That means that the vocabulary of T_2 has to be a subset of the vocabulary of T_1. The primary theory T_1 includes the secondary theory T_2. That is, T_2 is a special case of T_1. But this is not enough to establish the reduction between those theories. Other conditions must be achieved in order to fully reduce one theory to another.

3.3.2 Formal conditions for reduction

The formal conditions that relate the expressions of the reduced and the reducing theory are important elements for performing a proper reduction between theories and branches of science. This relation has to be logical, as Nagel stresses: «a reduction is effected when the experimental laws of the secondary science (and if it has an adequate theory, its theory as well) are shown to be the logical consequences of the theoretical assumptions (inclusive of the coordinating definitions) of the primary science»[27]. If this can be achieved, we have what is called a homogeneous reduction. On the contrary, if there are some terms in T_2 that are not present in T_1, we are dealing with a heterogeneous reduction, as we have said before. From a logical point of view, it is impossible to have a case of heterogeneous reduction, unless some formal requirements are introduced. These requirements are the conditions of *connectability* and *derivability*.

The condition of connectability states that some kind of relation has to be established between the new term present in the secondary science and the already existing terms of the primary science. There has to be some sort of connection in order to establish a relation between the two sciences if a reduction of one to the other is to be made. They cannot be incommensurables to each other. Besides connectability, derivability is also needed. Then, it is not enough for terms of the secondary science to be somehow linked to the ones of the primary

[27] E. NAGEL, *The Structure of Science*, 352.

science; this connection has to be logically derivable. Of course, derivability implies connectability, but not vice versa[28].

3.3.3 Non-formal conditions for reduction

Connectability and derivability are necessary but not sufficient conditions for reduction. It is also necessary that the theoretical assumptions of the primary science have some empirical evidence. Then, reduction is not a mere logical exercise based only on theoretical statements, but needs to be rooted in some empirical data. There are two reasons that explain why this empirical content is needed: one is that in every theory there are some observation statements that are not fully deducible from the fundamental theoretical postulates and, then, some empirical data is required; the other one is that the premises of science are not chosen arbitrarily, but they have to be stated as having some degree of certainty, and then they have to be empirically tested[29].

The degree of development of the theories at the time there is an attempt to reduce one to another is also a non-formal condition that has to be studied. The historical stage of a science, the point at which it is and the theoretical outcomes it has achieved are important issues that carry weight when reduction has to be performed. Let us see the example Nagel puts in order to understand this point:

> In particular, though contemporary thermodynamics is undoubtedly reducible to a statistical mechanics postdating 1866 (the year in which Boltzmann succeeded in giving a statistical interpretation for the second law of thermodynamics with the help of certain statistical hypotheses), that secondary science is not reducible to the mechanics of 1700. Similarly, certain parts of nineteenth-century chemistry (and perhaps the whole of this science) is reducible to post-1925 physics, but not the physics of a hundred years ago[30].

Not every secondary science is ready to be reduced to a primary science at any stage of its history. It depends on the scientific theory that characterizes

---

[28] Cf. E. Nagel, *The Structure of Science*, 353-354.

[29] Cf. E. Nagel, *The Structure of Science*, 358. Nagel admits that is difficult to establish how to provide this epistemological evidence. Nonetheless, what would be stressed here is that some empirical data is required as a non-formal condition for reduction.

[30] E. Nagel, *The Structure of Science*, 362.

the primary science. This point stresses the epistemological tenet we are dealing with. The question is not whether objects can be ontologically reduced to the most basal objects of reality. For epistemological reductionism it is not important what the world is made of, but just the epistemological rules that allow us to embrace it with some fundamental statements.

Nagel stresses the fact that sometimes the question about reduction is referred to properties instead of statements. This is, according to him, not the correct way to proceed. The question about the possibility of reduction of one science to another has nothing to do with the properties of the objects investigated by that science, but with the logical connections of the theories of each one. In his very words: «the supposition that, in order to reduce one science to another, some properties must be deduced from certain other properties or "natures" converts what is eminently a logical and empirical question into a hopelessly irresolvable speculative one»[31].

I think that this last statement of Nagel closes epistemological reductionism in a mere work of connecting ideas, regardless of the nature of the things those ideas represent. He has acknowledged the importance of some empirical data as a non-formal condition for reductionism, but in the end any data referred to the properties of the items studied is left out. It can be seen that focusing on the "nature" of the objects studied it is difficult to formalize them in order to fit them in a reductionistic scheme. There is too many empirical data in a given item to reduce it to just one of them and deal with it as a representative character to which all the rest can be reduced to.

My guess is that for an adequate epistemological approach to any system, it is not enough to consider the empirical data taken from the different items of the system. We omit something when we are focusing exclusively on the individual elements of the system or when we try to reduce the system to mere data that can be handle by just one fundamental science. The phenomenological richness of the system needs all-embracing categories that would preserve the particular features of the system; those features that are impossible to catch using only the most fundamental sciences.

---

[31] E. NAGEL, *The Structure of Science*, 365.

3.3.4 An example: the reduction of thermodynamics to statistical mechanics

Sooner or later, almost everyone who wants to investigate what reductionism is finds a reference that leads them to a paper where the case of the reduction of thermodynamics to statistical mechanics is cited. Many authors think that this is a clear example of a fully accomplished reduction. Others also sustain that we can talk of reduction, but only to a certain extent. There are overlapping areas between the two theories, but there are also specificities that seem to be irreducible. Nagel uses this example to illustrate the process of reduction of one science to another[32].

The study of thermodynamics has introduced some terms that are not present in classical mechanics. If we find concepts like "volume", "weight" and "pressure" in mechanics, new terms like "temperature", "heat" and "entropy" are coined by thermodynamics. Nagel himself states the key question: «By what reasoning is it apparently possible to drive statements containing such terms as "temperature", "heat", and "entropy" from a set of theoretical assumptions in which those terms do not appear?»[33]. The same question could be formulated as follows: Which are the conditions of connectability and derivability that allow T_2 (in this case thermodynamics) to be a logical consequence of T_1 (classical mechanics)?

Nagel focuses on some points of the argument, one of which is the need to introduce a statistical approach if we want to derive the Boyle-Charles' law for ideal gases form the principles of kinetic theory. Let us allow Nagel to explain with his own words the assumptions that are needed:

> Let the volume V of the gas be subdivided into a very large number of smaller volumes, whose dimensions are equal among themselves and yet large when compared with the diameter of the molecules; and also divide the maximum range of the velocities that the molecules may possess into a large number of equal intervals. Now associate with each small volume all possible velocity-intervals, and call each complex obtained by associating a volume with a velocity-interval a "phase-cell". The statistical assumption then is that the probability of a molecule's occupying an assigned phase-cell is the same for all molecules and is equal to the probability of a molecule's occupying any other phase-cell, and that (subject to certain qualifications

[32] Cf. E. Nagel, *The Structure of Science*, 338-345.

[33] E. Nagel, *The Structure of Science*, 343.

involving among other things the total energy of the system) the probability that one molecule occupies a phase-cell is independent of the occupation of that cell by any other molecule[34].

These assumptions leads to infer that pressure p at a given time is the average of the instantaneous collisions from the gas molecules to the walls of the recipient. Then, pressure p and mean kinetic energy E can be correlated with the formula $p = 2E/3V$. Giving the formula of the Boyle-Charles' law, $pV = kT$ (where k is a constant for a given gas and T its absolute temperature), both formulas can be linked as follows: $2E/3 = kT$, where the temperature is a function of the main kinetic energy of the gas. Then, we can say that Boyle-Charles' law of ideal gases is a logical consequence of the principles of classical mechanics if three assumptions are made: «an hypothesis about the molecular constitution of a gas, a statistical assumption concerning the motions of the molecules, and a postulate connecting the (experimental) notion of temperature with the mean kinetic energy of the molecules»[35].

3.3.5 Did we really achieve the reduction of thermodynamics to statistical mechanics?

Lawrence Sklar has reflected about the mechanism of reduction in general and about the paradigmatic reduction of thermodynamics to statistical mechanics in particular[36]. He is very conscious about the difficulty to fully achieve a truly reduction of one theory to another. The structure of the world is so complex that we have to be very careful when asserting that we have understood it with a handful of concepts that we think can explain everything. However, he does not want to eliminate reduction from the scientific scenario. His starting point is the conviction that things have to be studied deeper to critically evaluate if reduction could take place. It does not mean that the possibility of reduction is closed at all, but he just focuses on the difficulty of the question[37].

The first difficulty Sklar finds is not in the reduction of thermodynamics to

[34] E. NAGEL, *The Structure of Science*, 344.

[35] E. NAGEL, *The Structure of Science*, 345.

[36] Cf. L. SKLAR, «The Reduction (?) of Thermodynamics».

[37] Cf. L. SKLAR, «The Reduction (?) of Thermodynamics», 197-198.

statistical mechanics, but just in the very nature of the latter. In fact, «what is most peculiar about statistical mechanics, though, is its assumption of various probabilistic posits that, under most interpretations of the theory, fail to have their complete grounding in either the theory of the constitution of matter or in the underlying dynamics»[38]. So the first assumption that is difficult to justify is how classical mechanics has become statistical mechanics.

As a consequence of this assumption, statistical mechanics is a theory based on a probabilistic account, while thermodynamics is not. Of course, some assumptions can be made in thermodynamics to become statistical thermodynamics. The introduction of the probabilistic approach in thermodynamics makes this science easier to be derivable from statistical mechanics and, then, reduction could be more feasible[39].

Difficulties appear also when we try to establish connections between the concepts of both sciences. Indeed, if we consider that a gas occupying a certain volume has a given temperature, we will agree that this temperature can be assigned to different configurations of the system. A given value of temperature allows an array of multiple realizations of the system, a certain range of configurations of the gas. Then, the relation of the configuration of the system and the temperature it has is not a one-to-one relation. To avoid this difficulty derived from the complexity of the relations, statistical assumptions are made. Indeed, only when some magnitudes are treated in a probabilistic way, the bridging of the parameters of both sciences is less problematic[40].

It has been noted that one of the difficulties of the reduction of thermodynamics to statistical mechanics is the assumption of the statistical nature of classical mechanics. How can we justify a probabilistic approximation to mechanics? This range of probabilities should be located in the distribution of the initial conditions of non-equilibrium systems, which could not be fully stated. «The general consensus is that one cannot get the full non-equilibrium theory without making some basic posit about frequencies or proportions with which initial conditions are realized in the world when the system is prepared in a non-equilibrium condition»[41].

[38] L. SKLAR, «The Reduction (?) of Thermodynamics», 190.
[39] Cf. L. SKLAR, «The Reduction (?) of Thermodynamics»,193.
[40] Cf. L. SKLAR, «The Reduction (?) of Thermodynamics», 194-195.
[41] L. SKLAR, «The Reduction (?) of Thermodynamics», 200.

Initial conditions are something that we cannot fully characterize only by looking at how the system is constituted. The thermodynamical characterization of non-equilibrium systems needs to consider the range of the initial conditions in which the system is. These initial conditions cannot be fully established, they can only be framed into a range of probabilities. At least, one can hypothesize a plausible proposal of these initial conditions. Such a proposal can be sketched combining experimental data and the frame provided by thermodynamics. It is plausible to get a good characterization if the system is near equilibrium, but the enterprise becomes almost impossible when far from equilibrium systems are considered. Therefore, the reduction is no longer inter-theoretical reduction because, besides the theories that have to be related, experimental data is also needed to get a minimal approach to initial conditions. The problem is that there is no full access to these data, since initial conditions cannot be totally known. Sklar gets the consequence of this for the reductionist model:

> The fact that when the statistical mechanical treatments exists, and how they are to be framed when they do, is guided solely by the existing thermodynamics and experiment, and has no derivation from fundamental theory, makes a claim of reducibility of thermodynamics to statistical mechanics a dubious in a fundamental way[42].

Kenneth Friedman also reaches the same conclusion when he states: «For then the truth of the secondary theory depends essentially on certain facts. And the dependence on these facts would undermine the status of that theory, making many of its central claims true as a matter of fact, as opposed to as a matter of law»[43]. Friedman points out to another major difficulty to achieve reduction: «The fundamental reason is that most interesting thermodynamic processes are irreversible, in that the initial state cannot be recovered without some positive net expenditure of energy. By contrast, all mechanical processes are reversible»[44]. The irreversibility appears to affect thermodynamical processes, no mechanical ones. Furthermore, Friedman considers that irreversibility is an emergent property of thermodynamical processes.

[42] L. Sklar, «The Reduction (?) of Thermodynamics», 202.

[43] K. Friedman, «Is Intertheoretic Reduction Feasible?», 24.

[44] K. Friedman, «Is Intertheoretic Reduction Feasible?», 22-23.

Friedman also looks at what happens with homogeneous reduction. He thinks that, in these cases, there are not truly inter-theoretical reductions. In fact, those disciplines are too closely related to be considered a reduction between different theories. On the other hand, there is another different way for reduction to relate two theories that are so close in a homogeneous way. From that point of view, the secondary theory could be considered as a special case of the primary theory. If some special initial conditions are observed in the primary theory, then the secondary one can be deduced from the former. We can consider for example the relation between classical mechanics and special relativity, when the ratio of the relative velocities to the speed of light tends to zero[45]. Friedman concludes that homogeneous reduction does not exist. What is considered is another type of relation between theories; a relation based on an approximation when certain conditions are preserved[46].

4. **Reductionism and biology**

After having considered the types of reductionism, I have to rise the question of whether reductionist models could be applied to biology. This is, indeed, the question upon which this work is founded. Can reductionism offer a model in order to reduce biology to physics? Or, on the contrary, does biology continue to have certain characters that do not allow it to be reduced to physics? If so, which are those special characters that make biology resist such reduction?

In the recent history, an attempt has been made to reduce biology to chemistry and, eventually, to physics. The discover of DNA's molecular structure in 1953 by James Watson and Francis Crick [47], was a paramount issue that allowed to shed light into genes' chemical structure. It seemed as if we had reached the ultimate characterization of life. Imprecise concepts as "gene", "Mendelian laws", "dominance", and other ones that where coined by classical genetics, would be substituted by the molecular explanation that was at the bottom of

[45] Cf. K. Friedman, «Is Intertheoretic Reduction Feasible?», 25-26.

[46] Cf. K. Friedman, «Is Intertheoretic Reduction Feasible?», 28.

[47] Cf. J.D. Watson – F.H.C. Crick, «Molecular Structure of Nucleic Acids».

every single biological process[48]. Bernd-Olaf Küppers summarizes what is the working hypothesis of the reductionist programme applied to biology:

> The reductionistic research programme starts from the epistemological premises that there is no principal difference between non-living and living matter, and that the transition from the non-living to the living must be considered as a quasi-continuous one, in which no other principles are involved than the general principles of physics and chemistry. If there are certain limitations with regard to our physical understanding of living matter, then these are supposed to have a temporary but not a fundamental character[49].

The reductionist programme admits that biological phenomena are very complex, but this is just a *material* complexity, due to the increase of molecules and reactions that are in a cell. However, a deep investigation into the physical and chemical characteristics of all the molecules that are in a cell will reduce, sooner or later, biology to physics.

Nevertheless, as was seen afterwards, such reduction has not taken place; or at least not to the extension it was thought. There is a discussion about reductionism and anti-reductionism approaches in biology. The two extreme positions can be defined as follows: reductionism holds that molecular biology can explain everything that happens in biological systems. On the contrary, the more extreme position of the anti-reductionism argument defends the believe that higher levels of biological organization have explanatory arguments, which do not need to be accaountable for the molecular basis. As usually happens, the most interesting proposals are those that are somewhere in the middle of those two extreme claims.

David Hull showed that reduction of mendelian genetics to molecular genetics is not as easy as it was thought at first glance[50]. The first difficulty with which we come across is the fact that, a DNA molecule does not catch all the significance that had the concept of "gene"[51]. In fact, a DNA molecule, without

---

[48] For a detailed study about the role of reductionism in genetics, see S. Sarkar, *Genetics and Reductionism*.

[49] B.-O. Küppers, «Understanding Complexity», 241.

[50] Cf. D. Hull, «Reduction in Genetics».

[51] For an extended study about the concept of "gene", see P. Beurton – R. Falk – H.-J.

the context of the cell is an inert polymer. DNA needs to be embedded in a network of biochemical pathways to perform its role in its proper place and time. "Gene" is not only the stuff out of which the DNA chain is made, but also the other DNA and protein elements that regulate its expression, as well as other regulatory processes that modulate the function of the protein that comes out from that gene. As Mario Bunge says,

> undoubtedly, the most sensational advances in contemporary biology have been inspired by the thesis that organisms are nothing but bags of chemicals, whence biology is nothing but extremely complex chemistry. But this thesis, though heuristically enormously powerful, is not completely true. Chemistry only accounts for DNA chemistry: It tells us nothing about the biological functions of DNA –e.g., that it controls morphogenesis and protein synthesis. In other words, DNA does not perform any such functions when outside the cell[52].

Then, DNA has to be inserted in a cellular context, where reactions are spatially and temporally coordinated if the information it contains is to be activated to perform its role in the cellular metabolism. The importance not only of the metabolic framework of the cell, but also of the environment where the cell is, has its most critical relevance in biological processes, especially in those regarding development.

Another typical character of biology in general and of DNA in particular is the fact that "many-to-one" and "one-to-many" relations are established. That means that different genes are needed for a single function to be performed ("many-to-one"), and a single gene could be involved in different functions and, moreover, synthesize different proteins from the same DNA fragment ("one-to-many")[53]. It could also be seen in signalling transmission. The cell has many receptors anchored into the membrane in order to receive information from the environment. It has been seen that the same ligand could give different responses in the cell, depending on how many and of what type are the molecules responsible for the signal transmission inside the cell. Then, the same

Rheinberger, ed., *The Concept of the Gene.*

[52] M. Bunge, «The Power and Limits», 43.

[53] This could be achieved through mechanisms as differential splicing, frameshifting and posttranslational modifications of the synthesized protein.

outer molecule could rise to a different cellular response. On the other hand, we can also have the contrary phenomenon; different ligands that bind different receptors could activate the same cellular response.

In the "one-to-many" cases, the outcome is context-dependent. Then, the molecular nature of the DNA does not give a fully account of its function, since it depends on other molecules that cooperate with or regulate that very function. The "many-to-one" cases show us that cellular metabolism is an interconnected network of items that can perform everything that is needed for cell survival. There is no single item for performing one single key reaction or function. Here some authors talk of a "multiple realizability" argument, or some sort of redundancy in order to make sure that the function that has to be performed really is performed, either in one way or another.

Another difficulty emerges when we try to reduce biology to physics. Indeed, as it has been shown, epistemological reduction needs to reduce the laws of the reduced theory to those of the reducing theory. But, as Alexander Rosenberg states, there are no laws in biology. What were called Mendelian laws did not have the nomic features that physics laws have. Almost from the very beginning of Mendelian genetics, many exceptions to the regularities expected where reported (mitotic drive, linkage between adjacent genes, heredity linked to sex chromosomes, and so on). Then, «there are no laws of biology to be reduced to laws of molecular biology, and indeed that there are no laws of molecular biology, can be shown by the same considerations that explain why genes and DNA cannot satisfy reduction's criterion of connection»[54].

The concept of "law" is difficult to manage in biology. Biologists prefer to talk about "mechanisms" and "functions" rather than "laws" and "theories". It fits better with the phenomenon of life, its complexity, self-organizing and evolution. Of course that does not mean that biology is a lawless science. There are regularities, but the course of the biological phenomena is not tightly law-linked, as are the physical or chemical ones.

There is also the question of the boundary conditions. As we have seen, an explanation requires both laws and boundary conditions, but the notion of boundary conditions is different whether we consider simple or complex systems. Physical simple systems have boundary conditions that are contingent

[54] A. Rosenberg, «Reductionism (and Antireductionims) in Biology», 122.

quantities; they can be chosen arbitrarily. However, if we are dealing with complex systems, such as biological ones, those systems are very sensitive to boundary conditions, and their dynamics and properties strongly depend on what the values of this boundary conditions were[55].

The difficulty arises when biological systems have to be treated as physical ones. In fact, there is a very important difference between physical and biological systems; the formers can be characterized as equilibrium or, at least, near-equilibrium systems, while the latter are always far-from-equilibrium systems. The reductionistic approach tries to reduce the boundary conditions to the DNA sequence, but it does not take into account the fact that DNA is inside the cell, that it is itself a far-from-equilibrium system. Küppers explains it as follows:

> Up to now we have found no support for the hypothesis that the molecular boundary conditions of a living organism are a direct consequence of physical laws. Instead, we have a lot of experimental evidence for the assumption that all microstates in sequence space have the same probability of becoming realised under prevailing physical laws. Thus, since the total number of all possible microstates is extremely large, the probability of the realisation of a certain predefined microstate becomes vanishingly small. This in turn means that the selection and stabilisation of a predefined microstate, for example a microstate that carries biological information, cannot be explained within the framework of equilibrium physics[56].

This is a very important issue that biology has not yet answered. If all the microstates of a given system have the same probability but only a small fraction of them are achieved, which are the criteria used to narrow down the possibilities? An astonishing example of this great variety of microstates can be found when we study proteins. There are 20 amino acids used to synthesize proteins. Each site of the protein can be occupied, in theory, by any of the 20 amino acids. That means that the possible number of proteins *n*-length that can be synthesized is 20^n. If we consider that the average length of a protein is 300 amino acids, then a cell could make more than 10^{390} (20^{300}) proteins. It is

[55] Cf. B.-O. KÜPPERS, «Understanding Complexity», 251.
[56] B.-O. KÜPPERS, «Understanding Complexity», 253.

impossible to produce even a molecule of each kind, since there are not enough atoms in the universe to afford such synthesis[57].

A biologist would answer that natural selection is the one that has the job of selecting the proteins that are suitable to be synthesized. However, this is not a fully satisfying answer. From a certain point of view is difficult to ascribe to natural selection all the work of stretching the probabilities of the multiple possibilities of biological systems. There are moments in which the selection of a microstate cannot wait to be selected by natural selection to perform its role. I shall deal below with the question of what "selection" means and if this selection is always a "selection for" some trait or character.

Besides the question of natural selection, there is also the question of how the cell can react differently to the stimuli that come from the environment. We can think that the cell structures its own strategies to select the proper answer that would be more beneficial for it. Indeed, when a new stimulus enters the cell, it can virtually interact, physically, chemically or both, with all the molecules it would find in its way. But experience shows that does not happen at all. There is a spatial-temporal interaction among the proper molecules in order to translate and transport the stimulus from the outer membrane to the target molecules that have to be modified by this stimulus. Then, not all interactions are permitted, and only those that would help to canalize the stimulus into the proper direction will be allowed. This is something that is very clear when considering the mechanisms of cell communication. As Alberts *et al.* stress,

> in principle, the hundreds of signal molecules that an animal makes can be used in an almost unlimited number of signaling combinations so as to control the diverse behaviors of its cells in highly specific ways. Relatively small numbers of types of signal molecules and receptors are sufficient. The complexity lies in the ways in which cells respond to the combinations of signals that they receive[58].

Then, not all the possible combinations of the signaling elements are used. The cell selects somehow the ones that are useful. So it is not enough to make a list of the all-possible interactions of the elements to understand cell communication.The criterion and the means through which the cell selects some

[57] Cf. B. ALBERTS – *AL.*, *Molecular Biology of the Cell*, 136-137.
[58] B. ALBERTS – *AL.*, *Molecular Biology of the Cell*, 884.

solutions instead of others that are also possible is also to be found. Alberts *et al.* acknowledge that the challenge is to understand how the cell integrates all the information it receives.

The challenge of understanding how a cell integrates, processes, and reacts to the various inputs it receives is analogous in many ways to the challenge of understanding how the brain integrates and processes information to control behavior. In both cases, we need *more than* just a list of the components and connections in the system to understand how the process works[59].

My guess is that this "more than" has to do not with an additional stuff but with the way the cell deals informationally with its stuff. But before entering the informational account, let us turn now to consider in the next chapter the notion of emergence and see how it would help us to have a richer worldview of biology.

5. **Summing up**

I have shown that reduction is a methodological approach to nature that consists in dividing the items under study into their most fundamental parts. This procedure is very important in every scientific investigation. However, problems arise when this method is used as the only one that really tells us what the fundamental structure of the universe is. Then, reduction derives into reductionism, with ontological and epistemological tenets about how the world is and how we can know it.

The different kinds of reductionisms are not fully satisfactory and all of them present some problems when they try to draw a full picture of nature. There is the tacit assumption that sciences will be reduced to physics because the structure of everything is no more than the complex combination of the basilar items of the stuff out of which everything is made.

However, the reductionist paradigm cannot explain everything in biology. There are many characteristics of life that do not fit in a philosophical and scientific framework where absolutely everything can be traced back to the physical and chemical features of the most fundamental items that physics can identify. The multiple interactions inside the cell, the fact that all organisms are

[59] B. Alberts – al.,, *Molecular Biology of the Cell*, 886 (italics are mine).

far-from-equilibrium systems, the need to perform those reactions in the proper space and time, the impossibility to fully determine the boundary conditions of biological phenomena, and the need of an instance that can narrow down all the possibilities of combinations of a cell's items, makes the reductionistic approach in biology hardly attainable.

That is the reason why I am constrained to look for another way of facing biological phenomena; a way that would allow us to have a more complete explanation of phenomena that reductionism cannot cope with.

Chapter II

Emergence

1. What does emergence mean?

When we are in front of our table, we can perceive some traits such as its colour, shape, and touch. However, it could also be said that all these features are epiphenomenal, because what really makes a table be a table is the arrangement of the fundamental particles out of which all the rest is made. What can be said for a table could be also said for everything, including our mind. Science shows us that everything in our universe is made of the same pool of items. Non-living matter does not have different items from those that are present in living organisms. The stuff for making a stone is the same used to make our brain. The only difference between them is the diverse proportion of atoms and the organization they show. This fact opens the door to see the world from another perspective, where proportion and structure are important for the final result. Then it could be instructive to give some attention at what Mark Bickhard and Donald Campbell say:

> Perhaps phenomena such life and mind are somehow *emergent* out of lower level particles and processes. Perhaps they only exist insofar as those lower level particles and processes exist and occur, but they nevertheless have a reality of their own that comes into being, that emerges, when certain patterns or quantities or some other threshold criterion is satisfied. And, furthermore, perhaps, the reality they have *makes a difference*[1].

Whatever would emerge (entities, properties, processes, and so on), I guess that it should have some kind of causal powers, as will be shown. The novelty

[1] M.H. Bickhard – D.T. Campbell, «Emergence», 323.

arisen through emergence should have some causal impact; otherwise it would be just an epiphenomenon. Moreover, those causal powers must have the capacity to act upon the lower levels out of which they have emerged. However, it is worth noting that the concept of causal powers I am using here is wider than that of effectiveness or causal efficiency. Gennaro Auletta and colleagues distinguish those terms to make it clear that causal efficiency is not the only way to have causal incidence in a given system.

> *Effectiveness* means that the causal agent, in ideal condition, can positively give rise to a certain effect through interactions at the same ontological level, as when a ball moving with a certain speed is able, after collision, to set another ball in motion. *Power* here means having a role –although *not* an *effective* one– in the production of something. In other words, causes provided with causal power, but without effectiveness, influence the realization of a certain result. But even in ideal condition, they are not sufficient for producing that result since at least an effective factor must also be at work to actually achieve some production[2].

As the novelty emerged is said to be "the something more" that comes out from the sum of the parts, emergent phenomena could not be reconstructed appealing just to the traits of the components of the whole. Wholeness gives something more that is irreducible to the sum of the parts. Hence, we have stated briefly the main characteristics of emergence: novelty, irreducibility, the existence of different levels of reality and new causal powers of the emerged levels upon the lower ones.

The history of the question of emergence is one of the most interesting pages of the debate between philosophy and science of the 20th century. We can see how there has been an increasing awareness about the convenience to see nature from a holistic point of view. We are going to see how the vocabulary has been coined in response to the intuition that something has been lost when we dealt with nature just as something to be divided into parts for studying it.

To begin with, it would be useful to introduce the historical concepts of supervenience, weak emergence and strong emergence. These terms have been coined in order to point out at the irreducibility of natural systems and proces-

---

[2] G. Auletta – G.F.R. Ellis – L. Jaeger, «Top-down Causation by Information Control», 3.

ses. It would be shown that there are different conceptions about the extent to which emergence could be considered as something novel, non-predictable and non-deducible. The strongest form of emergence claims that a true emergent phenomenon entails that the emergence of properties or entities of the higher level have to have causal powers over the lower level. My aim is to show how this could be achieved without breaking the causal closure of the world. The notion of top-down causation implies a multi-layered structure of reality, where bottom-up and top-down causations are interconnected. Finally, this would lead to a reconsideration of the notion of causality.

2. **Historical landmarks**

One of the first precursors of the notion of emergence was probably Aristotle. He posits a principle, named *entelechy*, internal to every biological organism that directs its growth and actualizes the qualities that it has in potency. In this process everything is contained in the organism from the beginning, but has to be actualized by formal and final causes. This biological paradigm was still valid when Charles Darwin made his famous voyage in the *Beagle*[3]. Here we find the root of a non-material principle that directs matter to reach its proper form and perfection. Even though Aristotle's approach was a good intuition, it allowed the entrance of dualism in the explanation of the natural world, which entails many problems in establishing a scientific dialogue. Also Plotinus, with his doctrine of emanation, had some interesting insights for the concept of emergence.

> On Plotinus's view, the entire hierarchy of beings emerges out of the One through a process of emanation. This expansion was balanced by a movement of (at least some) finite things back up the ladder of derivation toward their ultimate source. The Neoplatonic model thus involved both downward movement of differentiation and causality and an upward movement of increasing perfection, diminishing distance from the Source, and (in principle) a final mystical reunification with the One[4].

The first three spheres that emerge out of the One are the intellectual (*nous*), the psychological (*psyche*) and the physical (*physica*). Plotinus explains

[3] Cf. P. CLAYTON, «Conceptual Foundations of Emergence Theory», 5.

[4] P. CLAYTON, «Conceptual Foundations of Emergence Theory», 5.

how further emergent phenomena lead to the apparition of new species. It is worth noting that for Plotinus everything emerges out from the One, which is the most perfect state. Then, different emanations and differentiations establish several levels of reality. On the contrary, in the conception of emergence defended in this work, the higher levels emerge from the lower ones, not vice versa. However, it must be admitted that, once the higher levels have been constituted, they could have some incidence over the lower ones.

The Aristotelian paradigm for science lasted until the so-called scientific revolution of the 17th century. This new investigation of the world was made through a reductionistic method, studying nature as if it were a machine that works with inexorable rules and where there is no room for exceptions, chance or imprecision. In the 19th century there were the first attempts to change the mechanistic worldview that scientific method had sketched. If we want to have a succinct idea about the emergent debate in the last part of the 19th century and during the 20th, it is very useful to divide the history of such debate in four phases, as Achim Stephan does[5].

The most important works of the first phase ar John Stuart Mill's *A System of Logic* (1843), Alexander Bain's *Logic* (1870) and George Henry Lewes' *Problems of Life and Mind* (1875). This first phase of the emergence debate was focused on the composition of causes. These authors distinguished between causes that are just a vectorial composition of precedent ones, and causes that have a real novelty compared to the ones out of which they emerge.

The second phase took place during the 1920's. The most important works of this period were: Samuel Alexander's *Space, Time, and Deity* (1920), Conway Lloyd Morgan's *Emergent Evolution* (1923) and C. D. Broad's *The Mind and its Place in Nature* (1925). Those authors belong to the so-called British Emergentism.

In the third period of the emergence debate, the very notion of emergence was strongly criticized[6]. We can place the start of this phase in 1926, when the "Sixth International Congress of Philosophy" took place. The Aristotelian

[5] Cf. A. Stephan, «Emergence».

[6] «During the mid-twentieth century, when philosophy in the English-language world was dominated by anti-metaphysical movements like logical positivism and various schools of linguistic analysis, emergentism was often trivialized if not ridiculed, and was largely ignored in mainstream philosophy», J. Kim, «Being Realistic About Emergence», 190.

Society also organized a symposium about emergence the same year, and published the contributions in the *Proceedings of the Aristotelian Society*[7]. Further articles were published in this period trying to clarify if it makes really sense to talk about emergence[8]. The end of this period arrived with the publication of two works that swept out emergence of the philosophical interest. Indeed, C. G. Hempel and P. Oppenheim's *Studies in the Logic of Explanation* (1948), and E. Nagel *The Structure of Science* (1961) coined a notion of emergence that was almost emptied of the suggesting intuitions that existed until then.

The fourth period began in the last 1970's and emergence was put again in the centre of philosophical consideration. It was the question about the mind/body relation that makes emergence an interesting concept to deal with again. Karl. R. Popper's *The Self and its Brain* (1977) and Mario Bunge's *Emergence and the Mind* (1977) were published in that period. Many other contributions can be found on the convenience of coining emergence for a better understanding of the connection between the mental and the physical[9].

These historical landmarks point at the necessity to re-elaborate the concept of causality. The reductionist approach is built in the conception of causality only from its efficient point of view. Indeed, «most contemporary criticisms to downward causation have deep roots (consciously or not) in this sort of assumption about causality in terms of causal explanatory exclusion»[10]. This exclusion refers to any kind of causal influence that is not identified with efficient causality. Those authors consider that only efficient causation has a true explanatory power. However, it is also possible to maintain the important role of

---

[7] Just to see how emergence became such a suspicious term, we quote here the conclusion of Leslie Mackenzie in this congress: «On the whole, it seems to me that the best use of the term [emergence] is to name provisionally cases where causation is accepted as a fact, but where the steps of their production are not yet capable of scientific description», W.L. MACKENZIE, «The Notion of Emergence», 68.

[8] Among them we can find, W.T. STACE, «Novelty, Indeterminism, and Emergence», P. HENLE, «The Status of Emergence», G. BERGMANN, «Holism, Historicism, and Emergence», A. PAP, «The Concept of Absolute Emergence», and C.W. BERENDA, «On Emergence and Prediction»,.

[9] For example, R.W. SPERRY, «Mind-brain Interaction», J.J.C. SMART, «Physicalism and Emergence», J. KIM, «Psychophysical Supervenience», and J. VAN CLEVE, «Mind--dust or Magic?».

[10] A. MORENO – J. UMEREZ, «Downward Causation at the Core», 107.

physical causation, as it is understood by the reductionist approach and, at the same time, reject its pretension of being the only way of characterizing causality. Moreno and Umerez defend a richer version of the concept of causality:

> Our argument in favor of a plural vision of causality lies in the necessity to postulate the existence of another type of causal link in biological systems (and in other kind of complex systems as well). This is not in contradiction with the principle of universality of physical causation, because anyway such new type of causal link requires complex underlying levels of physical organization[11].

3. Characterizing emergence

3.1 *Composition of causes: a first approach to emergence*

Stuart Mill was the first to introduce a difference among causality; more precisely, a difference on the composition of causes. There are causes that act together, and their resultant is the vectorial sum of their effects. Those causes follow "homophatic laws" or produce "homophatic effects". This model of composition is the one we find in mechanical processes, in efficient causes. On the other hand, there are causes that, when they interact, the effect is not the resultant of a vectorial sum; and that is why they are "heteropathic". Stuart Mill gives the example of chemistry, where the composition of two diverse substances give rise to a third one different from the former ones (for example, as happens with oxygen and hydrogen when they combine yielding water)[12].

Stuart Mill also points out that heteropathic laws rule heteropathic effects. So there are also new laws that emerge from the older ones; the new ones do not cancel the old ones, but supersede them at the new emergent level. Then, novelty is not only about new properties or new structures, but also there are laws that were not before emergence. Old laws are not violated; on the contrary, they are still doing their work, but new laws enter also in the emergent scenario to rule the new properties and structures[13]. In fact, is precisely in those new

---

[11] A. Moreno – J. Umerez, «Downward Causation at the Core», 107.

[12] Cf. A. Stephan, «Emergence», 28.

[13] Cf. B.P. McLaughlin, «The Rise and Fall», 60-61. As Stuart Mill states: «It is impossible to deduce all chemical and physiological truths from the laws or properties of simple substances or elementary agents, they may possible be deducible from laws which commence

structures where Bain identifies the new forces at play. It is the collocation of certain items what allows the appearance of new forces of nature. However, as was also stressed by Stuart Mill, the new forces cannot violate the principle of conservation of energy[14].

It was Lewes who used the term "emergence" for the first time in a technical sense. He calls emergent effects what comes out from Stuart Mill's heteropathic causes. Lewes, however, goes further and says that frequently there is no possibility of re-composing the process to individuate the role that each factor has had in the emergent effect[15]. As Stephan points out, Lewes would had an epistemological notion of emergence, since he suggests that someday the proess would be known which, until now, has had to be classified as emergent[16].

3.2 *Supervenience and weak emergence*

Alexander's *Space, Time and Deity*, was an attempt to clarify the mind-body problem through what afterwards was called "weak emergence". Alexander was interested in the question of the causal powers of the mental; how the mind can exert some sort of causality on a physical system such as the body. Matter acquires novel properties when it is organized in such a way that a higher order structure such as the mind could arise. Alexander says:

> The emergence of a new quality from any level of existence means that at that level there comes into being a certain constellation or collocation of the motions belonging to that level, and this collocation possesses a new quality distinctive of the

---

when these elementary agents are brought together into some mode-rate number of not very complex combinations. The Laws of Life will never be deducible from the mere laws of the ingredients, but the prodigiously complex Facts of Life may all be deducible from comparatively simple laws of life». Quoted in *Ibid.*, 61.

[14] Cf. B.P. McLaughlin, «The Rise and Fall», 63-64.

[15] «Thus, although each effect is the resultant of its components, the product of its factors, we cannot always trace the steps of the process, so as to see in the product the mode of operation of each factor. In this latter case, I propose to call the effect an emergent. It arises out of the combined agencies, but in a form which does not display the agents in action». G.H. Lewes, *Problems of Life and Mind*, Vol. II, 412; quoted in T. Stoeckler, «A Short History of Emergence», n.17.

[16] Cf. A. Stephan, «Emergence», 28.

higher complex. The quality and the constellation to which it belongs are at once new and expressible without residue in terms of the processes proper to the level from which they emerge[17].

It is worth noting that, although the emergence of a novelty that at first was not there is recognised, the explanation of that novelty is made by appealing to the characteristics of the lower level. The important issue for Alexander is that such novelty could not be predicted. Then, emergence is characterized as unpredictable, but it could be explanatory reduced to the level out of which emergence takes place[18]. Alexander is, hence, the first example of what is called weak emergence.

Supervenience, for its part, is a term that has its roots in the study of moral qualities and the relation between mental events and brain processes. In the field of morals, there were the works of G. E. Moore and R. M. Hare who, without using the word *supervenience*, established a relation between items of different levels of reality. Moore states that, «if a given thing possesses any kind of intrinsic value in a certain degree, then not only must that same thing possess it, under all circumstances, in the same degree, but also anything *exactly like it*, must, under all circumstances, possess it in exactly the same degree»[19]. Supervenience's conceptual weight can be synthesized by saying that there is no mental change without physical change. There is a correlation between mental events and the physical lower-events that are the basis that make the emergence of the mental possible. Kim summarizes it in two claims:

(1) One family of properties is "supervenient" upon another family of properties in

---

[17] Quoted in A. Stephan, «Emergence», 30.

[18] As Philip Clayton says: «weak emergence grants that evolution produces new structures and organizational patterns. We may *speak* of these structures as things in their own right; they may serve as irreducible components of our best explanations; and they may seem to function as causal agents. But the real or ultimate causal work is done at a lower level, presumably that of microphysics». P. Clayton, «Conceptual Foundations of Emergence Theory», 21. Stephan also stresses that there is a linguistic problem with Alexander's position. Alexander holds that there is true novelty, but it is described without residue in terms of the lower level. But is difficult to imagine how the lower level description can give account of the higher level when the latter one is defined as novel. Cf. A. Stephan, «Emergence», 31.

[19] Quoted in J. Kim, «Concepts of Supervenience», 154.

the sense that two things alike with respect to the second must be alike with respect to the first; but (2) there is no relationship of definability or entailment between the two families. These two theses often form a package: in fact, the main point of the talk of supervenience is to have a relationship of dependence or determination between two families of properties *without* property-property connections between families[20].

If it were possible to establish a property-to-property relation between the two families, the relation between them would be not of supervenience but of reduction. Supervenience tells us that the lower level furnishes a base from which novel properties could supervene. It establishes a connection between the levels, but this linkage cannot be done in a one-to-one fashion. However, the nature and relations of the lower level items determine the novelty of what supervenes. Although a clear relation between items of the two levels could not be traced, it is assumed that the lower ones micro-determinate what supervenes upon them. Let us see how Kim defines supervenience, a term to which he has devoted a great number of articles:

> To say that M supervenes on N_1, ..., N_n is to say that any system that has the base properties N_1, ..., N_n will necessarily have the supervenient property M. It is important to see that this is only a claim of determination or necessitation of one property by a set of properties, and that it says nothing about how M can be derived or deduced from the Ns, or about how the fact that something has M can be explained on the basis of the fact that it has Ns[21].

With this definition Kim places supervenience in an interesting point between reductive physicalism and any kind of dualism. In fact, he acknowledges that all reality is made of the same stuff, but at the same time there are higher-level properties that supervene on the intrinsic relations between the items of the lower level. Then, supervenience of new properties is a question of part/whole constraint; when certain items of the lower level are put in a certain constellation, forming certain structures, then some qualities arise as higher-level properties.

---

[20] J. KIM, «Supervenience and Nomological Incommensurables», 149-150.

[21] J. KIM, «Being Realistic About Emergence», 193.

We can see that, in the end, supervenience is claiming almost the same than weak emergence. These positions stress that there are new patterns in nature that arise out of lower levels; but, ultimately, it is physics that is doing all the work. Higher-level entities, and their aggregates, are just a kind of description without any ontological tenet at all. Higher levels do not have causal powers upon the lower ones. There is a non-derivability and non-deducibility relation a priori between both levels, but there is no causal influence of the higher upon the lower.

Supervenience, then, is a necessary condition but not a sufficient one to have a truly emergent phenomenon. Without causal powers, there is no true emergence. In fact, those phenomena that seem to be really emergent, as life and mind, are characterized by external relations, not only by the internal links established by the items of the lower levels allowing some sort of supervenience. Emergence of life and mind could not be understood without the respect they have to external conditions. Bickhard and Campbell explain the insufficiency of supervenience for understanding those phenomena:

> Supervenience is defined in terms of constituent particles and relations. It cannot handle external relations. But many critical phenomena, and important kinds of entities, are far-from-equilibrium, thus necessarily open, thus cannot be modelled without taking into account the external relations that maintain such far-from-equilibrium conditions, and the non-constancy of constituents that is involved in those open transactions[22].

Causality is the criterion for distinguishing supervenience from true emergence. As it has been shown, supervenience implies also novelty, but if this novelty does not entail some sort of causal influence from the higher upon the lower levels, we cannot speak of emergence. Then, emergence cannot be understood without top-down causation (or downward causation). The reality that emerges out of lower levels has to have causal powers upon those very levels. Emergence and top-down causation are two faces of the same coin.

The novel properties of supervenience cannot be predicted or derived from the basis of its ultimate constituents (at least *a priori*), but its novelty is only at the level of description. Ontological emergence, or strong emergence, for its

[22] M.H. Bickhard – D.T. Campbell, «Emergence», 341.

part, is not predictable or derivable from the lower level either, but has causal powers upon it. Furthermore, these causal powers cannot be reduced to the causal role of each single part of the whole[23].

Terrence Deacon proposes a classification of emergence, dividing it into first-, second-, and third-order emergence. I think that probably what he understands as first-order emergence can be identified with supervenience. Indeed, first-order emergent phenomena are those events ruled by higher-order thermodynamical processes. This type of emergence is the result of the interaction between molecules, leading to some properties that could not be ascribed to single items, like laminar flow, surface tension and viscosity. The new properties depend on the relatedness of the molecules and the proper conditions to achieve that relatedness. The emerged macro-properties are the result of the amplification of molecular interactions. Those interactions are aggregative, and the new properties supervene when that aggregations reach certain scale.

> Philosophers of science often refer to the dependence of higher-order properties on lower-order properties as "supervenience". Liquid properties supervene on these lower-order properties, including their interaction effects, and are therefore entirely determined by them. And yet we require a separate explanation for the fact that these properties are also to some extent independently converged upon despite a diversity of substrates[24].

This citation of Terrence Deacon gives us a lot information about the first-order emergence. We can see that this kind of emergence is characterized as supervenience. This means that we can trace a link between the micro-properties of the lower level and the macro-properties of the higher one. Hence, changes in the higher level are due to changes in the lower level. Therefore, macro-properties are of course novel, they are also unpredictable if we have not already seen such novelty, but they are not irreducible.

However, the most interesting thing is that different types of molecules could achieve those macro-properties of liquids, those patterns of organization. «So there are many possible ways that different micro-details of structure and

---

[23] Cf. M. SILBERSTEIN – J. MCGEEVER, «The Search for», 186.

[24] T.W. DEACON, «Emergence», 126-127.

interaction can converge to produce the same higher-order properties»[25]. This shows us that what supervenes is something that has to do with patterns of organization rather than with molecules of one type or other. The fact that supervenience is not an example of strong emergence could prevent us from paying attention at this fact. What really matters here is not the fact that we are in front of a weak example of emergence, but the confirmation that higher-level properties could supervene from lower levels composed by different kinds of molecules. The important point is the pattern, the macro-properties that we see as a product of the aggregation of the lower-level entities, despite the matter they are made of.

The macro-properties are not just a sum of the micro-properties of the lower-level items. The aggregative process that leads to the supervenience of the higher-level imply a certain selection of some characters, which are then amplified by the scale effect. Not all the details of lower-level items are relevant for the supervenience of the higher-level properties. Let us see how Deacon explains it with the example of water:

> Liquid water properties "supervene" on the properties of water molecules primarily because of *relational* features. In repeated microscopic interactions the specific unique features of individual molecules (e.g. their charge, geometry, orientation, momentum, internal vibration, etc.) distribute in such a way as to cancel one another in aggregate, thus leading to a higher ordered state. These astronomically many details cancel out, except for the average effect expressed globally via the relative linearity of the summed stochastic processes[26].

There is a selective cancellation of some details and the amplification of others. Not all the traits of the lower-level items are relevant to be a basis for the supervening properties. Robert Batterman, who is also interested in the emergence phenomena stressed by Deacon, also defends this. Indeed, Battermann says that this cancelling of the irrelevant data is heuristically useful in our exploration of emergence, and helps to leave out the irrelevant details that makes difficult for us to see the pattern that emerges from the lower level[27].

---

[25] T.W. Deacon, «Emergence», 127.

[26] T.W. Deacon, «Emergence», 127-128.

[27] Cf. R.W. Batterman, *The Devil in the Details*.

Emergent properties, then, are the result of a selective amplification of some details of lower-level items, in order to achieve certain regularities and stability at a higher level. These macro-properties could all fall into the category of form. The novelty of emergent processes is the appearance of new forms that have different properties from those of the lower-level items and that are not deducible *a priori* from the complete knowledge of the elements of the lower level, because we do not know the selective criterion that decides which details have to be selected and amplified. Deacon explains that this form has to do with distribution of characteristics, possibilities of interaction, boundary conditions and everything that could lead to a dimensional bias when scale is amplified[28].

This amplification phenomenon pervades homogenously through space and time. The new structure emerged does not perturb the dynamics of the amplification effect, but is a way to amplify this very amplification. There is no correlation between perturbation and the structure of the system. The emergent structure is shielded against perturbations because they form part of those details that are cancelled and then, they are not amplified.

3.3 *Strong emergence*

Conway Lloyd Morgan's *Emergent Evolution* (1920) is also a paramount work that gets into the notion of levels of reality, which is at the heart of emergence. Lloyd Morgan was a punctualist rather than a gradualist[29], from an evolutionary point of view; and this is a more adequate framework to give account of the novelty displayed in the evolutionary process. Then, he was closer to Alfred Russell Wallace than to Charles Darwin. However, he did not wait too long to take a different way from that of Wallace. Indeed, Wallace appealed to God's intervention in order to explain the novelty of each new level of evolution. Lloyd Morgan was uncomfortable with that approach, since this position conceived God just as "the God of the gaps". He agreed with Bergson about the fact that evolution creates novelty, as the French author defended in his work *Creative*

[28] Cf. T.W. DEACON, «Emergence», 128-129.

[29] Punctualism holds that evolution takes place through the accumulation of mutations that, in a relative small time, provoke significant changes in an organism's phenotype. Gradualism, on its side, says that changes in the organisms are done through many slightly intermediate states.

Evolution. But Lloyd Morgan did not agree that the motor of this creative capacity had to be an *élan vital*, or some other force out from nature[30].

Lloyd Morgan argued that novelty has to be explained by the different levels of organization that we can trace in nature. Each level has novel properties if compared to the immediately lower one. This novelty is due to the fact that the upper level is considered as a whole, with properties that cannot be individuated in the singular parts of the lower level. This conception of reality layered in different levels that have come to being through evolution have some consequence for our view of the world. As Lloyd Morgan says,

> it does imply (1) that there is increasing complexity in integral systems as new kinds of relatedness are successively supervenient; (2) that reality is, in this sense, in process of development; (3) that there is an ascending scale of what we may speak of as richness in reality; and (4) that the richest reality that we know lies at the apex of the pyramid of emergent evolution up to date[31].

The importance Lloyd Morgan gives to the concept of relatedness is very interesting. In fact, new levels of reality emerge from the lower ones because there are new relations between the already existing items. This relation is what makes the difference and allows the emergence of new characters.

A key point of Lloyd Morgan's work is that he conceives emergence at every step of the history of the universe and at every level of reality. Emergence is not something that appears only within the frontier of non-life and life, neither could it only be applied when we are dealing with the mind-brain problem. Those fields of investigation, of course, are challenging points to our understanding of the world and to our comprehension of what emergence is. However, emergent phenomena are something that happens at every level of reality, no

---

[30] Cf. P. CLAYTON, «Conceptual Foundations of Emergence Theory», 10-11. Lloyd Morgan explains it as follows: «Since it is pretty sure to be said that to speak of an emergent quality of life savours of vitalism, one should here parenthetically say, with due emphasis, that if vitalism connote anything of the nature of Entelechy of Elan –any insertion into physico-chemical evolution of an alien influence which must be invoked to explain the phenomena of life- then, so far from this being implied, it is explicitly rejected under the concept of emergent evolution». C. LLOYD MORGAN, *Emergent Evolution*, 12.

[31] C. LLOYD MORGAN, *Emergent Evolution*, 203.

matter how simple it is. It is something that has not only to do with the origin of life and the origin of mind, but with the very structure of the universe. It is a truly transversal phenomenon that allows us to catch the essential traits of our universe. Clayton stresses this particular, noting also that it prevents Lloyd Morgan to fall into vitalistic or dualistic conceptions.

> If emergence theories can point to emergent wholes only at the level of mind, they quickly fall into a crypto-dualism (or perhaps not-so-crypto one!); and if they locate emergent wholes only at the level of life, they run the risk of sliding into vitalism. Conversely, if significant whole-part influences can be established already within physical chemistry, they demonstrate that emergence is not identical with either vitalism or dualism[32].

I think that Clayton points out a very relevant issue. In fact, seeing emergence only when dealing with the mind/body problem leads to some difficulties, as has been shown with supervenience and weak emergence. I think that the problem has to be faced from its root, and that is the reason why my work is centred in biology, because if emergence appears in the simplest manifestations of life, then it would be easier to recognize it in the most complex and organized structures such as the brain.

It can be seen, then, that emergence begins to be a third way between vitalistic and mechanicistic theories. Vitalists claim that everything that characterizes living beings is due to some non-material component. On the contrary, mechanicists hold that every phenomenon could be explained through the composition of the items and the general physical laws that rule them. Emergence, for its part, admits that the whole universe is done with the same matter, but novelty could arise out of that matter; and the novel structures could have causal influences upon the lower-ones.

C. D. Broad adds another trait to emergent phenomena: the non-deducibility of the emergent properties[33]. The mechanistic approach holds that the

---

[32] P. CLAYTON, «Conceptual Foundations of Emergence Theory», 13. Harold Morowitz identifies up to twenty-eight steps of emergence. Cf. H.J. MOROWITZ, *The Emergence of Everything*.

[33] Non-deducibility has to be distinguished from non-predictability. Sometimes non-predictability has been linked to emergence, but it is misleading. In fact, «there can be novelty

behaviour of the whole is determined by how the lower level items are arranged and it could be deduced from the behaviour of the single items or from how they behave in other systems. On the contrary, the emergentist approach states that such deduction is not possible. Let us put it into Broad's words:

> Put in abstract terms the emergent theory asserts that there are certain wholes, composed (say) of constituents A, B, and C in relation R to each other; that all wholes composed of constituents of the same kind as A, B, and C in relations of the same kind as R have certain characteristic properties; that A, B, and C are capable of occurring in other kinds of complex where the relation is not of the same kind as R; and that the characteristic properties of the whole R(A, B, C) cannot, even in theory be deduced from the most complete knowledge of the properties of A, B, and C in isolation or in other wholes which are not of the form R(A, B, C). The mechanistic theory rejects the last clause of this assertion[34].

We cannot guess how the new laws of a hypothetical relation R would be; we have to wait to see such relation and then study which are the new laws that have emerged from that disposition of the items that compose the system.

Broad distinguishes between intra-ordinal and trans-ordinal laws. Intra-ordinal laws are the laws that operate within the same organizational level. The laws of chemistry are an example of intra-ordinal laws. Broad defines trans-ordinal law as «the law which asserts that all aggregates composed of such and such chemical substances in such and such proportions and relations have the power of reproduction would be an instance of a Trans-ordinal law»[35]. This example puts two different levels into relation: chemistry and the reproduction of organisms. Broad is saying that the laws that rule reproduction are not deducible from the laws of chemistry. He focuses on the non-deducibility of the emergent laws from the lower-level laws. However, as we shall see, the question does not end here. The non-deducibility of reproduction from chemistry comes from the fact that reproduction is a function that uses chemistry events according to

---

and predictability (e.g. in cases of the prediction of novel qualities of artefacts) and of course there may be no novelty but unpredictability, as in indeterministic sytems», A. Stephan, «Emergence», 32.

[34] Quoted in A. Beckermann, «Introduction», 17.

[35] Quoted in B.P. McLaughlin, «The Rise and Fall», 80.

some respect. Reproduction is an informational process that canalizes chemistry in order to fulfil its aim: the generation of another self.

Finally, another important contribution of this author to the conceptualization of emergence is the link that he establishes between the epistemological and the ontological approach. I have shown above (chapter I, section 3.3) that there is a discussion about the unity of science and the pertinence whether there should be special sciences or not. See how Broad tries to solve this question:

> On the emergent theory we have to reconcile ourselves too much less unity in the external world and much less intimate connexion between the various sciences. At best the external world and the various sciences that deal with it will form a kind of hierarchy. We might, if we liked, keep the view that there is only one fundamental kind of stuff. *But we should have to recognise aggregates of various orders*[36].

Broad thinks that the hierarchy of sciences reflects the hierarchy of reality, and each branch of science is devoted to study one level of organization of the world without being threatened by a possible reduction of all the branches to physics.

3.3.1 Second-order emergence

Terrence Deacon's second- and third-order emergence can be understood as strong emergence phenomena. It is in second-order emergence that there are some regularities that begin to have some causal impact in the dynamics of the system.

> The unpredictability of chaotic systems derives from the fact that interaction dynamics at lower levels become strongly affected by regularities emerging at higher levels of organization. This can produce a deviation-amplifying dynamic that propagates throughout the system. If perturbations of this type are incessant, bias comes to dominate over distributive tendencies[37].

Structural regularities on the higher level have a dynamical influence on the amplification process of the traits of lower-level items. So we are in front of

---

[36] Quoted in P. Clayton, «Conceptual Foundations of Emergence Theory», 9.
[37] T.W. Deacon, «Emergence», 130.

a top-down causation phenomenon. This top-down effect introduces a continuous bias in the system, preventing the returning of the system to an equilibrium state, where amplification takes place homogenously. The dynamical behaviour across time of this kind of systems is unpredictable, and this unpredictability «derives from the fact that regularities at lower levels have become strongly affected by regularities emerging at higher levels of organization»[38].

There is another characteristic that makes these second-order emergent phenomena different from those of first-order. As it has been shown, in first-order emergence, what matters is the pattern emerged, rather than the micro-structure of the lower-level. Indeed, the same pattern could be achieved by systems with different micro-configurations. But in the case of second-order phenomena, the nature of the elements that are selectively amplified is important, because they establish the unique boundary conditions of the system. Moreover, not only the kind of elements is important, but also the history of the system. In fact, initial conditions are critical to the future ongoing of the dynamical system. The conditions where the second-order process begins are not irrelevant. Systems behaving that way are ruled by non-linear equations, as happens for example with far-from-equilibrium thermodynamics. On the other hand, first-order emergence are phenomena that could be described with linear equations; that is why emergent phenomena of first-order could be fully explained appealing to the microstructure of the lower-level out of which it supervenes.

An example of this second-order emergence is autopoietic systems, those whose structure and behaviour depends on the autocatalytic reactions in which the elements of the system participate. «In autopoiesis the interaction dynamics of sets of different components is constrained both by configurational properties of the whole collection and by configuration symmetries and asymmetries that exist between the micro-configurations of the different classes of its components»[39].

Molecular autocatalytic cycles are an example of second-order emergence. Molecules are arrays of atoms tightly bound that have chemical entity and are able to interact with other molecules through strong or weak interactions. Each molecule has a variety of possibilities of interactions with other molecu-

---

[38] T.W. DEACON, «Three Levels of Emergent Phenomena», 101.

[39] T.W. DEACON, «Three Levels of Emergent Phenomena», 104.

les, depending on its chemical properties. An autocatalytic cycle privileges and amplifies some of this interactions in a way that a recurring reaction could take place, leading to the emergence of higher level properties and regularities. These regularities, for their part, act as structural constrains to further amplify the micro-constituents that have allowed its emergence. This is, in fact, what characterizes the molecular reactions that we can find in the different metabolic pathways of the cell.

Ultimately, the metabolic molecular dynamics that constitute living cells depends on autopoietic system dynamics constituted by many fully and partially autocatalytic sets of molecules. What do these examples of second-order emergent phenomena have in common? A kind of tangled hierarchy of causality where micro-configurational particularities can be amplified to determine macro-configurational regularities, and where these in turn further constrain and/or amplify subsequent micro-configurational regularities[40].

3.3.2 Third-order emergence

There is also a third-order emergence level that, besides the trait of self-organization and autopoiesis, also has evolvability. Deacon puts it as a separated level, but I think that this third-order emergence is a special case of the second one. Indeed, it could be seen as a case of strong downward causation that has the property of being capable of transmitting its second-order properties to another systems. That is why I think that third-order emergence is the second-order applied to biological systems. To have this evolvability that Deacon attributes to this kind of emergence, some sort of memory or informational instance is needed, such as a record of what happened in previous moments of the system and as a recurring instance to reuse those configurational patterns that played their role at a particular time in the system's life[41].

There is another difference between second-order and third-order emergence phenomena. The processes of the former are thermodynamically coupled, with exchange of matter and energy between structures; while in the latter, the amplification processes are maintained despite the separation in space and time that could be between different structures. The different traits of the system

[40] T.W. Deacon, «Three Levels of Emergent Phenomena», 104-105.

[41] Cf. T.W. Deacon, «Emergence», 137.

are correlated in space and time in order to maintain the unity of the whole. Memory is an essential character to preserve that unity. At that point I think that we can start to talk about a "self"; that is, a system whose characteristics persist in space and time and whose items are related to each other forming an integrated network of first- and second-order events.

Another important question is how these levels are related to each other. Elsewhere (section 3.3, this chapter) it has been said that there are some trans-ordinal laws that relate different levels of organization. The layered structure of the different levels helps us to imagine how top-down causation takes place. However, we have to think about these levels as something inclusive, allowing the co-existence of all of them in the same integrated system.

3.4 *A global characterization of emergence*

It has been shown that emergence has been differently characterized depending on the areas where it was necessary to introduce a non-reductive approach to the phenomena that were studied. The vocabulary of emergence has been forged in the interests of many different authors. Stuart Mill introduced the difference between homophatic and heterophatic effects; and he explained the difference between them with examples from the chemical world. Alexander, on his side, approaches the concept of emergence through the study of the mental. Even the field of the morals dealt with concepts that are related to that of emergence.

On the other hand there is the question on the strength of the emergent phenomena. Are they really new? Or can be ultimately reduced to more basilar phenomena? This issue introduced many classifications of emergence: supervenience, weak and strong emergence, the nature of the emerging qualities and the possibility to consider them as subjects of causal influence. I think, however, that we must not mix the causal powers with the concept of emergence, but consider them when we talk about top-down causation (see below). Emergence is a concept that allows us to think of the universe as a continuous process of generation of new structures and properties.

When we observe the universe, we can see that there has been a process of increasing complexity from the Big Bang to our present time. All along the history of the universe we can state the emergence of new and complex structures that integrate simpler and older ones. As Auletta states, the three main levels of reality that we can clearly individuate in the history of the universe are: the non-living, the living and the mental[42]. So we can identify two major transitions that can be classified as strong emergent phenomena. Of course, many intermediate levels can be also identified. From that point of view, emergence does not happen only in very singular moments of the history of the universe, but is something that has always been in process, since the beginning.

The historical appearance of these levels allows us to think of a genetic and dynamical process that leads to more and more complex structures in the universe. This genetic process is what I refer to with the concept of emergence. Emergence, then, is linked to the development of the universe. It is clear that in the beginning, the structures of the universe where simpler than they are now. Different kinds of atoms have been gradually synthesized in the core of stars. Then, life began initially in a very simple way, and then has been progressively leading to more and more complex structures and living beings since then. The appearance of man lead to a new emergent phenomenon, that of mind, which is rooted, in turn, in the previous non-living and living processes.

Then, my proposal is that we can talk of an emergent monism. All that exists in space and time are the basic particles and their aggregates that could be recognized by physics. There is only one kind of stuff in the world, and this stuff is the one recognized by the method of physics. This tenet prevents the introduction of dualisms and vitalisms in the emergence debate. There is only one kind of substance that constitutes the world, and that substance is the one characterized by physics. However, this does not forbid us to state that physics cannot be accountable for all the processes, events and items we find in the world. Indeed, the organization of physical matter reaches such levels of complexity and relations in order to allow novelty to arise[43]. Then, «we can understand the emergence of biological systems as both something new (and

---

[42] Cf. G. Auletta, *Cognitive Biology*.

[43] «Reality is ultimately composed of one basic kind of "stuff". Yet the concepts of physics are not sufficient to explain all the forms that this stuff takes –all the ways it comes to be structured, individuated, and causally efficacious. The one "stuff" apparently takes forms for which the explanations of physics, and thus the ontology of physics (or "physicalism" for short), are not adequate. We should not assume that the entities postulated by physics complete the inventory of what exists. Hence emergentists should be monists but do not need to be physicalists in the sense that physics dictates their ontology». P. Clayton, «Conceptual Foundations of Emergence Theory», 2.

therefore displaying properties and principles that are additional to) and in accordance with the laws of physics (as well as of chemistry)»[44].

Emergent monism acknowledges that there is just one kind of matter in the universe, but that matter can be structured in order to give rise to new structures and properties. Then, if novelty is not due to a different kind of stuff but to the rearrangement of the already existing matter, the emergence of that novelty has to be linked to the concept of information. Emergent systems deal with their items informationally; that is, they use them not according to the matter they are made of, but to the role they can play in the system.

The new structures emerged are new phenomenon, but they depend on what has previously emerged and on the nature of the most basilar elements of the universe. However, emergent properties are irreducible to, and unpredictable from, the lower-level phenomena from which they emerge. We cannot predict which properties would emerge from certain elements that are structured in a certain complex way. Once we notice the emergent phenomenon we can conclude that, given the same conditions and the same elements, we could expect the same phenomenon to emerge.

Once novelty has emerged, it can have some causal influence over the lower levels out of which novelty arose. That means that higher-level entities have causal powers over the lower-levels' items. Moreover, this claim implies the existence of different levels into which reality is divided. There are levels of organization where a whole in a certain level is, in turn, part of a whole in the next level.

> The doctrine of emergent evolution holds that, although the fundamental entities of this world and their properties are material, when material processes reach a certain level of complexity, genuinely novel and unpredictable properties emerge, and that this process of emergence is cumulative, generating a hierarchy of increasingly more complex novel properties. Thus, emergentism presents the world not only as an evolutionary process but also as a *layered structure* –a hierarchically organized system of levels of properties, each level emergent from and dependent on the one below[45].

In the following section I shall show how these causal powers act upon the lower levels. This would compel us to revisit our concept of causality and to

---

[44] G. Auletta, *Cognitive Biology*, 60.

[45] J. Kim, «"Downward Causation" in Emergentism», 121-122.

explore the new forms of influence between different organizational structures. We shall see that, to understand downward causation in biology, notions such as "information" and "function" must be introduced, as will be done in the next chapters.

4. **Top-down causation**

In the previous section we have focused on the concept of emergence and saw how the proper arrangement and structure of matter makes a difference regarding the parts taken separately. In this section we want to go further and show how the emergent levels have causal powers and can influence the lower ones. This will led us to reshape our conception of causality and discover that there are other kinds of causes besides the efficient and the material ones. When talking about top-down causation, there is always a certain suspect about what sort of cause we are talking about. Auletta and colleagues points out that difficulty:

> The majority of scientists think that top-down causation means to act directly on a lower ontological or less complex level of reality without the intermediary of causes acting at this same level. Indeed, a majority of both top-down causation supporters and detractors assume that it substitutes for the specific modes of action at lower ontological levels. It is easy to see that in this way, the principle of the closure of the physical world from the point of view of action and interaction would be violated in most cases, and at least in those where actions according to goals are involved[46].

It is very important to make it clear that top-down causation is not a cause that substitutes the classical causes, nor is it a cause that can have a direct influence from one level onto another. Top-down causation does not violate the physical closure of the world. We do not need to postulate any special entity with spooky properties that allows causality to jump between levels and to have incidence bypassing the physical laws. Top-down causation uses the causes already existing in each level and establishes a relation and a mutual influence between them. Davies clarifies that

[46] G. AULETTA – G.F.R. ELLIS – L. JAEGER, «Top-down Causation by Information Control», 3.

> top-down talk refers not to vitalistic augmentation of known forces, but rather to the system harnessing existing forces for its own ends. The problem is to understand how this harnessing happens, not at the level of individual intermolecular interactions, but overall – as a coherent project. It appears that once a system is sufficiently complex, then new top-down rules of causation *emerge*[47].

Howard H. Pattee has stressed that there is a difficulty in understanding the concept of top-down causation. This difficulty is linked to what we have already pointed out about the independency with which the different levels of organization are studied by the special sciences. It is not easy to see the causal link that emerges when all this levels are interacting together if the only way we have to study them is by separating them[48].

Again we have the necessity of a whole approach to nature and to study it with categories that can be applied to all levels of organization. Top-down causation is one of those categories, since the causal relation between different levels of organization can be established, whatever the levels under consideration were. Then, it is not a privileged level that has this particular causal power, but each level can act upon the one immediate below and can be influenced by the one immediate above.

4.1 *Levels of emergence*

Before entering the characterization of top-down causation, let us look more carefully to the layered structure of nature. When nature is studied in depth, different levels of organization are discovered. Indeed, we can see that there is a scale of organization and complexity from the most fundamental physical particles to the whole universe. A classification of these levels of organization could be as follows: sub-atomic particles, atoms, molecules, cells, organisms, communities, ecosystems and, finally, the universe. However, this classification is just phenomenological. It can give the impression that it is only at the end,

---

[47] P. Davies, «The Physics of Downward Causation», 48.

[48] «Downward causation is a difficult concept to define precisely because it describes the collective, concurrent, distributed behavior at the system level where control is usually impractical, rather than at the parts level where focal control is possible. Downward causation is ubiquitous and occurs continuously at all levels, but it is usually ignored simply because it is not under our control». H.H. Pattee, «Causation, Control, and the Evolution», 70.

when we consider the universe as a whole, that everything fits as a coherent structure. From that point of view, it could be thought that the universe is the entity that steers the organization of the other levels of organization. I do not want to give the impression that I am postulating the universe as a sort of agent that determines the dynamics of emergence and the structure of the levels of reality. Science has established the layered structure of the universe according to the criterion of size. However, the emergence of novelty has little to do with size, but with complexity; that is, the relations established between different items, both at the same and at different levels.

It is easy to notice that every level of organization has its proper sciences that study it. For example, the molecular level is studied by biochemistry and molecular biology, while botany, zoology and microbiology are interested in the organism level.

Hence, there is certain correspondence between the ontological layered structure of nature and the epistemological approach that sciences took to each level. Special sciences try to delve into a certain level of organization and investigate how things work on that level. This approach has been used since the scientific revolution of the 17th century. New sciences appeared and they distributed among them the levels of organization each one has to focus on. The problem is that those different levels were seen as separated compartments that can work on their own. No attention was paid to the linkage between them. It was tacitly assumed that if the different levels worked when studied separately, things should also match when they were put together.

The concept of emergence, with that of levels of organization, allows us to have a more complete picture of how things work in nature. With this conceptual framework it is feasible to think of an interconnected network of levels of organization that influence each other. Lower levels are conditions of possibility of higher ones and these higher ones can, in turn, influence the behaviour of the lower ones. There is a causal relation between these different levels. We do not have to think about the lower levels as something that is contained in the following levels, as if lower level items lose their properties and characteristics when they are embedded in a higher level of organization. If it were so, there would only be a bottom-up relation between them. What I want to stress here is that even a single atom allows new properties to emerge when it is placed in a new context provided by higher levels of organization. Emergence is not a process of construction of reality from the smaller elements to the bigger ones; it is a process of emergence of novelty due to the new relatedness that is established among the different levels of reality. The point to understand emergence is not only a question of scale but also of complexity. New properties emerge because of the complexity of higher levels, which provide the surrounding conditions where new relations can be established with lower level items. The novelty emerged is not the result of an additive process of certain items constructing a bigger structure, but a new relation established between different levels of reality, between items with different degree of complexity. The key of novelty is relatedness, not the mere addition of items.

Conway Lloyd Morgan postulated the existence of such levels of organization when he developed a primary version of what was later known as top-down causation. Lloyd Morgan was interested in the relation between those levels. It is clear that the lower levels are the condition of possibility of the higher ones, so some sort of bottom-up causation is at work. But what about the higher levels? Have they any causal influence over the lower ones? They should, if we do not want emergence to be an epiphenomenon. Lloyd Morgan says that «when some new kind of relatedness is supervenient (say at the level of life), the way in which the physical events which are involved run their course is different in virtue of its presence – different from what it would have been if life had been absent»[49].

Higher levels have causal influence in the way processes at lower levels take place. There is no violation of lower level laws, but the presence of higher levels make the laws rule in a new manner. New levels of organization harness the lower level events. For example, consider the role of calcium in inorganic and organic chemistry. In the non-living world we can find calcium mostly as forming the calcium carbonate compound, found in some types of rocks. On the other hand, even if it could also be found as a structural element in the bones of vertebrates, the role of calcium in the living world has properties that are not seen in the inorganic world. For example, calcium is an essential partner in muscle contraction; it is also one of the most important second messengers involved in cellular signal transduction and plays a role in cell adhesion mechanisms[50].

It has been said that the action of downward causation by the higher-level is to restrict or narrow the possibilities of the lower-level items. I think it is true,

---

[49] C. Lloyd Morgan, *Emergent Evolution*, 16.

[50] Cf. B. Alberts– al., *Molecular Biology of the Cell*, 660; 912-916; and 1135-1136, respectively.

but i retain that a more appropriate formulation could be found. In fact, what the higher-level does is not just a restriction of possibilities, but a canalization of the lower-level traits that have special significance for the higher level. Then, the aim of downward causation is not restriction but selection of possibilities of the lower level (and to do that, of course, we need to somehow restrict or constrain them, as we shall see).

For example, in the case of calcium and its role in muscle contraction, there have to be some conditions to fruitfully use calcium properties within the other partners involved in muscle contraction: the cytoplasm has to have particular properties to have calcium in the form Ca^{2+} and prevent its combination with other molecules; the structure of the endoplasmic reticulum is also needed, since calcium is stored there until it is released in response to a proper signal. So we can see that some particular environment has to be preserved and also the spatial separation and the timing of the events have also great importance to canalize calcium into the network of events responsible for muscle contraction. Then, the role that higher-level entities can play over lower ones is context-dependent. The insertion in a different context makes calcium acquire a new relatedness. The context of the cell, with all its structures and properties, provides calcium with the surrounding conditions to perform a different role if compared with the role it had in the non-living world. Being in a different system, in relation with different partners, allows novelty to emerge; not because something different has been introduced, but because a new context and relation has been provided. Higher levels canalize lower-level properties giving them the proper boundary conditions for the possibility of novel events.

That is the reason why the function of calcium for the living beings is not reducible to the chemistry of that element, because it does not depend on its properties, but on the context into which it is inserted. A new context and the interaction with new partners is what make novelty emerge. This characterization of emergence allows for the opening of the door to an actualization of the Aristotelian formal causality.

4.2 *Weak downward causation*

In this form of top-down causation, the higher level is seen as a structural instance that organizes the form into which the lower-level items have to be arranged. The higher level gives a pattern, a way of ordering the lower level items. There is a novelty in the upper level regarding the lower one. However, although this novelty is unexpected, it is deducible from the conditions of the lower level. One way for representing weak top-down causation is using phase-space models, as Claus Emmeche suggests:

> Phase space maps all the possible states of a system into a space defined by a set of dimensions, each of them corresponding to a parameter of the system. Through a continuous change in these parameters, any change in the system will be modelled by a trajectory in the phase space. Classical conservative mechanical systems will result in one distinct trajectory through the phase space, but various dampened, thermodynamic systems lose energy all along and may approach the same behavior as systems with other initial conditions. An *attractor* is the name of a set of points in the phase space in which trajectories with many different initial conditions end[51].

Stuart Kauffman compares the dynamics of the attractor to that of water moving from mountains to lakes and valleys. Lakes are point attractors where the water will end if it has the opportunity to move due to a perturbation. Peaks are very unstable points where, if there is water, just a little perturbation is needed in order to abandon that point and go down towards a more stable state. When water is allowed to have a dynamical behaviour, it will move towards the phase-space of the mountain's shape reaching more and more stable states[52].

We should not think of an attractor exerting some strange force over the items of the system. It is just a structure capable of adjusting its own stability between a certain range of conditions. There is a window of tolerance within which the attractor can "buffer" the effect of the perturbation and recover the stability of the system. There can be such strong perturbations to which the system cannot respond to and, consequently, the attractor would be disrupted.

I think that this kind of downward causation could be linked to supervenience, since there is no higher level influence in a strong sense, but just a phase-space through which the system can adopt different configurations based on the energetic balance.

---

[51] C. Emmeche – S. Koppe – F. Stjernfelt, «Levels, Emergence, and Three Versions», 26-27.

[52] Cf. S.A. Kauffman, *The Origins of Order*, 175-177.

4.3 *Strong downward causation*

Emmeche talks about strong downward when «a given entity or process on a given level may causally inflict changes or effects on entities or processes on a lower level»[53]. As I have said, these changes cannot violate the closure of the physical world; which means that each level cannot directly produce a causal efficient alteration in another one.

One of the difficulties with strong downward causation is that it seems as if firstly, the system has to be assembled with its parts, and only after that has been achieved, would it have some causal role from above. It can be thought as a temporal sequence: initially there is the bottom-up causation of the fundamental items of the system; and then, once we have the whole, it can begin to act downwards. From a biological point of view we may think that physics and chemistry do their work when a biological system is being made. And then, there is a biological influence over the physics and chemistry of the items that have allowed the biological system to emerge.

It is important to notice that the biological system is not first constructed physico-chemically and then it can exert some downward effect biologically. The point is that the biological system is constituted by physico-chemical items that are organized in a certain structure constrained by the boundary conditions imposed by the system. Moreover, the progressive constitution of the biological systems, which is reached through a spatial-temporal sequence of events known as development, means that each step of such development leads to new structures that will exert new constraints for the further elements that have to be introduced in the next developmental stages. Then, even if we cannot talk about a temporal precedence of the whole with respect to the parts, there is a logical priority of the former over the latter[54].

Another characteristic of strong downward causation is that its effect is not deducible from the initial conditions of the system and the rules governing the lower level items. There is an intrinsic non-deducibility. David Chalmers compares strong and weak downward causation from that point of view:

---

[53] C. EMMECHE – S. KOPPE – F. STJERNFELT, «Levels, Emergence, and Three Versions», 18.

[54] Cf. C. EMMECHE – S. KOPPE – F. STJERNFELT, «Levels, Emergence, and Three Versions», 21-23.

> To be clear, one should distinguish *strong* downward causation from *weak* downward causation. With strong downward causation, the causal impact of a high-level phenomenon on low-level processes is not deducible even in principle from initial conditions and lower-level laws. With weak downward causation, the causal impact of the high-level phenomenon is deducible in principle, but is nevertheless unexpected. As with strong and weak emergence, both strong and weak downward causation are interesting in their own right. But strong downward causation would have more radical consequences for our understanding of nature[55].

I think, however, that the classification of top-down causation into weak and strong is not of great help. The classification is normally proposed according to the unexpectedness of the emerging event and its non-deducibility. But the question about the novelty is something that has to do with the concept of emergence. On the other hand, when we talk about how this novelty exerts some causal influence over the systems, we have to introduce the concept of top-down causation.

The novelty that has emerged has to have causal powers over the lower levels out of which this novelty has arisen. That is the reason why, to my mind, it is only worth speaking of emergence and top-down causation in their strong version. Assuming the role of top-down causation as a causal power on its own, the category of causality can be revisited.

5. **Causality revisited**

I think that the notion of top-down causation opens new possibilities to rethink causality. It is not here the place to explain and comment the four causes according to Aristotle in full, but just to show that they are not enough to account for the richness of the concept of cause I am dealing with. Aristotle distinguished four causes: the material cause, the one out of which things get started; the efficient cause, which is the principle of changing or rest; the formal cause, which makes an object be that object; and the final cause, the end for which a thing is done[56].

It is well known that modern sciences only pay attention to efficient cau-

---

[55] D.J. Chalmers, «Strong and Weak Emergence», 249.

[56] Cf. Aristoteles, *Metafisica*, 983a24-30.

sation, and left out the other three causes. This is due to the deterministic vision that has pervaded sciences since the 17th century. That is the reason why thinking about causality was linked to the notion of determinism. But this is not how things go. I prefer the definition of causality offered by Auletta, who says that causality is what *concurs* to produce a certain effect[57]. It is worth noting that the verb used is to *concur*, not to *determine*. This is a broader framework for the notion of cause and gives the opportunity to deal with causality in a more realistic approach. Indeed, it is very difficult, at least in biology, to ascribe a certain effect to a given cause. We can find many examples where a cause could lead to multiple effects and where an effect could be the result of many concurring causes. Causal relations are not always one-to-one, but many-to-one and one-to-many.

Let us follow Auletta in his characterization of causes and see how this new approach to causality fits much better with the phenomenology of the living world and helps top-down causation to find its place in it. Firstly we can identify three types of main causes: a) formal causes, which are constraints imposed from higher levels upon lower ones; b) material causes, such as conditions of possibility out of which higher levels could emerge without violating lower level laws but also not being reduced to those levels; and c) dynamic causes, which is causality understood as determining a certain effect[58]. The effectiveness of the dynamic causes is due to a physico-chemical interaction among the items at stake. We can think of a system of balls exchanging kinetic energy through their physical interaction, or of a chemical system where the dynamic causes imply the exchange of matter and energy when chemical bonds are formed and disrupted.

This classification allows us to see the causes that operate at each level of organization we are considering. Formal causes are top-down causes, material causes have a bottom-up direction, and dynamic causes act at the same level. Only dynamic causes can have an efficient effect because they act at the same level and respect the causal closure of the world. If we want to have an effect in a given level, the only way to achieve it is through dynamic causation, because it is that kind of cause that havs an effective exchange of energy and momentum

[57] Cf. G. Auletta, «How Many Causes Are There?», 44.
[58] Cf. G. Auletta, «How Many Causes Are There?», 45.

between the items that are involved. However, this causation can be influenced from the bottom (furnishing the conditions of possibility of the material cause) and from the top (constraining what is emerging from the lower level)[59].

Dynamic causation is then what does all the work, at least if we understand work as the classical efficient cause. But there are other kinds of causes besides the efficient one that could also be classified as dynamic. One of them is circular causality, which involves feedback circuits, allowing separate items to become members of the same network and be regulated as a whole. This kind of causality can determine self-increasing processes. An example of circular causality in biology could be any cyclical metabolic pathway, such as the Krebs cycle. The different metabolites that are produced and consumed in the cycle are circularly related, and the concentration of some of them regulates key points of the cycle in order to activate or stop it[60].

Finally, inside the dynamic causes we can also find final causes, which can be teleonomical, teleological and intentional. Teleonomical final causes are found in systems that are dynamically conveyed to a local attractor. The system can follow different paths, but it is canalized towards those that led to those attractors. It is worth noting that the attractor is nothing out of the system, or separated from it. The attractor is the system conveying its elements to a final state. The system is not only a formal cause, but can also regulate the evolution of the elements in achieving the final state. These attractors canalize the system through the different possibilities of the phase-space.

Teleology, for its part, goes a step further and consists in selecting from the environment the information that is useful for the system. At this point items are no longer considered according to its physical and chemical properties, but are informationally treated. Here the system selects a certain input according to some standards that have to be reached. Then, inputs are compared with that standard, which is the survival of the system. In teleology, a system controls another system in order to select the inputs that are useful for maintaining the former[61].

Intentionality, in the end, is a form of final cause present only when two

---

[59] Cf. G. Auletta, «How Many Causes Are There?», 45.

[60] Cf. J.M. Berg – J.L. Tymoczko – L. Stryer, *Biochemistry*, 475-501.

[61] Cf. G. Auletta, «How Many Causes Are There?», 46.

individuals can share intentions about a third item. Until now, this can only be ascribed to humans. «It is interesting to notice that material causes are bottom up while formal causes are top down. Of the dynamic causes, efficient, circular, and teleonomic causes are at the same level, while teleologic and intentional causes are both top down»[62]. Intentionality will not be dealt with in this work, since I am interested in how emergence and top-down causation take place at the molecular and cellular level.

The point here is to make clear that the only cause that can have an efficient incidence at a given level is circular causality, because it is the only one that acts within a certain level of organization. Then, the question is how the remaining causes can influence circular causality and then, indirectly, exert some kind of influence over the items of a certain level. These causes play their role as constraints.

5.1 *Constraints*

The concept of constraint was first used in physical mechanics. The motion of a pendulum and the rolling of a ball are constrained by the geometrical conditions of the environment in which the motion takes place. It is not a force what constrains the elements of the system, but the way they are related and connected. «The orderly context in which the components are unified and embedded constrains them. Constraints are therefore relational properties that parts acquire in virtue of being unified – not just aggregated – into a systematic whole»[63]. The initial conditions of a system constrain its future dynamics.

Alicia Juarrero points out an important issue in the problem of constraints. Indeed, to constrain a system means to lower the degrees of freedom of that system, since constraints narrow down the possibilities of the system. Then, if constraints are acting in nature it would be expected that organisms would be more and more constrained and, therefore, less able to perform a wide variety of actions. However, this is not what happens in nature; quite the opposite happens. Consequently, «some constraints must therefore not only reduce the number of alternatives: they must simultaneously create new possibilites»[64].

---

[62] G. Auletta, «How Many Causes Are There?», 47.

[63] A. Juarrero, *Dynamics in Action*, 133.

[64] A. Juarrero, *Dynamics in Action*, 133.

As I will show below in Shannon's theory of communication, the equiprobability of alternatives in a message (or a system) allows us to have all the potentiality of the combinations that can be made. But precisely because the combinations are equiprobable, they cannot carry any message. «Constraining the number of ways in which the various parts of a system can be arranged reduces randomness by altering the equiprobable distribution of signals, thereby enabling potential information to become actual information»[65].

Juarrero distinguishes two kinds of constraints: context-free and context-dependent. She compares context-free constraints as the constraint imposed by a piston over a certain gas. The gas is compressed to adopt a non-random distribution of its molecules, since they are confined to the limits defined by the piston, and cannot readopt the range of configuration to which they had access when the volume was greater. If we want to give a biological example, we can say that the cellular membrane is the most important context-free constraint. It puts together everything that is inside it, facilitating the interaction of all the molecules that are there. The membrane physically constrains all the molecules to be together in a common space, allowing them to easily perform their reactions because of their spatial proximity.

> As soon as the membrane has been created, it acts as a macroscopic constraint, harnessing the motion of the microscopic components within it, in a very simple but important way: avoiding diffusion, and hence, increasing reaction rates (and even allowing them). In other words, the membrane should be considered as a functional structure[66].

However, we cannot talk of a greater degree of organization in gas. «If nature rely solely on context-free constraints to create form, matter might clump and agglomerate, but differentiation and complexity would not exist»[67]. Then, we need a kind of constraints that curtails randomness, but at the same time has to deal with certain possibilities in order to generate novelty. To achieve that, context-sensitive constraints are required.

The need for context-sensitive constraints is clear when we think of lan-

[65] A. Juarrero, *Dynamics in Action*, 134.

[66] A. Moreno – J. Umerez, «Downward Causation at the Core», 112.

[67] A. Juarrero, *Dynamics in Action*, 136.

guage. When we receive a given letter from a message, depending on the language in which it is written, there would be a certain probability of guessing the next letter. For example, in English, it is highly probable that if we read a "q", the next letter would be a "u". The probability of having a "u" is constrained by the fact that there has been previously a "q". This interdependence allows languages to increase their possibilities of communication. «Without contextual constraints on sounds and scribbles, communication would be limited to a few grunts, shouts, wails, and so forth that would be severely restricted in what and how much they could express»[68].

In nature, complex dynamical systems emerge when the items of that system are mutually related and impose certain constraints. The behavior of the system depends not only on the spatial context given by the rest of items, but also on the temporal context where the action takes place. That is why feedback loops are so important in the dynamics of the system. A feedback loop allows integrating past events into the present state of the system. And the present state of the system can have a backward influence on the former steps. Then, those kinds of systems are context-sensitive to their very history, because the feedback circuits embody past events and give the actual system its boundary conditions depending on those past events[69].

Juarrero further divides context-sensitive constraints into first- and second-order contextual constraints. First-order contextual constraints operate at the same level of organization. Second-order contextual constraints emerge as constraints acting upon lower levels. Constraints reorganize the lower level, opening it up as a pool of alternatives to which the emergent structure can access. The structure imposes constraints restricting the degrees of freedom of the items. Then, the flow of matter and energy cannot take all the possible ways[70].

> The constraints that wholes impose on their parts are restrictive insofar as they reduce the number of ways in which the parts can be arranged, and conservative in the sense that they are in the service of the whole. But they are also creative in a different, functional sense: those previously independent parts are now components of a larger system and as such have acquired new functional roles. The newly cre-

[68] A. JUARRERO, *Dynamics in Action*, 138.
[69] Cf. A. JUARRERO, *Dynamics in Action*, 139-140.
[70] Cf. A. JUARRERO, *Dynamics in Action*, 142-143.

> ated overall system, too, has greater potential than the independent, uncorrelated components[71].

I think it would be useful to clarify Juarrero's last quotation. Indeed, it could seem as if the parts of the whole, once they are integrated in the system, have acquired a new function. But things do not go exactly like that. What happens is that a given part, when inserted into a system, is no longer the same part that it was when considered alone. It is the new relation acquired with the partners of the system that allows a new function to emerge. And this novelty is irreducible to the parts of the system because what has emerged is precisely a new relation between the items, not just a sum of their properties.

Then, constraints can harness the organization of the system in order to make room for new functionalities. It is a creative way to arrange and select some conformations of the system in order to achieve certain functional characters that otherwise would not be reached. We begin to see that the constraining activity of the higher levels upon the lower ones is not something blind or without an aim. On the contrary, this constraining has to be somehow linked to the functional needs of the system.

The emergence of new levels of organization imposes certain constraints over the lower ones. This allows the system to take control of its dynamics. When we think about physical laws, we have in mind their universality and inexorability. Laws always behave in the same way. However, focusing on biology we can see that life has local control over certain events. We can say that biology imposes certain constrains over physics in order to "harness" physical laws and link them to the needs of life to be maintained[72]. There is not a violation of laws, but a use of them in such a way that are canalized for the sake of the living being.

Moreno and Umerez have also reflected about the role of constraints in biological systems. These constraints do not violate physical laws at all. Again, the question is to notice that laws are deterministic, but they are not the unique source of causality.

> Physical laws are just able to determine very basic and general features within the realm of the possible. But they still leave plenty of room for equally possible ins-

[71] A. JUARRERO, *Dynamics in Action*, 144.

[72] Cf. H.H. PATTEE, «The Necessity of Biosemiotics», 116-117.

> tantiations which might be actualized further due to factors ranging from mere chance to any kind of selective process, passing through weaker types of constraining influence. Therefore, depending on the system under analysis, it may well happen that we are able to give a physical causal explanation which determines only a range of possibilities which need to be supplemented with an additional causal explanation at a different level[73].

Michael Polanyi says that, when we talk about life, we can see that «its structure is that of a boundary condition harnessing the physical-chemical substances within the organism in the service of physiological functions»[74]. Then the organism, as a higher-level structure, has a causal influence over the elements of the lower levels. We can say that the organism exerts some control over the lower-level elements that constitute it. However, «the control of a system by irreducible boundary conditions does not *interfere* with the laws of physics and chemistry»[75].

Küppers shows also that the impossibility to state the boundary conditions in biological systems makes them irreducible to physics and chemistry.

> Thus a selective self-organisation of the microstates in sequence space seems only to be possible under the conditions of non-equilibrium processes, in the course of which an a priori indeterminable number of microstates is narrowed down to a few

---

[73] A. MORENO – J. UMEREZ, «Downward Causation at the Core», 104-105. It is also noticed by El-Hani and Pereira: «Biological systems act as constraining conditions for the relational properties of their components. They are now controlled, in the sense that they cannot enter in any kind of relation, but rather their relational properties are spatially and temporally restricted by the organization of the higher-level system. A cell, for instance, causes its components to have a much more ordered distribution and function in time and space than they would have in its absence». C.N. EL-HANI – A.M. PEREIRA, «Higher-level Descriptions», 121.

[74] M. POLANYI, «Life's Irreducible Structure», 1309.

[75] M. POLANYI, «Life's Irreducible Structure», 1310. Davies formulates it as follows: «We do not need to discuss *two sorts of forces* even though we do need to invoke two aspects in the causation story. The molecule's motion is caused by the push and pull of neighbours, *in the context of* their own global, systematic motion. Thus a full account of causation demands appeal to (i) local forces, and (ii) *contextual information* about the global circumstances. Typically the latter will enter the solution of the problem in the form of constraints or boundary conditions». P. DAVIES, «The Physics of Downward Causation», 38.

> biologically relevant ones. The present results of the new paradigm of self-organisation show unambiguously that the process of molecular self-organisation is essentially subject to certain principles of selection and optimisation, and these can be reduced completely to the known laws of physics. However, the results also indicate an inherent limitation of the reductionistic research programme. Thus, although the existence of specific boundary conditions can be completely explained as a general phenomenon within the framework of physics, it is not possible to deduce physically their detailed structure. This has been revealed by a through analysis of evolutionary dynamics in sequence space. The fine structure of biological boundary conditions reflects the historical uniqueness of the underlying evolutionary processes, which, by definition, cannot be described by natural laws[76].

Having arrived at this point of my work, it should be clear that I want to propose a philosophical framework that does not want to go against physicalism, nor to be reduced to it. I feel comfortable with the expression "emergent monism", because it points out that the novelty that emerges is not contradictory to the maintenance of the physical laws that rule the most fundamental items of the universe. If they were, I would be defending a philosophical approach that wants to bypass the chemical and physical laws. Then, I would have to postulate some other items that follow different rules and that play their role without taking into account the laws that govern their basilar elements. This is exactly what I want to avoid. However, I do not want to go to the other extreme position, which sees physicalism as the only real force at play. From that point of view, new patterns of organization and new properties would be seen just as an epiphenomenon; that could be useful from an epistemological point of view but, in the end, what is doing all the work is just physics. Novelty can be acknowledged, but it does not perform any real causal work.

Then, if emergence does not produce a novelty that can override physicalism, but what emerges cannot be reduce to physics, how does it work? Robert Van Gulick offers an interesting approach:

> One possible solution would focus on the respect in which higher-order patterns might involve the selective activation of lower-order causal powers. Mircroproperties that were causally irrelevant in most configurations, for example because

[76] B.-O. KÜPPERS, «Understanding Complexity», 255.

> their random actions cancelled out and had no significant overall effect, might exert a powerful causal influence in a small range of cases involving higher-level patterns that brought those micro-powers into a coherent mode of action making a major difference to the overall operation of the system[77].

Van Gulick proposes a model where there is a selective activation of causal powers. Lower levels would have a certain causal capacity to perform different acts. What the higher level would do is to select certain causal powers offered by the lower level. Then, «higher-order properties act by the *selective activation* of physical powers not by their *alteration*»[78].

I think, however, that Van Gulick's approach is right, but the notion of "causal powers" is rather vague and can lead to think only of efficient causation. El-Hani and Pereira, for their part, do not identify novelty with the emergence of a new efficient causation. Indeed, they think of the novelty emerged from the lower levels as some form of constraints that organize in space and time the traits of the lower level items and their way of interaction. This novelty has to be understood «as a kind of formal and functional causation, that is, as a new mode of coordinating and controlling the relational properties of the components that comes into being as a higher-level system appears in the evolutionary course»[79].

In this case the causal powers that emerge are not efficient ones, but formal and functional. What these new causal powers do is to coordinate and to control the properties of the lower level items, in order to be activated when they are needed. However, this model is not fully satisfactory, since we do not know exactly what is selected and which criteria are used to select one or other property to be activated. Further philosophical concepts, proposed in the next part of this work, are needed in order to clarify how this selection takes place.

On the other hand the notion of "selective activation" can be misleading. Indeed, higher levels do not directly activate some properties of the lower levels that have been dormant until then. Higher levels provide a contextual environment that introduces a new relatedness between the already existing elements. Then, rather than an activation, it would be better to talk of a new context in which a given element establishes new relations and is canalized by higher levels.

---

[77] R. Van Gulick, «Reduction, Emergence, and the Mind/Body», 64-65.

[78] R. Van Gulick, «Who's in Charge Here?», 83.

[79] C.N. El-Hani – A.M. Pereira, «Higher-level Descriptions», 122.

Summing up, it can be said that there is a dynamic interplay between random happenings in the lower levels and formal constraints from above that integrate those novel events in order to lead the system to a further stable state. Then, there is a constant dynamics of the system, which from that point of view could be seen as a transient complex of interrelations that integrates novelty through the constraining of the already existing structures[80].

5.2 *Top-down causation in biology*

At the end of this chapter, I want to stress the fact that top-down causation is a very important epistemological and ontological category that can be of great help to study biological systems. Indeed, the cell is a complex network where all the elements within it are interconnected. We can identify different levels of organization in the cell, which can be taken as the ontological levels I have been talking about (see 4.1). The free atoms that are found in solution in the cytoplasm constitute the most basilar level of the cell. In the next order of complexity we find the molecules, formed by the bonding of atoms in a certain way. We can find mainly four classes of molecules in a cell: proteins, nucleic acids, carbohydrates and lipids. Those molecules can be of many different kinds, varying their complexity and structure. We can find also some molecules that are a compound of two different ones, as it happens with the proteins to which a lipidic molecule is attached.

The interactions and relation between molecules could be even more complex. In the case of the ribosome, which is the organelle where protein synthesis takes place, there is an assembly of protein and ribonucleic acid (RNA), yielding a complex structure that recognizes either the messenger RNA (mRNA) that has the coded information to be translated into protein, and the transfer RNA's (tRNA's) that carry the proper amino acids that are coded by the mRNA (Figure II.1). So here we can see an example of a multi-molecular assemblage that puts many kinds of items into relation. Moreover, is precisely that way of assembling that makes the ribosomal properties to emerge, in order to be able for interacting with the other partners that are essential for protein synthesis.

---

[80] Cf. G. Auletta, *Integrated Cognitive Strategies*, 140.

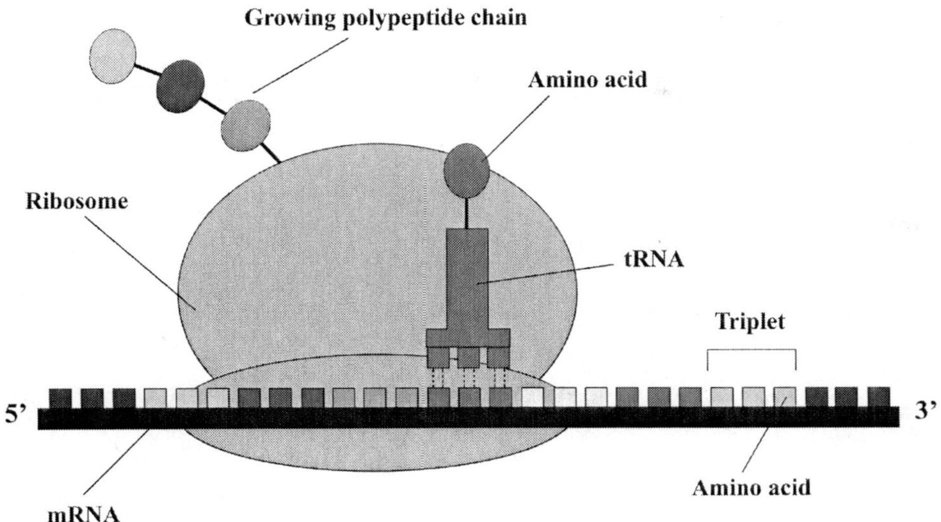

Figure II.1: The ribosome reads the nucleotides of the mRNA molecule. Each group of three nucleotides is a triplet and codes for a single specific amino acid, which enters the ribosome attached to a tRNA molecule. The ribosome provides the context where a tRNA can specifically read a triplet of mRNA and then its attached amino acid can be bound to the growing polypeptide, which grows one amino acid each step.

Groups of molecules could form bigger structures than the ribosome, as such as the cellular membrane, the mitochondrion (where there is the energetic conversion of the cell) and the endoplasmic reticulum (very important for protein maturation) to name a few. Those structures exert physical constraints over the lower level items. For example, the mitochondrion establishes an ionic gradient between the cytoplasm and its inner space. This makes the mitochondrion the place where the energy of lipids and carbohydrates is converted into ATP, the energetic storage molecule that the cell uses when it deploys reactions that need energy.

We can also consider the cellular level as a further ontological level, where all the previous ones are included and related among them in order to allow new properties to emerge, if compared with those levels taken separately. For example, only the cell as a whole can integrate the energy needed to produce a protein, the molecular machinery to synthesize it, the signals in order to drive it to its proper place, and the necessary elements to regulate it in order to deploy its function in the cell properly. The cell as a whole canalizes all these elements, which pertain to different levels of organization, in order to achieve the result that is important for it.

From that point of view, we have to think in those levels of organization not as something perfectly aisled. Indeed, higher levels are the way lower level items can be in such a tight relation in order to concur to causal efficiency, which in biology means that there is rupture or formation of chemical bonds. The ribosome is a structure that canalizes mRNA and amino acids in order to put them close to, step by step, make the polypeptide chain. The information of the mRNA dictates the order in which amino acids have to be added. In the end, the ribosome establishes chemical bonds between amino acids, but it has to do it according to the information given by the mRNA; and then, to perform it properly, it needs to have a certain three-dimensional structure, not only to accommodate the molecules in order to be closer, but also to be regulated by other proteins and then be controlled and integrated into the whole metabolic network of the cell. The ribosome can be seen as a higher-level structure that exerts top-down causation over lower level items (mRNA and amino acids). It does not violate the closure of the molecular level, but just provides the contextual conditions in order to allow certain molecular reactions to take place.

We can see the issue from the point of view of emergence and from the perspective of top-down causation (Figure II.2). The ribosome is a structure that emerged at a certain moment in evolution. It is the result of the interaction and coordination of certain proteins and RNA molecules, which in turn have come from more basilar units such as molecules and atoms. A certain combination of those molecules has led to the emergence of the ribosome as we know it. So a certain combination of proteins and RNA macromolecules was the molecular level out of which emerged a new relation among them that yielded an organelle that can translate strings of mRNA in strings of amino acids. But this complex structure, in turn, can exert a top-down effect over mRNA molecules and the amino acids that match with mRNA's coded information. Then, the ribosome is a higher-level structure that has a top-down effect over the molecular level.

We can see that the ribosome can bind macromolecular structures like tRNA and mRNA. This binding is done in a way that allows the nucleotides of tRNA to establish hydrogen bonds with those of mRNA; and this specific recognition establishes the order through which amino acids are linked by the peptid bond to make the protein. In the end, higher-level instances facilitate the encounter

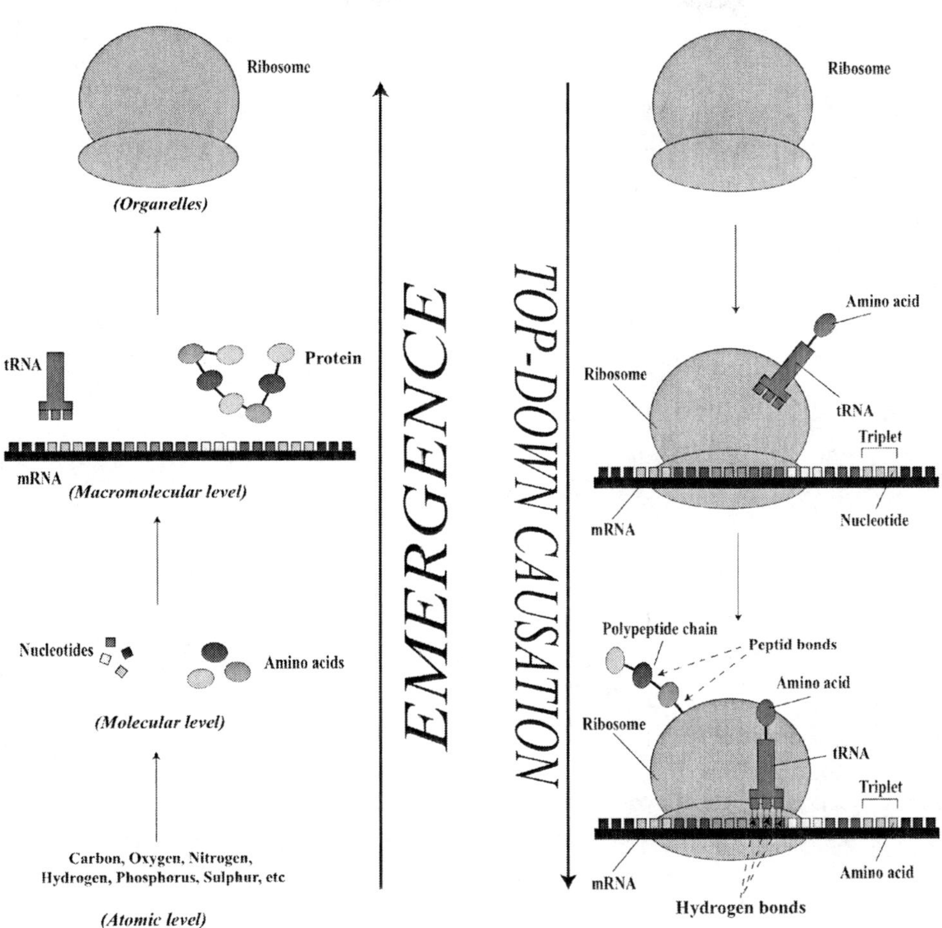

Figure II.2: Emergence has to do with the genetic process of constitution of the differents levels of reality. Here we have only focussed on the atomic level, the molecular, the macromolecular and the organelles, which are structures composed of different kinds of molecules. Top-down causation, for its part, is the causal power higher levels exert in order to put lower level items into certain contexts to deploy a certain reaction.

between the proper partners in order to allow reactions at molecular level to take place.

That is the reason why Moreno and Umerez say that the novelty that characterizes life can be explained in relational terms. That means that we cannot explain a character of a living system without making reference to all the rest of traits of that very system. We cannot understand the function of a given protein if we do not know how its gene is regulated, where it is synthesized, where it will perform its role in the cell and with which partners it would interact to control its own activity or that of other proteins. So to give a full explanation to the functional account of a given protein, we have to answer that the *cause* of that protein is the cell as a whole[81]. Davies expresses it as follows:

> Top-down talk refers not to vitalistic augmentation of known forces, but rather to the system harnessing existing forces for its own ends. The problem is to understand how this harnessing happens, not at the level of individual intermolecular interactions, but overall – as a coherent project. It appears that once a system is sufficiently complex, then new top-down rules of causation *emerge*. Physicists would like to know whether these rules can ultimately be derived from the underlying laws of physics or must augment them. Thus a living cell commandeers chemical pathways and intermolecular organization to implement the plan encoded in its genome. The cell has room for this supra-molecular coordination because it is an open system, so its dynamical behaviour is not determined from within the system. But openness to the environment merely explains *why* there may be room for top-down causation; it tells us nothing about *how* that causation works[82].

My aim in this work is to propose an answer to the question on how causation works, as asked by Davies. But if this new kind of causation works through the harnessing or canalization of certain lower level causal powers, this question has to be preceded by another one and would lead us into a further one. We have to see *what* is selected, *how* it is selected and *why* (or *for what*) it is selected. This is what we are going to deal with in the next part, but we can advance that information is what is selected, semiosis establishes how it has to be selected and functions are the criteria to decide why it is selected.

6. **Summing up**

It has been shown that the notion of emergence has its own history both in philosophy and science. The authors presented agree with the fact that in nature there is something new that comes from the organization of lower levels. This novelty could be seen just as something unexpected, but ultimately reducible to

---

[81] Cf. A. Moreno – J. Umerez, «Downward Causation at the Core», 110.

[82] P. Davies, «The Physics of Downward Causation», 48.

the lower levels, or as something really novel, not only unpredictable, but also with no way to establish a relation of derivability that would allow us to trace back all the novel characters to the features of the lower level items.

Emergence bets for something more, for features that appear when certain degrees of complexity are reached and critical thresholds are crossed. This novelty does not add any strange substance to the system, but exerts its causal influence and allows the system to make a difference, to do things that otherwise it were not be prepared to do.

Emergence of novelty implies a top-down causal effect upon the lower levels out of which emergence has come. This top-down causality does not violate the closure of the physical world, since its effect is exerted through constraints and selection of relevant information useful for the system.

SECOND PART

INFORMATIONAL PARADIGM

Chapter III

Information

1. The concept of information in biology

The concept of information does not have a very long history in science. It is a term that suggests something that has more to do with our everyday life than with the vocabulary of philosophy or science. The notion of information is present in many areas of our knowledge and we use it currently in many situations. We talk about the information read in the newspaper, the information stored in the libraries or in the strings of "0" and "1"s that make our computers work. Nonetheless, for not so many years, information has been pervading scientific literature from quantum mechanics to physics, chemistry, biology, neuroscience, ecology, and even to philosophy and theology[1].

It seems as if information strives to become a general category that could help us in our effort to unify our knowledge of reality. The fact that it is applied to a great variety of topics makes information a transversal category to many fields of knowledge. If information is something that could be predicated of items at different levels of organization and complexity, then it is worth considering it as a paramount epistemological category.

On the other hand, the promiscuous application of this concept in many fields could lead to misunderstandings and wrong applications of that concept from one issue of interest to another. My aim is not to study the concept of information and its use in all the fields mentioned above. I am going to focus my attention on the importance the concept of information has for molecular biology. Indeed, the informational approach in molecular biology is a new perspective through which we can study old problems from a different point of view. Information shall

---

[1] To get introduced to the concept of information as a common framework to understand the different levels of reality, see P. Davies – N.H. Gregersen, ed., *Infromation and the Nature of Reality*.

offer a new philosophical framework to scientists in order to better understand how life works. Of course, the term "information" has been present in molecular biology since the beginning of this discipline but, as it shall be seen, the historical framework in which biology coined the informational vocabulary exerted some influence in how those terms were understood.

The concept of information in biology is something rather elusive, difficult to conceptualize and to comprehend with the reductionistic tools with which biological disciplines have developed. However, if we consider what has been said in the last chapter about emergence, information could be seen as the *more* that emerges from the sum of the parts of the system; it is an emergent item, linked to matter, of course, but not to *that* matter. Information is something that emerges out from matter, but it refers to the relations established between different items despite the matter they are made of. Furthermore, information has causal incidence over the matter out of which it has emerged. Information, then, would not only help us to understand emergence, but would also be a key concept to better understand top-down causation. As it has been said, there is no strong emergence until what has emerged has causal powers over the lower levels out of which something emerges, and information could play an important role in pointing out which powers are at play and how they are canalized.

The introduction of the concept of information in biology in the way we are going to propose can help to comprehend the commonalities that characterize life and the principles that steer its organization. Information is not just something placed in the genes that have to be switched "on" or "off". The informational approach wants to widen the field of information to everything within the cell, not only its genetic material. Furthermore, the environment in which the cell is embedded also has an informational role for it. It could be indeed, a new paradigm to understand the whole biology from a different point of view.

2. **Information as a paradigm**

2.1 *Shannon's model of communication*

Let us begin considering what information is from the point of view of Shannon's mathematical theory of communication[2]. Claude Shannon was an

[2] Cf. C.E. SHANNON, «The Mathematical Theory of Communication».

engineer at the Bell Company and worked in the improving of communications. His model was focused on the problem of transmitting information from a source to a receiver. The source transmits information, which is conveyed by a channel in order to reach the receiver. During the transmission of information, some noise can interfere with the signal and, therefore, the receiver would not get exactly what had been sent by the source. Shannon was interested in reducing that noise. Thus, the aim of this model is to improve the accuracy of message's transmission and to assure that during its passage through the channel no information is being lost.

Shannon was aware that this concept of information has nothing to do with the meaning of the message sent by the sender. The model is not interested in the meaning of what is sent, but in maintaining the proper conditions during its transmission through the channel in order to avoid any lose of information. He was very conscious of the limits of that model, and he never tried to apply it to other fields outside of telecommunication[3].

It is worth noting that in this model of communication, what has to be transmitted is selected from the source. The message is clearly defined from the beginning, and it is that very message that has to be preserved from the interference of noise in order to reach the receiver as it has been sent by the source. Warren Weaver, who wrote a book with Shannon entitled *The Mathematical Theory of Communication* to make Shannon's model accessible to a wider audience, distinguishes three levels of the communication problem. In the first level there is the problem of accuracy, of a correct transmission of the message without lose of information; it is focused at a syntactical level, regarding only the order of the items in the message. The semantic problem arises in the second level, and is interested in how the items of the message convey its meaning. The third level has to do with pragmatics, the way in which the received message

[3] «The fundamental problem of communication is that of reproducing at one point either exactly or approximately a message selected at another point. Frequently the messages have *meaning*; that is they refer to or are correlated according to some system with certain physical or conceptual entities. These semantics aspects of communication are irrelevant to the engineering problem. The significant aspect is that the actual message is one *selected from a set* of possible messages». C.E. SHANNON, «The Mathematical Theory of Communication», 379.

alters the receiver[4]. Shannon and Weaver's work focused only in the first level, the syntactical one.

To understand the pertinence of this model, it has to be said that Shannon's work is historically framed a few years after World War II. He not only worked for the Bell Company, but also for the MIT (Massachusetts Institute of Technology), which have military applications to improve weapons and communications, including cryptography[5]. That is why he was interested in a model of communication whose main aim was the transmission of a message with high fidelity. This could have been the difference between victory and defeat at war.

Another aspect in which the model focuses on is that of the quantification of information. R. V. Hartley, who was also an engineer at Bell Labs, established a logarithmic law for information:

$$H = K \log s^n$$

H is the amount of information, K is a constant, n the number of kinds of items with which the message is made and s the size of a given message. Then, s^n is the number of possible messages of length s made with the different items n. As Shannon's model establishes that the source selects the message according to their relative probability, he uses Hartley's equation in this particular form:

$$H_n = -\sum p_i \log p_i$$

H_n is the information content, i's are the possible events, and the p_i's the probabilities related to those very events. Then, the information content tells us about the amount of messages a certain source can have; it is a measure of the capability of permutation of the items in order to generate different messages of a certain length. The amount of information is intrinsic to the characteristics of the source. Then, we have also here a measure of the *potential* information a

[4] Cf. L.E. KAY, *Who Wrote the Book*, 98.

[5] Shannon wrote also an article about encrypted communication: C.E. SHANNON, «Communication Theory of Secrecy Systems». To have a brief historical background about the importance of Shannon's work for military investigation is worth reading L.E. KAY, *Who Wrote the Book*, 91-102.

source can generate and eventually sent. The sent message is just a fraction of the whole amount of *potential* alternative messages that the source can generate. What is sent is the *actual* information that has been chosen from the source to be transmitted. Auletta has formulated it as the information accessibility principle: «The whole information potentially contained in a system may only be partially accessed»[6]. He characterizes this principle when dealing with quantum mechanics, but also opens the possibility to apply this principle to any system that deals with information.

2.2 *Information in thermodynamics*

It is not difficult to see that Shannon's formula for the quantification of information has elements that are formally similar to that for Boltzmann's entropy. Let us see then if Shannon's concept of information has some relation with entropy. To univocally characterize a physical system, a certain amount of information is needed. Information is a quantitative magnitude that establishes the necessary instructions to define a certain system. Information is required to univocally characterize the actual arrangement of the system that has been chosen from the whole potential set of possible arrangements. Werner Loewenstein explains the relation between information and the possible configurations of a system:

> In general, then, we compute the information inherent in any given arrangement of matter (or energy) from the number of choices we must make to arrive at that particular arrangement among all equally possible ones. We see intuitively that the more arrangements are possible, the more information is needed to get to that particular arrangement[7].

However, information is a dimensionless magnitude. Nonetheless, we can use entropy as a partner of information. If information is a measure of how the system is ordered, entropy measures the degree of disorder of such a system; that is, the amount of possible configurations a system can have. The entropy of a system increases with the disorder, increasing also the number of possible

[6] G. AULETTA, *Cognitive Biology*, 37.
[7] W.R. LOEWENSTEIN, *The Touchstone of Life*, 7.

configurations. A crystal of salt (NaCl) has low entropy, and that entropy increases when the same crystal is dissolved in water, since the Na^+ and Cl^- ions could adopt more configurations in solution than when they are fixed in a crystallization lattice.

From a physical point of view, information is a measure of the order of a system. If the items of the system can have many configurations, this means that the system is a *macrostate* that can adopt several *microstates*. Consequently, more information is needed to characterize all the possible configurations of the system. For example, the amount of information to characterize a crystal is less than the needed to characterize a gas. The atoms in a crystal have just one possible position in the lattice, which strictly defines their relations with the surrounding atoms. On the other hand, one would need more information to characterize a system such as a gas, whose atoms can adopt many spatial configurations at a given macrostate characterized by a certain pressure, temperature and volume.

Then, we can see that the information of a system has to do not only with a given configuration of the system, but also with all the possible arrangements the system may have. In the case of the crystal, the information of that very crystal is the same as the all-possible configurations, since a given crystal can only have one configuration, and then entropy would be zero. In the case of a salt solution, the number of possible configurations of the atoms in interaction with water is considerable higher, despite in a certain moment we see just one of these possible configurations. In that case, the actual information is just a fraction of the possible information.

Boltzmann establishes the relation between entropy and the number of possible configurations of a system in his famous equation:

$$S = k \log P$$

S stands for entropy, P is the number of possible configurations and k is a universal constant known as Boltzmann's constant. Then, we can see the relation between Shannon's concept of information and Boltzmann's entropy. Information refers to the amount of data needed to fully characterize a given system in a certain state. Entropy, on its side, gives us the rest of the alternative possible combinations of these data. In Terrence Deacon's words, «Shannon's measure of the potential information conveyed by a given message received via a given communication channel is inseparable from the range of signals that could have been received but were not»[8]. Shannon's quantification of information is a measure of the potential information of a system, the whole group of microstates that can be actualized from a given macrostate.

2.3 *The inadequacy of Shannon's information for biology*

The concept of information entered biology as a straightforward application of Shannon's model into the sciences of life. As we have said, Shannon thought that this model of communication could only be applied to the transmission of information between a source and a receiver. However, it was too suggestive to be left out when information was becoming an important concept in the biological field. It was at the end of the first half of 20th century when this conceptual framework was established in biology, making information a part of the biological vocabulary in a very precise sense, as Lily Kay remembers:

> The picture changed radically at the end of the 1940's; World War II and the subsequent militarization of science and culture in the cold war played a significant role in this shift. Several leaders of the information revolution had a major impact on the biological sciences and the social sciences, including molecular biology (still within the protein paradigm of heredity). Though the mathematical aspects of these works did not influence the technical content and experimental agenda of molecular genetics (as many information theorists had predicted) the discursive framework – the information discourse – that it stimulated did endure[9].

Applied to the molecular biology domain, Shannon's theory tells us about the all-possible ways to arrange a given message to be transmitted from one biological system to another or even inside the same biological system. However, as has been said above, Shannon's model does not consider the meaning of the message that is being sent from the sender to the receiver. Consequently, the application of this model to molecular biology would not ask about the meaning of what is transmitted. The important think is to univocally establish the information and to send it faithfully. It does not matter if the message has a meaningful content for the receiver or not. The point is just to transmit it without error from one place to another.

---

[8] T.W. DEACON, «What Is Missing», 159.

[9] L.E. KAY, *Who Wrote the Book*, 329.

One of the questions that arise when Shannon's paradigm is applied to biology is: who is the sender and who is the receiver? In Shannon's model it is clearly established, but it is difficult to see how this takes place in the cell. In the frame of military applications, it is assumed that the senders and receivers of the messages are human beings who have some previous knowledge about the apparatus used in transmission, and they have certain expectancy about what they are going to receive. The model does not say anything about the meaning of what is transmitted, but the fact is that both the sender and the receiver have some background about what is to be transmitted. It is easy to see that this model of communication cannot be applied to the cell. Biological systems such as cells do not have human intentionality. When talking about Shannon's model of information, Deacon says that this model «left us with a deflationary theory of information, from which content, reference, and significance are excluded and irrelevant»[10].

Susan Oyama also stresses that the communication model cannot be applied directly to molecular biology. There are some similarities, of course, but there are also some traits that are rather different.

> What corresponds to the source, for instance, that selects symbols from a set and encodes the message? One might argue that the zygote represents a source whose DNA is assembled at fertilization, or that the process of reproduction, or the species, or the gene pool is the source. The channel might be ontogeny itself. Where communication theorists deal with relatively closed systems, however, in which information may be degraded but not increased, ontogeny is an open one in which energy is dissipated but complexity often increases. The engineer's message can only be harmed or lost by transmission, while the ontogenetic "message" is created in transmission[11].

It seems as if all the process of information transmission were the development of the organism, the sender corresponds with the genetic and environmental load, with which it begins to develop, the channel would be the developmental process itself, and the receiver is the mature form of the individual. If we see the problem form an evolutionary point of view, the source would be the individual

---

[10] T.W. DEACON, «What Is Missing», 158.

[11] S. OYAMA, *The Ontogeny of Information*, 76.

and the receiver its offspring. The scheme is the same. Oyama warns against this straightforward application of that informational model because it does not take into account the fact that information, in biological systems, has an ontogeny. Biological systems do not just transmit information from one place to another, but they create their own messages depending on which content has to be transmitted.

Biologists, are aware of the fact that genes do not carry information only in Shannon's style. Genes also carry an instructional content, but also the application of this concept to genetic information is misleading in many cases. Instruction is thought to be the only causality genes can afford. From that point of view genes are not descriptions of the traits that are expressed on the phenotype, but the instructions that, once executed, lead to a certain phenotype. That is why Peter Godfrey-Smith warns that seeing things from this perspective, «if genes and phenotype do not match, what we have is a case of unfulfilled instructions rather than inaccurate descriptions»[12]. Genes are seen only as the instructions or the blueprint of the biological system. But the notion of "unfulfilled instructions" leads us to think of a reference link between genes and the final traits. From that point of view, there is a semiotic relation among them. We cannot trace a one-to-one correspondence between genes and traits; the relation between them is more complex, as it will be shown.

Furthermore, another very important difference is that Shannon's model selects the message at the beginning of the communicational act, which only consists in reducing the noise that could alter the message during its transmission through the channel. The most important question is the reliability of the message that has been chosen at the start of the transmission event. However, the cell is not prepared to react to every single change that takes place inside or outside its membrane. It would be impossible to process all the information the cell receives from outside. Indeed, what the cell does is to select those stimuli that are useful to it beforehand. One way of doing that, for example, is to put specific proteins in the outer membrane in order to bind certain molecules. Then, the cell is constantly scanning its environment, looking for molecules that could bring information of interest. Then, the important aspect of the information is not whether all the information of the environment is faithfully transmitted to

[12] P. GODFREY-SMITH, «Information in Biology», 107.

the cell; what really matters is that the cell is previously prepared to identify the relevant information according to its own criteria. So the selection is done at the end of the process, not at the beginning.

John Maynard Smith thinks that there are strong similarities between the translation of DNA into protein and the transmission of a human message by Morse code. The application, nevertheless, is not straightforward, since in Morse code the codification of the initial message and its interpretation by the receiver needs human beings. They are needed not only for the emission and the reception of the message, but also for designing the apparatus that codes and decodes the message. However, for Maynard Smith the question is not difficult to solve, since «where an engineer sees design, a biologist sees natural selection»[13]. Natural selection has produced the variety of possible DNA messages that are at the source, and has selected those that have "meaning", those that in some way favoured the survival of the organism. Moreover, Maynard Smith states that the relation selected between the message and the function to which is linked, is because of intentionality. Of course he does not say that we have to postulate that genes, proteins or cells have a sort of mind; in the case of molecular biology, intentionality should only be predicated from natural selection[14]. Finally, Maynard Smith says that the application of Shannon's model to genetics poses some problems, not with the transmission of information, but with its meaning. As he states, «in biology, the question is: How does genetic information specify form and function?»[15]. It is worth noting the effort Maynard Smith makes to deal with the concept of information in biology, but I think that it is misleading to talk about natural selection as having some sort of intentionality. He also tries to distinguish the very act of transmission from the importance of what is

---

[13] J. Maynard Smith, «The Concept of Information in Biology», 179.

[14] Cf. J. Maynard Smith, «The Concept of Information in Biology», 189-190. Maynard Smith explains the intentionality of natural selection as follows: «How, then, can a genome be said to have intentionality? I have argued that the genome is as it is because of millions of years of selection, favoring those genomes that cause the development of organisms to be able to survive in a given environment. As a result, the genome has the base sequence it does because it generates an adapted organism. It is in this sense that genomes have intentionality. Intelligent design and natural selection produce similar results». J. Maynard Smith, «The Concept of Information in Biology», 193-194.

[15] J. Maynard Smith, «The Concept of Information in Biology», 181.

transmitted; but this leads him to talk about *meaning*, a concept that I would not use in biology (at least at the cellular level).

I think that the Shannon's model cannot be applied straightforwardly to biology. It is needed to find another way to solve the problem. One of the main questions is how to think of a cell without using concepts that are devoted to human intentionality. Biologists have used "informational talk" for many years and this has led them to use anthropomorphic terms when they are referring to biological systems simpler than humans. My guess is that a broader concept of information has to be used in biology to avoid such misunderstandings.

2.4 *A new theoretical framework*

I do not want to reject Shannon's model, but just to propose a wider framework where the whole universe, and especially biological phenomena, can be better understood from an informationally point of view. Auletta points out the central issue of this informational model:

> The classical theory of information has supported the idea that information is a two-term process, namely, a process in which we have an input and a corresponding output. This is due to the circumstance that it has been formulated as a communication theory for controlled exchanges of information, so that, in general, we assume a determined output given a certain input. When, on the contrary, we deal with a general theory of information acquisition, we are always dealing with at least three terms: An unknown (not controlled) parameter, some data, and a certain information selection[16].

Then, information is not a question regarding just a sender and a receiver, but is something that involves three partners (Figure III.1). The processor constantly offers informational variations that change with time and is the starting point of the whole process. The processor provides the raw data, the huge amount of information that the system receives and generates. There is also a regulator, whose structure contributes to determine which information is acquired by the system; that is, which fraction of the whole potential information is accessible at a given time. The decider does the final act of detection, selecting a piece of

[16] G. AULETTA, *Cognitive Biology*, 47.

information and making the whole process irreversible. Then, the difference with Shannon's model is that this wider informational paradigm is based on a triad (processor, regulator and decider) rather than in a dyad (sender and receiver), and in the fact that the selection event takes place at the end of the process, not at the begining[17].

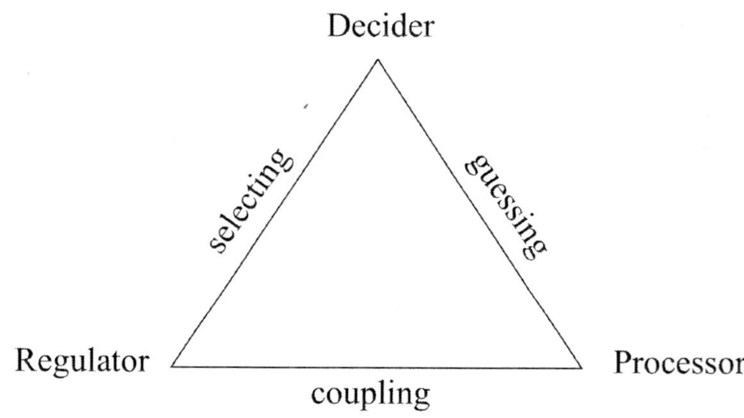

Figure III.1: Triadic structure of any dealing with information system. The processor provides the source of information, the regulator's structure allows some of this information to be actualized and the decider selects the information that can tell something about the processor.

The coupling between the regulator and the processor allows information to be acquired by the system. The decider selects the information that comes from the regulator. The information selected from the regulator is coupled with the raw data that come from the processor. Then, the decider, because of its linkage with the regulator, can make a guess about the processor. Even if one of the elements is know as the decider, it is not a model that has to do with information regarding intentional knowledge. This is a model that can be applied to any system. Then,

> any time that a decider (even randomly) selects and eventually stores some information – which, through an appropriate coupling, reveals something about another

[17] Cf. G. AULETTA, *Cognitive Biology*, 48.

> system – we have something that, at a pure physical level and without any reference or goal-directed action, bears some structural relation to what we are authorized, at another level and in another context, to call a true guess[18].

Shannon's model of sender and receiver is then inserted in a wider framework. Systems that deal with information are not characterized by the classical dyadic scheme of sender and receiver, but by a three-partner relation established between the processor (source of variety), the regulator (contributing to determine how information is acquired) and the decider (selecting at the end of the process which information is acquired and eventually stored). This richer approximation to the concept of information would be very challenging to the aim of characterizing life as a phenomenon that is not reducible to physics and chemistry.

3. **Information in biology**

My interest is focused on the concept of information applied to molecular biology. It can be seen that in many books of biology and biochemistry the concept of information plays an important role. It is said that DNA is the molecule that stores information, that the development is the proper activation in space and time of this information contained in the genes, and that reproduction is the passing of all this information to the next generation. Inside the cell there is a network of signalling elements that drives information from one place to another, and also integrates the information that comes from outside, leading to a proper response to those stimuli. Studying the cell, we can say that life is characterized by the processes that allow to express the information stored in the DNA during the development, by the integration of the inter- and extracellular stimuli that it receives and by the transmission of information to the next generation.

It was with the discovery of DNA's structure in 1953 by James Watson and Francis Crick that the term information became popular in biology. In their famous article, Watson and Crick not only elucidated the three-dimensional structure of the DNA molecule, but they also confirmed its role as information carrier and also suggested a mechanism to transmit this information to the

[18] G. AULETTA, *Cognitive Biology*, 48-49.

next generation[19]. Hence, since this paramount discovery, there is a theoretical framework that states that genes are the locus of information, that their proper expression is essential for the cell, and that their accurate transmission to the next generation is crucial for the life of the descendants. The molecular biology of the DNA is then thought of as a process of expression and transmission of information. Genes are informational stores that have to be activated at specific times and places, and their activation or repression leads to the synthesis of proteins, which are the products of those genes.

Information flows from DNA to RNA to protein, as it is established by the "central dogma" of molecular biology. The stored information in DNA is activated and translated into RNA. This RNA is then read by the ribosome in order to translate the information encoded by the nucleotidic sequence into the polypeptide chain. This can be done thanks to the genetic code, especially to the adaptor molecules that link each three nucleotides of DNA (and RNA) with one amino acid. Information is translated from the language of the genes into the language of the proteins (Figure III.2).

The expression of the genetic information has many editing processes, which make sure that information is flowing correctly and properly expressed. One of these editing processes is splicing, which cuts off introns from the RNA molecule, linking the remaining exons. Introns are pieces of RNA that do not have to be translated into protein. That is why splicing takes place before the translation of RNA into protein by the ribosome. When the polypeptide chain is synthesized, we do not have yet a mature protein. It has to pass through a process of maturation: folding, modification of some amino acids and, eventually, the association with other polypeptides in order to form a multi domain protein. At the end of the editing process we can say that the linear information of the DNA has been expressed in the three-dimensional shape of the protein, and the returning path from protein to DNA is impossible[20].

From that point of view, the expression of a gene and its translation into

[19] Cf. J.D. Watson – F.H.C. Crick, «Molecular Structure of Nucleic Acids» and J.D. Watson – F.H.C. Crick, «Genetical Implications of the Structure».

[20] Some kind of virus, called retrovirus, can synthesize DNA from RNA, but this does not suppose a violation of the "central dogma", but an exception to the Weismann's barrier, as it is explained in G. Auletta, *Cognitive Biology*, 302-304.

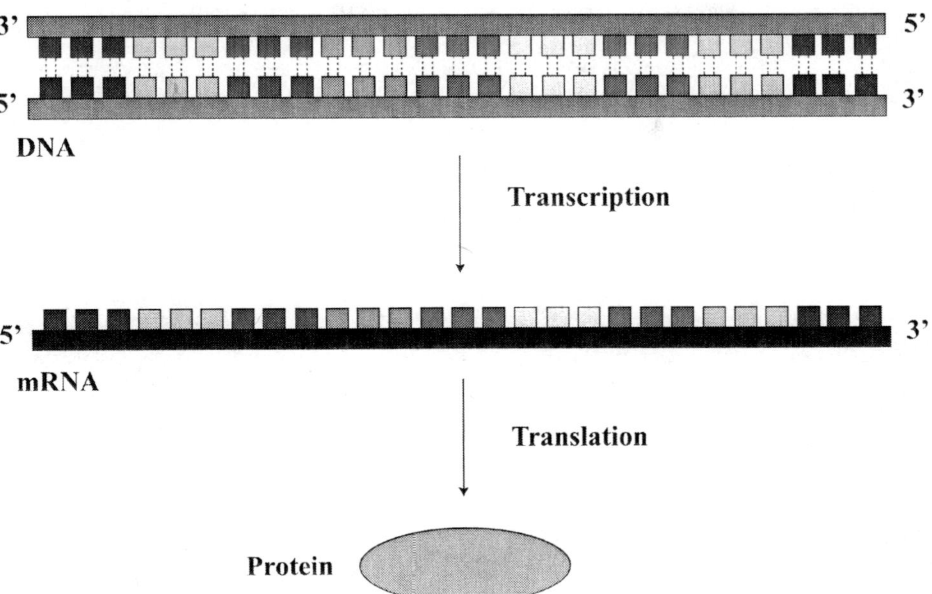

Figure III.2: The "central dogma" of molecular biology. Information in a cell flows from the molecule of DNA double stranded to a mRNA copy single-stranded. This process is known as transcription and consists simply in the copy of the linear information contained in the DNA strand into a mRNA molecule. Ribosomes and other partners can get access to mRNA and translate it into a protein. In this last step there is no more codified information. The linear information conveyed by mRNA has been translated in a three-dimensional protein.

a protein has been seen as a communication event between the sender and the receiver. Moreover, the informational paradigm in biology according to Shannon's model is not only used for the question of genetic information and its expression into protein. There is also another source of information for the cell beyond its own genes: the environment. The environment could be seen as a source of information for the cell, which sends it a great variety of stimuli, both physical and chemical. Furthermore, considering the chemicals present in cell's environment, we could individuate ions, sugars, proteins, lipid molecules, complex inorganic molecules and many others.

According to Shannon's model, the environment would be constantly bombarding the cell with huge amounts of information. The cell would be as an incredible processor of all the information generated in the environment, which acts as a source of variety. However, the cell does not process all the information it receives from the environment, but that would be the conclusion if we take the environment as a source of information and the cell as the receiver. Then, the cell does not behave as it was expected according to Shannon, since it selects to which stimuli it has to be more sensitive.

Finally, the sender-receiver scheme is also used to understand the transmission of information from one generation to the next, being the former the sender and the latter the receiver. Then, we see that the conceptual framework provided by Shannon's model is used to embrace both ontogeny and phylogeny, and explain them from an informational point of view.

Summing up, when Shannon's model is applied to biology, we have that the flow of information takes place between the genetic system and the cell (ontogeny), between the environment and the cell (epigeny), and between an organism and its offspring (phylogeny). The information model that was originally thought up for a technical problem has been applied straightforwardly to biology. As we are about to see, that leads to some difficulties when considering biology under this point of view.

In the next chapter I shall develop further the informational paradigm that has been sketched in section 2.4. Before doing that, I will expose different approaches that have been made to the concept of information in biology. This will be helpful to highlight the insufficiency of the sender-receiver model and to point out the necessity of a different paradigm to deal with the concept of information in biology.

3.1 *Genocentrism*

The informational paradigm in biology has been focused in the gene as the key concept[21]. Genes are the informational store of the cell, and they play the critical role in causing the characteristics of the cell. It seems as if the information stored in the genes (genotype) is the sender that has the variety of messages to be transmitted to the receiver, that in this case is the molecule into which a gene is translated, generating a certain trait (phenotype). Then, we can see that Shannon's model of communication has been adopted in molecular biology. Genes are the source of variety that stores the potential information. This information is eventually activated and translated into protein, which

[21] For an historical account of the concept of gene, see P. Beurton – R. Falk – H.-J. Rheinberger, ed., *The Concept of the Gene*.

is seen as the receiver or final product of the informational message sent by genes. As Godfrey-Smith states, what was initially applied to the transmission of information from DNA to protein was extended far beyond the process of protein expression, leading to a general informational paradigm in biology, which Godfrey-Smith summarizes in the following affirmations:

(i) The description of the whole-organism phenotypic traits (including complex behavioral traits) as specified or coded for by informational contained genes;
(ii) The treatment of many causal processes within cells, and perhaps of the whole-organism developmental sequence, in terms of the execution of a program stored in the genes;
(iii) The idea that genes themselves, for the purpose of evolutionary theorizing, should be seen as, in some sense, "made" of information. From this point of view, information becomes a fundamental ingredient in the biological world[22].

This is the more intuitive aspect of genetic information. The genes carry the information needed for the synthesis of the polypeptide chains they code for. To have a certain polypeptide chain, the genetic sequence has to be univocally established; otherwise the gene would codify to a different polypeptide chain. If the sequence is altered, we have different information and, then, a different polypeptide chain would result. That is the reason why it is important to make sure that the information stored in the gene would be faithfully translated into protein, and no noise must alter the transmission process from DNA to protein. Then, Godfrey-Smith says that «when a biologist introduces information in this sense to a description of gene action or other processes, she is not introducing some new and special kind of relation or property. She is just adopting a particular quantitative framework for describing ordinary correlations and causal connections»[23]. The concept of macrostate and microstate used when the relation between information and entropy was explained can be applied here (section 2.2). Indeed, «Shannon information represents the potential information contained in the macrostate, that is, the greatest amount of knowledge that can result from a full knowledge of the microstate»[24].

[22] P. GODFREY-SMITH, «Information in Biology», 103-104.
[23] P. GODFREY-SMITH, «Information in Biology», 106.
[24] B.-O. KÜPPERS, *Information and the Origin of Life*, 40.

Let us to critically study, in depth, this approximation to the concept of information in biology from the point of view of the centrality of the gene as the sender of a message that reaches its ultimate receiver when a protein is synthesized.

3.1.1 The "coding for" paradigm

Godfrey-Smith has pointed out that the only thing a gene codes for is a polypeptide chain. Reading an mRNA string according to the rules of the genetic code only leads to a sequence of amino acids put together according to that code. That is the only meaning for coding: to change the lineal sequence of nucleotides for an equivalent string of amino acids following the rules of the genetic code. If a given polypeptide chain needs the help of other proteins to reach its proper folding, this is something that is not codified by the sequence of the gene that has led to that polypeptide chain. Moreover, there are many proteins that, once they are folded, will be chemically modified by other proteins. These modifications are not codified by the gene, but are due to the activity of proteins that are codified by other genes. Then, «it is a mistake to take the concept of genetic coding and extend the reach of the relationship, past amino acid sequences, to complex traits of whole organisms»[25].

Hence, knowing what is coded is only a part of the story. We cannot infer the whole causal chain of events that would lead to an organism if we exclusively look at the polypeptide chains that are coded by genes. There are many other elements that are not coded and play important roles in the ontogeny of the organism. This calls into question the notion of gene as the unique source of biological information for the cell. Moreover, genes alone can do nothing, since they need the cooperation of the cell as whole in order to properly express the proteins they code for.

Godfrey-Smith agrees that the term "genetic coding" can be used in biology; but it has to be applied to a particular aspect of biology, which is the translation of messenger RNA (mRNA) into protein. It is misleading to use that concept in other fields of biology. At the same time, it would not be of great help to

[25] P. GODFREY-SMITH, «Explanatory Symmetries», S328.

distinguish between traits that came from a genetic coding process and traits that do not[26].

It is worth noting that one thing is to say that a gene *codes for* a certain polypeptide chain and a different one is to say that a gene *causes* a certain trait. «According to the standard framework, both genes and environmental conditions *cause* traits, but only genes *code for* them»[27]. I find the distinction established by Auletta explaining the concept of cause very useful (Chapter II, section 5). He says that a cause is something that *concurs* to some effect, and is not always something that *determines* a certain effect. Biology also needs a broader framework that can lead to a conception of genes not causing traits in a deterministic manner, but concurring to them.

There are some positions that are not enthused about dichotomising the factors that take part in the developmental process. Hence, those positions would not accept the division of traits into coded and not coded. «So the questions addressed here are as follows: Does the concept of genetic coding make a real contribution to our understanding of biological systems? And if so, which is the nature of this contribution?»[28].

As Godfrey-Smith says, «the "genetic code" is, strictly speaking, the rule linking RNA base triplets with amino acids. This "interpretation" of the RNA determines the "interpretation" of the DNA from which the mRNA was derived»[29]. Then, we talk of "genetic coding" because we have the following characteristics: as nucleic acids and amino acids are different kind of chemicals, a non-trivial rule must exist to link them; there are combinatorial rules that specify for an amino acid every three nucleotides, and the triplet codes for the same amino acid despite the neighbouring triplets it may have; finally, there is nothing in the chemistry of a certain amino acid that obliges it to be linked to a certain triplet. It can be said then, that the link is arbitrary[30].

The problem is that the concept of "genetic coding" has been used far beyond translation. Then one can think that a certain biochemical pathway is genetically coded, when what is really genetically coded is just the primary

---

[26] Cf. P. Godfrey-Smith, «On the Theoretical Role».
[27] P. Godfrey-Smith, «On the Theoretical Role», 27.
[28] P. Godfrey-Smith, «On the Theoretical Role», 29.
[29] P. Godfrey-Smith, «On the Theoretical Role», 32.
[30] Cf. P. Godfrey-Smith, «On the Theoretical Role», 32-33.

structure of the proteins that are involved in that pathway. But there are a lot of other elements that are not genetically coded: the post-translational modifications of the proteins, the cofactors that bind the proteins for their proper functioning, regulatory molecules of non-protein nature, the spatial-temporal disposal of the elements for improving a pathway's yield, the membranes where they are attached, etc. All these elements can, on their part, be the result of a coding process, but the interaction established by two or more elements that are the result of a coding process is not coded. The interaction follows other rules from that of the coding processes.

This does not mean that genes have no causal role over phenotypic traits. They do have, indeed, but this long-range causal role is different from that of coding. Genes code for polypeptide chains, and the mature proteins are inserted in a metabolic web, where could play causal roles, but those are not coded by genes[31]. To this regard Kim Sterelny asks,

> can we actually find discrete, fairly autonomous physical mechanisms that are capable of mapping genotype differences onto phenotype differences? The mechanism that map genotype differences onto protein differences have been identified, but we cannot just assume that such mechanisms must exist for the phenotype as a whole[32].

The point is that for the polypeptide chain to have its full causal incidence in the cell's metabolism, the cell as a whole is also needed. A given protein, isolated from the rest of the cell, does nothing. It needs a metabolic network to have energy for its synthesis. A system of inner and outer membranes in the cell is also needed in order to sort out the protein to its final place. And finally, it is also needed a genetic system that will code for other proteins involved in the protein control. Then, it is impossible to fully characterize a protein's causal role without taking into account the three systems that conform the cell: the metabolic, the genetic and the selecting systems (the membrane and the different compartments that it encloses).

---

[31] Cf. P. Godfrey-Smith, «On the Theoretical Role», 35.

[32] K. Sterelny, «The "Genetic Program" Program», 200.

3.1.2 The arbitrariness of the genetic code

The concept of arbitrariness is an important one to understand the informational role of the genetic code (Figure III.3), but also to shed some light on the concept of information that I want to introduce. Roughly speaking, we could say that the genetic code is arbitrary because the assignments between codons and amino acids could have been different from what they actually are. «It could be argued that arbitrariness expresses the claim that a code with a different set of assignments would have been functionally equivalent to the "canonical" code»[33]. Indeed, this is what Crick states when he characterizes the actual genetic code as a "frozen accident"[34].

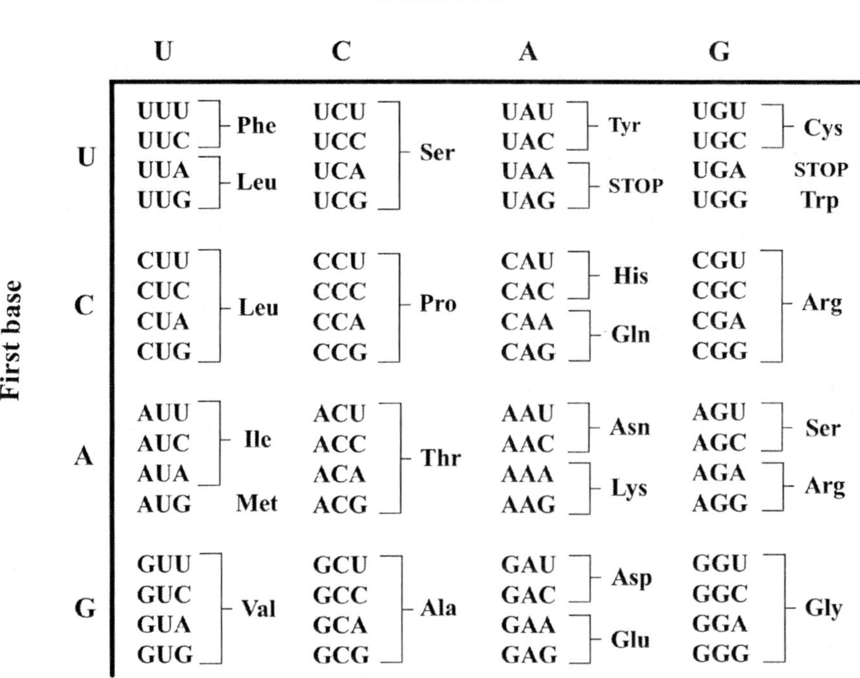

First base \ Second base	U	C	A	G
U	UUU, UUC – Phe; UUA, UUG – Leu	UCU, UCC, UCA, UCG – Ser	UAU, UAC – Tyr; UAA, UAG – STOP	UGU, UGC – Cys; UGA – STOP; UGG – Trp
C	CUU, CUC, CUA, CUG – Leu	CCU, CCC, CCA, CCG – Pro	CAU, CAC – His; CAA, CAG – Gln	CGU, CGC, CGA, CGG – Arg
A	AUU, AUC, AUA – Ile; AUG – Met	ACU, ACC, ACA, ACG – Thr	AAU, AAC – Asn; AAA, AAG – Lys	AGU, AGC – Ser; AGA, AGG – Arg
G	GUU, GUC, GUA, GUG – Val	GCU, GCC, GCA, GCG – Ala	GAU, GAC – Asp; GAA, GAG – Glu	GGU, GGC, GGA, GGG – Gly

Figure III.3: The genetic code. Every three letters of the mRNA string codify for an amino acid. The fact that a given triplet codes for a certain amino acid is not a question of chemical necessity, but just the ascription that was made when the genetic code became *frozen*.

[33] U.E. STEGMANN, «The Arbitrariness of the Genetic Code», 208.

[34] «The controversial proposal was that fixation at an early stage was not caused by selec-

However, Crick thought that the code was "frozen" at an initial stage, and the expansion of the genetic code, including further amino acids, would have been done through a process where natural selection took its part, and then arbitrariness would be left out in this further phase of the code. This fact opened the door to theories that call the arbitrariness of the formation of the genetic code into question. However, admitting that the "frozen accident" *freezes* the whole genetic code at once, there are some arguments that challenge that hypothesis. Indeed, the genetic code could have been the result of selection, history, chemistry, or a combination of the three[35]. However, the arbitrariness we are looking for is not that of the total chemical independence *during* the formation of the genetic code. What matters here is the fact that the chemical nature of DNA does not impose chemical restrictions in the sequence of the polypeptide chain it codes for.

Arbitrariness, nonetheless, has been defended because of the indirectness of the interactions between codons and amino acids. Since codons do not interact with amino acids to be linked in the genetic code, indirectness has been taken as a characteristic for arbitrariness. However, this is misleading, since «any remote chemical effect of a cause would then be related arbitrarily to its cause»[36]. Stegmann explains it with a clear example:

> Suppose there was a chemical principle to the effect that the stereochemical properties of binding sites within molecules like tRNAs [transfer RNA] and aminoacyl-tRNA-synthetases depend on each other. Then, codons would still interact indirectly with amino acids but, since both codons and amino acids would need to fit to the binding sites, the principle relating intramolecular binding sites would eventually determine the kind of structure a codon must have in order to specify a particular

tion and was not predetermined by stereochemical constraints. Crick proposed that codon and amino acids did not interact directly with each other even during earliest code evolution. Without the specificity of direct interactions, the chemical properties of codons or amino acids did not restrict or determine the set of assignments which then could become fixed». U.E. Stegmann, «The Arbitrariness of the Genetic Code», 207.

[35] Cf. R.D. Knight – S.J. Freeland – L.F. Landweber, «Selection, History and Chemistry».

[36] U.E. Stegmann, «The Arbitrariness of the Genetic Code», 209.

amino acid. Therefore, the structural relation between codons and amino acids would be chemically necessary[37].

There has to be some structural relations between codons and amino acids, of course, but these relations do not have to be established with chemical necessity. Consider the example of the codon CAC and the amino acid histidine (His). There are some specifications that have to be fulfilled in order to make CAC to codify His. CAC must be complementary to the tRNA anti-codon GUG (Figure III.4.B). Histidine, for its part, must have some properties that allow it to be recognised by the aminoacyl-tRNA-synthetase that links it to the proper tRNA; the one that has GUG in its anti-codon arm (Figure III.4.A). So we can see that certain relations must be preserved in order to have a link between the CAC triplet in the DNA and the His in the polypeptide chain.

However, there is no chemical necessity or constraint that forces it to be that way. Then,

> there is no chemical reason stating a dependence between the binding sites within molecules. Thus, to say that there is no chemical reason why CAC codes for histidine but not for glycine is to claim the following: there are chemical properties of CAC (and histidine) which are required for CAC to specify histidine, but they are not required in virtue of a chemical principle; and, therefore, it is chemically possible that CAC could specify for glycine or any other amino acid[38].

Indeed, it could have happened that an aminoacyl-tRNA synthetase could have recognized the tRNA with the GUG anti-codon and link it to the amino acid glycine instead of histidine. Then, when that tRNA would be paired with the codon CAC, it would be glycine what would be added to the polypeptide chain instead of histidine.

It also can be defended that the arbitrariness of the genetic code comes from the fact that its first assignments were made by chance; but chance is not even necessary for arbitrariness between DNA codons and amino acids. «If the code's assignments had become fixed by selection [due to a random process],

[37] U.E. STEGMANN, «The Arbitrariness of the Genetic Code», 216.
[38] U.E. STEGMANN, «The Arbitrariness of the Genetic Code», 215.

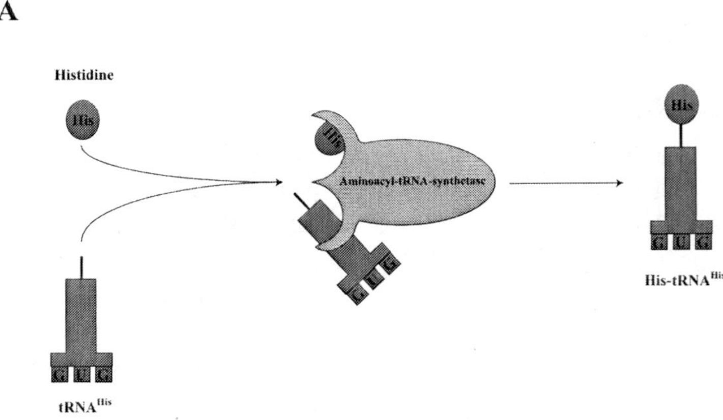

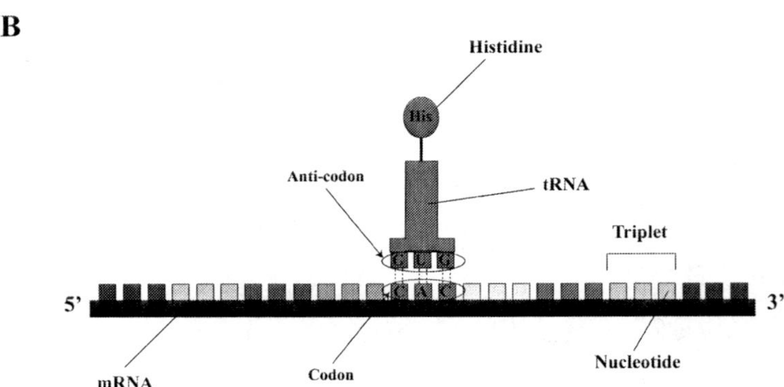

Figure III.4: The arbitrariness of the genetic code. A) Every amino acid is linked with a tRNA. This linkage is done through a family of enzymes named Aminoacyl-tRNA-synthetases. Every member of this family links an amino acid with a specific tRNA. B) The anti-codon of the tRNA fits with the codon of the mRNA molecule that has to be translated into protein. Then, a given codon fits with its specific tRNA anti-codon, which in turn carries a specific amino acid. But there are no chemical restrictions among the partners, and the ascription of amino acids and codons could have been otherwise.

then there would not have been a chemical principle making the binding sites of tRNAs depend on each other. The structural relations between codons and amino acids would still be chemically arbitrary»[39].

Finally, there is also the perception that arbitrariness has to do with evo-

[39] U.E. Stegmann, «The Arbitrariness of the Genetic Code», 216.

lutionary contingency. But contingency is not also sufficient for arbitrariness. Consider the molecule of DNA. There is a chemically necessary relation between the two chains, since they are complementary. Complementarity is not a contingent fact if DNA has to do the work it does. However, the fact that we found thymine instead of uracil (found in mRNA) in DNA molecules is something contingent, since uracil can also form hydrogen bonds with adenine. Then, the pairing A-T is an evolutionary contingent, but chemically necessary. Then, evolutionary contingency is not sufficient for chemical arbitrariness[40].

However, it is one thing to admit that the genetic code is arbitrary, and another to say that, as a consequence, it can establish semiotic relations. We have seen that this conventional relation is found in the genetic code, but there is no such arbitrariness in the other informational processes in which information is involved: replication and transcription. These two processes are chemically necessary. So here arbitrariness is not needed to make information pass through different generations. Stegmann concludes that «some of the processes expected to involve semantic information are certainly not chemically arbitrary and, therefore, chemical arbitrariness is not a necessary condition for a semantic relation»[41]. I agree with Stegmann only if the word "semantic" is changed to "semiotic". We cannot talk about semantics at molecular level. Semantics is a property that we only find in humans, because we only are capable of abstract thinking and symbolism, the conditions of possibility to ascribe a meaning to a certain item.

The question is that the arbitrariness of the genetic code is not initially to having some semiotic relevance, but to be able to have a great variety of possible combinations without chemical restrictions over the neighbouring nucleotides or amino acids. Once the polypeptide chain has been coded according to the arbitrary genetic code, there are many other molecules that would ascribe to that protein a semiotic relation with the cell as a whole that goes beyond the codification of that very protein.

Jacques Monod says that the binding of two molecules that have some functional consequences is based in gratuity (in French *gratuité*). This means that the link to the function that those molecules are related does not depend on

---

[40] Cf. U.E. Stegmann, «The Arbitrariness of the Genetic Code», 217.

[41] U.E. Stegmann, «The Arbitrariness of the Genetic Code», 219.

the chemistry of the molecules. There is no necessary connection between the chemical nature of the molecules and the functional role in which they are involved. Maynard Smith says that the absence of necessity between the chemical structure and the functional role of a molecule is due to its symbolic nature. I would rather use, once again, the term "semiotic" instead of "symbolic", as has been explained above. That means that there is something arbitrary in the assignment of a certain function to a particular chemical item.

Of course, we can say that there is a correlation between the activation of a gene and the appearance of a certain trait, between the activation of a certain cell signal cascade and the alteration of the molecules that depend on that signalling. «But when we say that, we are saying no more than what we are saying when we say that there is an informational connection between smoke and fire, or between tree rings and a tree's age»[42]. The point is that the source-receiver informational framework is not enough to apply the concept of information in biology. We need a three-sided informational paradigm rather than a sender-receiver model. Arbitrariness implies the linkage between a sign and its reference, despite its chemical structure, that can be selected by a third partner, as I shall show.

3.1.3 Is the "informational talk" in biology a metaphor?

Some investigators asked themselves if the metaphor of the DNA as a book where all the information is stored should be applied to everything in molecular biology. As has been stressed above, the informational paradigm from the field of human communications has been applied to biolog. However, the concept of information is not univocal, and it is necessary to specify its meaning to avoid misunderstandings. When people think of information, what immediately comes to mind is carrying something from one place to another. Furthermore, information can be seen that way if there is a receiver that can understand what the sender is communicating.

Indeed, biological vocabulary is full of terms used in human communication. We talk about the *transcription* from DNA to RNA and the *translation* from RNA into protein. The expression *genetic code* is also broadly known and makes reference to the Morse code. Furthermore, when we characterize how the

[42] P. Godfrey-Smith, «Information in Biology», 106.

cell expresses the information contained in its DNA to the final protein, terms as *messenger*, *editing*, *proofreading* and *correct* are used.

Many people think that this "informational talk" is just a metaphor to talk about what happens within the cell, an anthropomorphic way to understand cell metabolism. It is believed that this metaphor will last until we discover the ultimate mechanisms that could clarify how the cell really works. When we have a complete knowledge of what really happens in the cell, we will drop those information-loaded concepts. This is the aim of the reductionistic programme, to reduce biology to physics or even to a more fundamental level, explaining everything that happens in biology with the vocabulary of such grounding theory.

I think that the question is not to abandon the "information talk" as we progressively shed more light over the molecular mechanisms of the cell. The point is that the informational paradigm that has been used in biology does not tell the entire story. It was based on dyadic relations between the source and the receiver; relations that are adequate when we are talking about human communication and technology, but when this paradigm is used to understand the flow of information in a cell, it cannot solve the issue. It has the main problem that it is difficult to identify the sender and the receiver, and it seems that the channel is the same cell, which is also sending and receiving. Moreover, we cannot talk about the cell in the same way that we talk about a human mind. Then, biologists think that, because of these difficulties, it is needed to analyse the molecular mechanisms of the cellular processes in order to make things clear and abandon the "informational talk". The talk about information is due to the incompleteness of the picture, and biologists need to see the deterministic mechanisms of everything to leave out that metaphorical language.

As I shall stress in the next chapter, I propose an informational paradigm inspired by the work of Charles Sanders Peirce to deal with these issues. The problem is not the informational talk in biology, but the informational paradigm that is used to talk about information in biology. According to Peirce, the relations between items are not dyadic but triadic, which makes the informational paradigm richer and allows informational talk not to be just a metaphor but a very interesting tool to approach to the complexity of the cell.

Let us now see the informational approaches in biology that are not centred in the gene as the unique source of information. There has been a growing concern that also the environment plays an important role in determining which information is expressed, when and how.

3.2 *Genes versus environment*

Much has been written about the relative importance that genes and the environment have in influencing the phenotype of the organism. There are authors that give an almost exclusive importance to the genes, while others thought that it is the environment that does the major part of the work. Susan Oyama showed that thinking about the genes and the environment as two different sources of information that compete to shape the phenotype is misleading. It is not correct to say that a trait is genetically or environmentally determined. On the other hand, we have Richard Dawkins' view, which firstly focused in the central role of the gene as a selfish item that only looks for its own perpetuation[43]. However, the same Dawkins has had to review his position, since he noticed that the environment also plays an important role for the organism.

The discovery of the hereditary material, its structure and its organization has led to a genocentric point of view regarding biological information. The gene is the molecule responsible for passing information from one generation to another. The fidelity during the copying process ensures that the descendant would have the same genetic equipment. Besides the importance the gene has for the process of transmission of the inherited traits from one generation to the following one, it also has a paramount role in reconstructing an individual from anew, starting from the single cell out of which a full adult would arise. This developmental process is also guided by genes, and depends on their fine-tuned regulation in space and time in order to have the proper cellular types in the right developmental phase and place emerge. It was thought that genes were the cause of everything in the organism.

However, the genes alone cannot do anything. The DNA is a rather inert molecule that has to be activated to perform its role. The activation or repression of the genetic material depends on its interaction with external molecules and on its binding with internal metabolites of the cell. Then, the genetic material is not something that actualizes its information blindly to what happens inside the cell or outside it. The expression of the genes is modulated by the products of the metabolism of the very cell and by the information that the cell receives from the environment.

This has led to consider that traits that characterize an individual depend

[43] Cf. R. DAWKINS, *The Selfish Gene*.

both on genetic and environmental inputs. Genes and environment contribute to the developmental plan of the organism. However, new approaches suggest that no difference should be made between genes and environment as if they were two different sources out of which the organism gets the necessary information to its developmental plan. Some authors talk about replicators and other ways to face the question, avoiding any language that could seem as if they were talking about the genetic and the environmental input separately. Of course the other extreme position is also represented. There are authors that give very little importance to genetic material, and they think that the environment is what really drives both the development of an individual and the evolution of the species.

Genetics tries to leave out the discourse of innate and acquired traits, and focus its attention on the influence that both genetic and environmental factors may have on each trait[44]. Griffiths also thinks that too much importance has been given to the role of genes in development. Everybody agrees that the environment plays its role in developmental processes, of course; but the conception remains that genes are still a special kind of cause. Griffiths tries to reconcile positions stating what he calls the "parity thesis": «Any defensible definition of information in developmental biology is equally applicable to genetic and non-genetic causal factors in development»[45].

I think, however, that even if this clarification of the roles of the genes and the environment is useful, it does not touch the important question I want to underline. Saying that genes are not special causes and that the environment has to be taken into account to fully understand information transmission in biological systems, is only a way to acknowledge that the source of information is more complex than was first thought. The informational paradigm according to Shannon has not changed; only the partners involved are more sophisticated.

[44] «The division of traits into innate and acquired has been replaced by attempts to determine the relative influence of genetic and environmental factors on each trait», P. GRIFFITHS – R. GRAY, «Developmental Systems and Evolutionary Explanation», 277.

[45] P.E. GRIFFITHS, «Genetic Information», 396.

3.2.1 The extended-replicator model

Dawkins was first focused on the centrality of the gene as the main actor of biological systems. Then, he shifted his position from a genocentric perspective to another one where there is an acknowledgment of other traits beyond the genes that could influence the phenotype. Griffiths and Gray are interested also in levelling the contribution that genes and environment do to the replication of the developmental system.

Kim Sterelny, Kelly Smith, and Michael Dickison try to reflect on the gene as a representation, as something that has a semantic meaning for the biological system, which they call "replicator". As if the semantic properties of genes were something that allow them to represent the biological function to which they contribute when their information is instructed. It seems as if what is selected are the genes (or other factors) that have a role in representing a certain function. That means that the representational information of the gene is strictly related to a certain function. They also argue that environmental factors have a similar developmental role as genes, and they have also semantic properties; that is, they represent the outcomes they are supposed to produce.

Sterelny, Smith and Dickison retain the centrality of the gene, but they also acknowledge that other interactors are needed for the proper function of the replicator. As the authors acknowledge, this model arises from the recognition that, despite the importance of the gene, more and more non-genetic factors are needed[46]. The extended replicator theory has some commonalities with Developmental Systems Theory (see below): genes are not enough for development, inheritance is something that goes from phenotype to phenotype, and evolution has not only to do with competition[47].

These authors even disagree with the "genetic programme" theory, which sees genes as the blueprint that dictates phenotypic features. Information is something that not depends only on a genetic blueprint.

> A signal carries information about a source to the extent that characteristics of the signal co-vary with features of the source. So the genome carries information about a phenotype just so long as features of the genome co-vary with phenotypic characters.

---

[46] Cf. K. Sterelny – K.C. Smith – M. Dickison, «The Extended Replicator», 378.
[47] Cf. K. Sterelny – K.C. Smith – M. Dickison, «The Extended Replicator», 379-380.

> Phenotypic plasticity means that this co-variation is far from perfect. It improves, of course, if we hold the environment fixed. Aspects of the genome will co-vary well with traits *in an environment*[48].

If we hold the genome fixed, we can trace correlations between the environment and the phenotype. Then, we can see correlations between any given element of the developmental system with the developed phenotype if the remaining elements are held constant. There is no strict correlation between genes and developmental outcomes, since other environmental factors have to be taken into account. However, it can be said that genes represent some outcomes, because it is not the correlation that matters, but the linkage to the functions the organism has to perform. The different elements of the replicator can be seen as representations of the functions that have to be achieved. And the correlation among them can only be understood when the replicator is taken as a whole[49].

It has to be said that this replicator model shifts from genocentrism to a new way of facing biological information that is more in accordance with what emergence claims. The role of the genes cannot be seen as the only source of information in biological systems. Genes interact with many others informational components, and the notion of replicator comprehends the need to relate all this items to each other to have the full picture.

However, I think that the problem here is the notion of representation. What does mean that the genome represents developmental outcomes? How can a gene represent a function? Does this mean that other non-genetic factors have nothing to do with functions? My position is not that genes represent developmental outcomes or functions, but that genes are signs that stand for a function in some respect. An icon establishes the correlation between the sign and the referent, as we shall see below.

Another objection to this "extended-replicator" model is that it is difficult to establish the frontier between individuals. Is not easy to see where a single organism begins and where it ends, since there are informational sources that are not contained in it but that play some role in the developmental system. This has lead to the proposal of another type of natural kinds, another type of interactional units.

[48] K. STERELNY – K.C. SMITH – M. DICKISON, «The Extended Replicator», 385.

[49] Cf. K. STERELNY – K.C. SMITH – M. DICKISON, «The Extended Replicator», 387.

The natural kinds are replicators, interactors and lineages. Genes are paradigm replicators, but not quite the only ones: in asexual or genetically homogeneous populations (e.g. cheetahs) much larger units –chromosomes, genomes or even the organism itself may qualify. Organisms are the paradigm interactors but not the only ones[50].

3.2.2 Genetic programme

Very often the genome is seen as the programme that, once it begins, will ordain matter as is established by its blueprint. Genes are just the software that makes the hardware work. In the genetic programme analogy, the hardware would be everything in the cell that is not a gene. Susan Oyama shows the deficiency of this approach:

> The point of the blueprint analogy, though, does not seem to be to *illuminate* developmental processes, but rather to *assume* them and, in celebrating their regularity, to impute cognitive functions to the genes. How these functions are exercised is left unclear in this type of metaphor, except that the genetic plan is seen in some peculiar way to carry itself out, generating all the necessary steps in the necessary sequence[51].

Oyama says that what was firstly predicated by vitalism about the form or the spirit, is now predicated of the gene. Genes are invested with special qualities that would help to put order into the chaos of matter that has to be structured in order to become a system. But Oyama rejects this vision. There is no need to invoke strange properties to genes in order to have a developmental system. She upholds that the very system constrains and controls the way it functions, without spooky entities that would be needed to drive the developmental process. Moreover, she also says that genes "mean" different things depending on the tissues they are expressed, at the time they are activated, and to which stimulus it responds[52]. Let us see how she introduces this new approach that sees organisms as developmental systems.

[50] K. Sterelny – K.C. Smith – M. Dickison, «The Extended Replicator», 393.
[51] S. Oyama, *The Ontogeny of Information*, 54-55.
[52] Cf. S. Oyama, *The Ontogeny of Information*, 62.

3.2.3 Developmental systems theory

One of the paramount books that explains and defends the developmental system theory (DST) is Susan Oyama's *The Ontogeny of Information*[53]. DST wants to overcome the dichotomy between genes and environment. Genes are a developmental resource as are environmental factors. Indeed, the maternal heredity of the cytoplasm's zygote is important for the development of the already started organism. Genes, of course, have an important role, but no more important than the role of other factors. DST, then, does not point out the difference between innate and acquired traits; neither does it stress the differences between genotype and phenotype. It focuses on developmental systems as the real main characters of evolution. Oyama says that «information "in the genes" or "in the environment" is not biologically relevant until it participates in phenotypic processes. It becomes meaningful in the organism only as it is constituted as "information" by its developmental system. The result is not *more* information but *significant* information»[54]. Griffiths has seen the implications that this approach has for evolution:

> We suggest an etiological solution: the developmental system consists of the resources that produce the developmental outcomes that are stably replicated in that lineage. By adopting this definition, we bring out the radical implications of the new approach to development for the theory of evolution. The developmental system goes far beyond the traditional phenotype, yet all its elements are parts of the evolutionary process[55].

DST blurs the frontier between genes and environment. What really matter for development and for evolution are the elements implicated in replicating the organism from one generation to another, no matter where those factors come from. Oyama does not assign a special informational role to genes, as if they were the central source of information. The important question is that a developmental process takes place, in which there are developmental sources that feed the process with information. Some of these sources are genetic and

---

[53] Cf. S. OYAMA, *The Ontogeny of Information*.

[54] S. OYAMA, *The Ontogeny of Information*, 16.

[55] P. GRIFFITHS – R. GRAY, «Developmental Systems and Evolutionary Expla-nation», 278.

others are non genetic. «There is no room for any distinction between some resources that contain the information and others that "read it" or "provide the material conditions for its realization"»[56].

Hence, DST claims for symmetry when talking about genes and environmental factors; they have the same informational role. Then, what is inherited is the developmental processes and contexts of the past generations. What is transmitted is a developmental system that has been modelled by the informational input received by the genetic factors of the previous generations and by the environment that was at that time. We can say that «species-typical traits are reconstructed in the next generation by the interaction of the same sorts of developmental resources that were present in earlier generations»[57]. Then, information has its own ontogenetical process, as is stressed in the title of Oyama's book.

> A developmental system is a collection of interacting entities, some inside an organism's skin and some outside it. The interactions characteristic of a developmental system give rise to a reliably recurring *life cycle*: A set of interactants found in early ontogeny causally create conditions under which later developmental stages appear, and these later stages also create the interactants that will give rise to a new generation. Every factor which recurs reliably across generations, as a consequence of the causal interactions involved in the life cycle, is inherited and is part of the developmental system itself[58].

One of the weak points of DST is that it is difficult to trace the limits of a given developmental system. We can broaden the environmental source as far as we want, and we can go back in time to characterize the inheritance we have received as far as we want also. This is what Sterelny has named the Elvis Presley problem: «Elvis Presley is part of my developmental system, being as he was causally relevant to the development of my musical sensibilities, such as they

---

[56] P. Griffiths – R. Gray, «Developmental Systems and Evolutionary Expla-nation», 284.

[57] P. Griffiths – R. Gray, «Developmental Systems and Evolutionary Expla-nation», 284-285.

[58] P. Godfrey-Smith, «Explanatory Symmetries», S325.

are. Yet surely there is no system, no sequence, no biologically meaningful unit, that includes me and Elvis»[59].

«So the theory is interested in those developmental resources whose presence in each generation is responsible for the characteristics that are stably replicated in that lineage»[60]. What matters is developmental outcomes that have a role in replicating the developmental system. These developmental outcomes have an evolutionary history, but this must not lead us to return to the distinction between acquired and innate characters. The important thing is that developmental outcomes or individual outcomes are not distinguishable. Analyzing a trait we cannot tell if it has appeared due to a certain evolutionary history (such as the thumb) or as a contingent characteristic of the individual (as having an scar on the thumb). The key is not the distinction of the traits themselves, but their pattern of explanation[61]. DST points out the elements (whatever their origin may be) that, generation over generation, replicate the developmental system. There are genetic and environmental traits that can be involved in that replication; but there could be environmental or genetic traits that are not involved at all in such replication cycles.

Since it is difficult to establish the limits of a developmental system, Griffiths and Gray prefer to talk of developmental processes rather than developmental systems.

> The developmental process is a series of events which initiates new cycles of itself. We conceive of an evolving lineage as a series of cycles of a developmental process, where tokens of the cycle are connected by the fact that one cycle is initiated as a causal consequence of one or more previous cycles, and where small changes are introduced into the characteristic cycle as ancestral cycles initiate descendant cycles. [...]. The things that interact with the organism in developmental interactions are developmental resources. Some of the resources are products of earlier stages of the process, others are products of earlier cycles of the process, others exist indepen-

[59] Quoted in P. GRIFFITHS – R. GRAY, «Developmental Systems and Evolutionary Explanation», 286.

[60] P. GRIFFITHS – R. GRAY, «Developmental Systems and Evolutionary Explanation», 286.

[61] Cf. P. GRIFFITHS – R. GRAY, «Developmental Systems and Evolutionary Explanation», 286-287.

dently of the process. These distinctions, while real, do not bear on the type of role which the entity plays in the developmental process[62].

Then, we can define the developmental system as the sum of items that are involved in the developmental process. These items are developmental resources, which can have diverse origin. Then, every developmental process has an evolutionary explanation, but the developmental resources could have an evolutionary explanation or not[63].

This point of view is very interesting, but puts some difficulties to my informational approach. For example, DST focuses on eliminating the distinction between genetic and environmental traits, but stresses the developmental system as the process by which the information is passed from one generation to another, what it is really doing is to widening Shannon's informational vision from ontogeny to phylogeny. DST does not really proposes a new informational paradigm for biology; it just points out that all the biological information has the same quality, no matter where it comes from.

The DST approach leaves out the vision that makes genes the directors of development and considers the rest of the factors as raw material with which genes work. Genetic and non-genetic factors have the same importance for DST. This approach helps us to have a different position regarding the meaning of biological information. From a genocentric point of view, information flows from the gene to its expression into traits. The developmental process is just the channel that establishes the bridge between genotype and phenotype. Then, a certain gene is always sending the same message, the same type of information. If it is finally not expressed in a trait, is because something has prevented it. It seems as if genes have an intrinsic meaning. But, as Oyama points out, «if the same message can be interpreted in several ways, it is harder to consider the several meanings as inherent in the message»[64]. It is interesting that the message can be differently interpreted depending on the environment it is expressed. The message carried by the gene should not be considered as invariant information

[62] P. Griffiths – R. Gray, «Developmental Systems and Evolutionary Explanation», 291.

[63] Cf. P. Griffiths – R. Gray, «Developmental Systems and Evolutionary Explanation», 291.

[64] S. Oyama, *The Ontogeny of Information*, 23.

that is exactly the same wherever and whenever it is expressed, but as a factor that can be tuned by the environment in which it is expressed. This issue shall be shown in the case studied in the third part of this work.

Let us return briefly to the "coding for" debate. Oyama tries to counterbalance the importance of genes stressing the important role that non-genetic factors play in development. But she goes further and intends to give a new insight into biological causality. She states that in biological causality there are many influences that come from different sources. The final result of all these mutual influences depends on the system as a whole, which somehow modulates the final influence of one item in the overall phenotype. The effect of a given factor depends on its inner chemical properties, but also in the way it enters in the complex combinations with the rest of elements at play[65].

Information is something that is not just expressed and that is all; it is not something that is densely packed and it has just to be unpacked to see its effects. Information has its own ontogenetic process. Oyama leaves out all attempts to rebirth the homunculoid solution. It seems as if genetics has put in the gene what previously was said of the homunculi. From that point of view, development is just the unfolding of what is already at the first stages of development, the construction of a given blueprint established in the genetic programme.

Oyama points out the fact that information is not transmitted only in the genes, but it has to be reconstructed at each generation. Then, what is transmitted is the information of the developmental system, which is in the genes but also in the environment. There is a great shift here in the conception of what development is. Development is not just the construction of an organism from its blueprint, but a dynamical event that needs the reciprocal interaction of the parts involved. She stresses the importance of the developmental system, and the fact that it includes traits that until now have been considered as accidental. Oyama's proposal construes us to think on the organism not only as something that exists within its skin, but also as something that is intrinsically linked to factors that are located out of its skin[66].

The problem is that she does not offer a way to understand *how* those mutual dependent interactions take place. Information is something that happens, I

[65] Cf. S. Oyama, *The Ontogeny of Information*, 17-18.
[66] Cf. S. Oyama, *The Ontogeny of Information*, 26.

agree with that; but the question is how these interactions take place among the elements in order to fulfil the proper developmental aim. Oyama has proposed a framework that prevents us from thinking of information as something already pre-existing in biological systems; since from that point of view, development would be just the unfolding of what is already defined. Of course, the cell cannot decide what to change in the sense we commonly understand "decision". The problem, again, is to apply an anthropological category as "decision" to a biological system that cannot decide anything in the way we do. However, not deciding like humans does not mean being committed to determinism[67].

This allows us to understand change, not in a deterministic manner, but as a dynamics that takes into account the previous history of the cell and the informational role of the elements that are present in the system.

> Change, then, is best thought of not as the result of a dose of form and animation from some causal agent, but rather as a system alteration jointly determined by contemporary influences and by the state of the system, which state represents the synthesis of earlier interactions. The functions of the gene or of any other influence can be understood only in relation to the system in which they are involved. The biological relevance of any influence, and therefore the very "information" it conveys, is jointly determined, frequently in a statistically interactive, not an additive, manner, by that influence and the system state it influences[68].

Some authors express their critics to the DST model. One criticism is that it seems as if everything is linked to everything, as if a sort of universal holism pervades life. This affirmation could be true from a certain point of view, but the strength with which it is affirmed by DST commit the model to deal equally with all causes without any difference. Some distinction of the causal relevance of the factors involved in the developmental process is needed. There is also the difficulty to explain how those informational interactions take place, since if all information is equiprobable and has the same importance (as occurs in Shannon's model), is difficult to explain life informationally. The other problem

---

[67] Cf. S. Oyama, *The Ontogeny of Information*, 36.
[68] S. Oyama, *The Ontogeny of Information*, 37-38.

is that the frontier between organisms is not clear, since the model is focused in developmental systems, not individuals[69].

3.2.4 Modern History

Jonathan Michael Kaplan and Massimo Pigliucci also face the question of whether it makes any sense to talk of a gene coding for a phenotype or not. «We can properly speak, we will argue, of a gene being "for" some trait only when the gene was maintained by natural selection in the recent evolutionary history of the organism by its causal association with the trait in question»[70]. In some cases there would be a gene that codes for a trait; but in many others the picture would be that genes are associated with particular traits in many different ways. In the latter sense, what defines a certain trait is not the presence or absence of a single gene, but the cooperation of a group of genes involved in that trait and its interaction with the environment. The authors acknowledge that «there will always be relatively few genes that are known to be "for" particular traits; however, if these have been of great evolutionary significance, a language that encourages a program of looking for them is still to be recommended»[71].

Two extreme positions can be found in the literature. Dawkins' position says that it would be enough to have a statistical association between genes and phenotypic traits to see that those genes are for these traits. On the other hand there is the DST proposal, which does not want to privilege the gene over the rest of factors that play a role in the developmental process. However, Kaplan and Pigliucci think that there are cases where natural selection acts on genes, and then is worth keeping the "genes for traits" talk[72].

These authors propose a functional approach to deal with the "gene for" question. They adopt Godfrey-Smith's "Modern History" approach, which defines biological functions roughly as the dispositions or effects a trait has and

[69] Cf. K. STERELNY – K.C. SMITH – M. DICKISON, «The Extended Replicator», 382-384.

[70] J.M. KAPLAN – M. PIGLIUCCI, «Genes "for" Phenotypes», 190.

[71] J.M. KAPLAN – M. PIGLIUCCI, «Genes "for" Phenotypes», 191.

[72] «The position that it is impossible to find genes for traits is untenable not because the gene should hold a central place in our thinking, and not because we are old-fashioned adaptionists who want to make every trait out to have been the product of natural selection acting on genes. Rather, the problem is that there are cases where the locution seems to us entirely natural». J.M. KAPLAN – M. PIGLIUCCI, «Genes "for" Phenotypes», 193.

that explains its recent selection. Sterelny and Kitcher alert against the fact that in a "gene for" framework, one can be tempted to admit that a certain character can be shaped by natural selection without taking into account other traits of the organism[73]. However, the Modern History proposal of talking about "genes for" is not such simplistic[74].

There would be many cases in which it is not possible to ascribe a certain trait to a single gene[75]. Genes that have complex epistatic[76] effects do not code straightforwardly for a trait, but their interaction makes the derived trait as an emergent property of the interacting process. The question is not to stop talking about "genes for", but to establish a criterion to make proper use of the expression, knowing that possibly it could be a relation of linkage between a single gene with a single trait. The criteria to be known for a proper "gene for" talk are the following:

a) That the gene is statistically associated with the trait.
b) The most important aspects of the biochemical and developmental pathway form the gene to the trait.
c) That the trait in question really was the likely subject of natural selection in the species' recent evolutionary history.
d) That the maintenance of the gene in the species in its recent evolutionary history can actually be the result of (c)[77].

[73] Cf. K. STERELNY – P. KITCHER, «The Return of the Gene».

[74] «The first thing to notice is that no statistical correlation, no matter how strong, will be enough on the Modern History view to talk of there being a gene for some trait. Indeed, it will turn out that no single kind of evidence will ever be enough. Merely knowing something about the evolutionary history, or the population genetics, or the developmental pathways, or any other isolated bit of biological information, will not suffice. Some combinations of these will be required». J.M. KAPLAN – M. PIGLIUCCI, «Genes "for" Phenotypes», 194.

[75] «Only rarely is a molecular gene causally associated with a single phenotype. There is a many-many relation between genes and phenotypes: each gene contributes products that act to produce many different phenotypic features, and each phenotype relies causally on the action of many different genes». N. SHEA, «Representation in the Genome», 319.

[76] As Futuyma explains, epistasis is «an effect of the interaction between two or more gene loci on the phenotype or fitness whereby their joint effect differs from the sum of the loci taken separately». D.J. FUTUYMA, *Evolution*, G-3.

[77] J.M. KAPLAN – M. PIGLIUCCI, «Genes "for" Phenotypes», 195.

It is interesting to note that the concept of epistasis allows me to question the communication theory model and try to introduce an alternative framework. Shannon's model fits very well with the concept that one gene codifies for one trait. Then, the ontogenetic process would be just to make sure that each gene has the appropriate channel and the less noisy interferences to achieve the transmission of its message to the protein for whom it codifies. But in epistasis this does not happens at all. On the contrary, genes are expected to interact with each other in order to express a certain trait. In these cases, the trait is the emergent product of gene interaction. The question here is not to reduce the noise, because the noise (for example, another gene involved in the epistatic interaction) could be part of the essential items to fulfil the proper expression of the trait. Then, the point is not the faithful transmission of the message of a single gene, isolated from any possible interaction, but the way all those genes interact to achieve the trait. What matters here is the respect, the way, the modality, in which the interactions are made.

Another interesting example is pleiotropy, «a phenotypic effect of a gene on more than one character»[78]. The example reported by Kaplan and Pigliucci is the very well known case of the Hb^s mutation of the globin gene. This mutation diminishes the capability of human hemoglobin to exchange oxygen and carbon dioxide with blood. Furthermore, this mutation alters the shape of red blood cells, preventing the malaria agent from binding to its surface and its consequent reproduction. «The alterations in the shape of the red cells and the consequent resistance to malaria are therefore not what the Hb^s allele was originally "for". But –depending on the environmental conditions- the mutation actually does confer a selective advantage in malaria-infested areas»[79]. We see here that the "for" of the gene changes within the respect under which that gene is considered[80].

Kaplan and Pigliucci report a very interesting example that helps to under-

[78] D.J. FUTUYMA, *Evolution*, G-7.

[79] J.M. KAPLAN – M. PIGLIUCCI, «Genes "for" Phenotypes», 197.

[80] «The Modern History view of genes being for phenotypes is a powerful tool for thinking through the way in which evolution can work to change the biological significance of various genes. On this view, genes that were for a particular phenotype can, in time, become for other phenotypes entirely, or indeed, for nothing at all (e.g., they may become pseudogenes). Similarly, genes that were not in fact "for" anything in particular (at least at the gross phe-

stand the complexity of the "gene for" approach[81]. They take the example of the *phyB* gene of the weedy crucifer plant *Arabidopsis thaliana*. This gene codes for a family of light-harvesting molecules named phytocromes. The molecule (PHYB) is sensitive to the ratio of red to far-red (R:FR) wavelengths of light. The molecule can adopt two different configurations, depending on the R:FR ratio (Figure III.5). This ratio is an indicator of competition for *A. thaliana*. Too much vegetation could be a serious threat for *A. thaliana*, since the surrounding plants would absorb the red fraction of light (the one that is energetically profitable for photosynthesis), and would return the far-red portion. This fact would affect the relative amount of the two PHYB configurations, leading to the switch of different growing strategies to face such competition. These strategies would be mainly the suppression of branching, the acceleration of flowering time, seed germination, cell elongation, cotyledon expansion, seedling appearance and leaf production. Furthermore, it has to be considered that PHYB interacts also with other photoreceptors that are sensitive to blue light, increasing then the complexity of the whole signalling.

This particular example shows us how difficult it is to answer the question "what is the *phyB* gene, for?". Here is where problems become harder to solve.

> If we use the studies associating mutations at the phytochrome B locus with their phenotypic effects, we are led to conclude that the gene is for the control of germination, cell elongation, leaf production, flowering time, apical dominance, branching pattern, and reproductive output –to say the least. One problem with this sweeping generalization is that in fact a lot of other genes also seem to be "for" the same characters, with only partial (and sometimes contrasting) overlap with the effects of *phyB*[82].

On the other hand, there is the entire dependence of the activation of all these signals on the environment. The incidence of the R:FR light ratio on the configurations of PHYB starts all the signalling pathways in which the

---

notypic level) can, over time, gain that sort of biological significance». J.M. Kaplan – M. Pigliucci, «Genes "for" Phenotypes», 202.

[81] Cf. J.M. Kaplan – M. Pigliucci, «Genes "for" Phenotypes», 205-207.

[82] J.M. Kaplan – M. Pigliucci, «Genes "for" Phenotypes», 206.

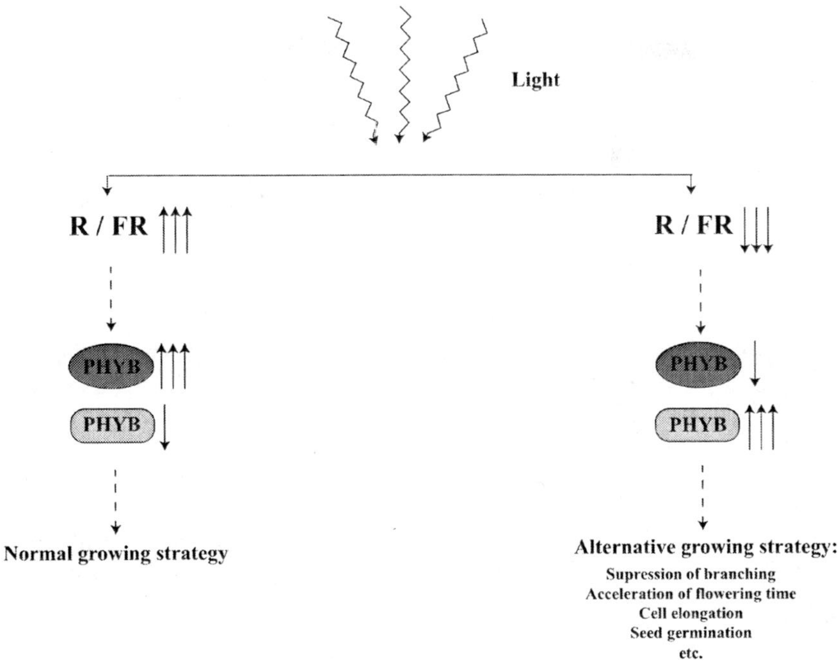

Figure III.5: Different growing strategies in *A. thaliana* depending on light availability. If the ratio of the red / far-red light spectrum is high, that means that the plant gets enough red light. Then, the two possible configurations of the protein PHYB are also balanced towards the normal one. On the contrary, if the ratio red / far-red is low, this is a sign that other plants in the surroundings are absorbing red light, and reflecting far-red light. Then, the other conformation of PHYB becomes predominant and an alternative growing strategy starts in order to compete for the light.

phytochrome is involved. Then, we can see that «genes do not do anything by themselves, but only in relation to their environment, both the external one, and the one provided by the actions of other genes»[83].

But things can be even more complicated. If we consider the evolutionary history of *phyB*, we can see that it pertains to a very ancient family of photoreceptors that were present before the existence of flowering plants. So initially, those kinds of photoreceptors were not associated to flowering time. This family of photoreceptors has been maintained through evolution because it was, and

[83] J.M. KAPLAN – M. PIGLIUCCI, «Genes "for" Phenotypes», 207.

still is, a gene family "for" photoreception. However, their role has been changed over time. They still codify for proteins that react to light, but this "reacting to light" character has been used in different respects depending on the organisms in which it is expressed[84].

It can be said that *phyB* acquires different respects in its evolutionary history. Is as if the self (decider) links the sensitive reaction of *phyB* to light (the sign) to different pathways related to the availability of light (object). But it is worth noting that this is a sign of the presence of competitors. I guess that the sign could be different (for example, the oxygen:carbon dioxide ratio), but the plant's reaction to competition would be maintained; perhaps through different pathways, but is something that the plant must be aware of if it wants to survive.

3.3 *The semiotic approach*

Shannon's model was not thought up to deal with the semantic content of the information that is transmitted. That is the reason why Griffiths distinguishes two concepts of information: causal information and intentional information[85]. Causal information derives from Shannon's mathematical theory of communication, as has been seen above. The causal concept of information establishes a correlation between the source and the sender; a correlation established through a channel[86].

On the other hand, intentional information has to do with human thoughts and their intentionality. Griffiths notes that one of the characteristics of intentional information is that it could be false; there is a capacity to misrepresent. There has been an attempt to apply the intentional vocabulary to genetic information, as we have seen before with Maynard Smith's account. «Genetic information is usually described as if it made sense to speak of a

[84] Cf. J.M. Kaplan – M. Pigliucci, «Genes "for" Phenotypes», 207.

[85] Cf. P.E. Griffiths, «Genetic Information», 396-398.

[86] «There is a channel between two systems when the state of one is systematically causally related to the other, so that the state of the sender can be discovered by observing the state of the receiver. The causal information content of a signal is simply the state of affairs with which it reliably correlates at the other end of the channel». P.E. Griffiths, «Genetic Information», 397.

phenotype misinterpreting the message in the genes and hence appears to be intentional information»[87].

I think that the concept of intentionality cannot be applied to molecular biology; neither that of semantics. Intentionality implies the use of abstract thinking and a symbolic method of communication, which the cell certainly does not have. My guess is that to differently deal with all the information the cell has to cope with, the concept of semiotics is very much appropriate. I shall devote the following chapter to this concept. But let us now to expose some attempts that try to go through a richer concept of information in biological systems.

3.3.1 The importance of the linking between the source and the receiver

Eva Jablonka[88] focuses her interest on the biological concept of information also; particularly in the semiotic aspect of information in biological systems. When those systems are studied, it is clear that Shannon's model of information quantification is not enough, since this model leaves out concepts such as function and meaning; terms that are of paramount importance in biology. That is why she tries to shed some light on the common features of different processes of information transmission[89].

First of all, the message sent by the source has some reaction in the receiver. The information that is transmitted acts upon the receiver, generating a response to the received message. Secondly, the reaction of the receiver depends on the

[87] P.E. GRIFFITHS, «Genetic Information», 397. Teleosemantics is another concept used to talk about this semantic aspect of information: «Maynard Smith proposes to analyze the intentionality of genetic information using teleosemantics, the philosophical program of reducing meaning to biological function (teleology) and then reducing teleology to natural selection. Whilst there is considerable controversy about whether such reductions can be successfully carried out, teleosemantics remains one of the more popular programs for naturalizing intentionality». P.E. GRIFFITHS, «Genetic Information», 399.

[88] Cf. E. JABLONKA, «Information: its Interpretation».

[89] «We have here a motley collection made up of environmental cues (the appearance of the sky, day length), man-made instructions (train timetable, recipe, software), evolved biological signals (alarm call, chemical signal), and hereditary material (a very special type of evolved biological signal). What do they have in common?». E. JABLONKA, «Information: its Interpretation», 580.

way the source is organized, not on the energy that is traded in the transmission event or on the chemical composition of the items involved. In third place, the reaction is a response that is beneficial from an evolutionary point of view. Finally, variations in the form of the source led to variations in the form of the response[90].

It is worth mentioning the importance of the evolutionary history of the source-response connexion. As in the Modern History approach, the information transmission depends on the previous history of the items involved in such particular transmission. The link between the sender and receiver is the point where evolution works on. Jablonka's approach does not differentiate genetic and non-genetic traits, but locates the key elements not on the traits themselves, but in the way those traits are linked to the whole causal web of the cell.

> I shall argue that when thinking about biologically relevant information, a functional-evolutionary perspective is necessary, but the focus should be neither on the evolution of the signal "carrying" the information, nor on the evolution of the final specific response. Rather, it should be on the evolution of the system mediating between the two –on the interpreting system of the receiver. The definition I suggest accommodates environmental cues as potential informational cues, and makes it easier to thing about non-genetic information and its transmission through non-genetic inheritance systems[91].

Then, what is subjected to evolution is neither the nature of the signal nor that of the receiver, but the connexion made between them and the benefits that that connexion yields to the system. What matters here is the informational connection that links the signal with the response of the organism. This is a very interesting approach, since Jablonka tries to catch the informational aspects of the question despite the physical and chemical nature of the elements involved. However, there is still something missing, since there is nothing that can tell us about the criteria of these links. She makes a step further, but maintaining Shannon's framework of source and receiver. Let us see Jablonka's definition of biological information:

---

[90] Cf. E. Jablonka, «Information: its Interpretation», 580-581.

[91] E. Jablonka, «Information: its Interpretation», 581-582.

> A source – an entity or a process – can be said to have information when a receiver system reacts to this source in a special way. The reaction of the receiver to the source has to be such that the reaction can actually or potentially change the state of the receiver in a (usually) functional manner. Moreover, there must be a consistent relation between variations in the form of the source and the corresponding changes in the receiver[92].

The term "form" must be understood as the organization in space and time of the features of the source. Jablonka stresses that the object of natural selection is neither the input coming from the source nor the output from the receiver, but the interpretative process to which they are linked. What is selected is the way in which the relation between sender and receiver is coupled. A consequence of this is that the organism has, through the linking process, a very important role in the ontogeny of information. Jablonka states that it is the receiver who constructs information; the receiver shapes the source when the input coming from it is linked to some response in a certain manner.

Focusing on the link between the input and the output, Jablonka's model also admits the gratuity (also called semioticity) of the signal, as Monod says. This means that there is no necessary connection between the "form" of the signal and the function it triggers in the receiver. «This gratuity is an inevitable consequence of the evolved nature of any biological signalling system: if evolution had been different, the biologically produced signal associated with a given adaptive response could, and almost certainly would, have been different»[93].

3.3.2 The tinkering strategy

Nicholas Shea focuses also in the semantic properties of DNA[94]. He agrees that the relation between the order of the bases on a DNA strand and the polypeptide chain it codes is a coding relation that is established despite the semantic aspect of information. Then, as it has been said above, genes only code for the primary structure of proteins, not to full phenotypes. Since a protein

---

[92] E. JABLONKA, «Information: its Interpretation», 582.
[93] E. JABLONKA, «Information: its Interpretation», 586.
[94] Cf. N. SHEA, «Representation in the Genome».

has to interact with the rest of components of the cell, there is a many-to-many relation when we consider proteins and phenotypic traits[95]. From that point of view, there is no preconceived plan out of which the organism is fully determined. The developmental process is not the execution of an engineer blueprint. Both phylogeny and ontogeny are the result of developmental processes that take into account genetic and non-genetic factors.

Shea introduces a new intuition when he explains that, both in ontogeny and phylogeny it does not seem as if the best way to explain it were to appeal to a transmission model of information. As we have seen, this approach is very focused in the faithfulness of the information transmission, as if the message should be clear from the beginning and then transmitted without error. This leads to think about life as a system that is very clear about what it wants in a certain situation and follows the path to achieve it. But seeing the oddness of some evolutionary paths followed by life, notices Shea, it is difficult to admit that those solutions where the result of a well established plan and a very fine-tuned design. Indeed, he says that we have to think about life, and especially about evolution, as a tinkering process rather than a designed or fixed one[96].

Shea distinguishes two aspects of the genes, two senses of genetic information. The *developmental properties* of a gene are established by its capacity to give rise to certain proteins. The contribution a gene gives to development is the protein it codes for. Coding and correlation are developmental properties of the gene. There is, however, a further characteristic of genes: their *selectional properties*, which are linked to *teleofunctions*. Each gene could have many effects when interacting with the whole web of the genetic and metabolic system. The

[95] Cf. N. SHEA, «Representation in the Genome», 314.

[96] Susan Oyama agrees also with this issue: «I am suggesting that *both* processes [phylogeny and ontogeny] show the contingent quality of tinkering, in the sense not of randomness or disorder but rather of subtle and opportunistic dependence on particular conditions and materials. This accounts for both their vulnerability and their power, as well as their "ingenuity". Rather than the directedness of planned activity, it is such inspired tinkering that characterizes life processes, the marvelous results notwithstanding. In the case of normal development, however, the scraps and bits of twine are all at hand, and the qualities». S. OYAMA, *The Ontogeny of Information*, 46.

gene is selected for one of its phenotypic effects, and this selection is made because of its functionality for the organisms[97].

Of course we can say that genes carry information in Shannon's sense, but this kind of information does not make the difference that is needed when we talk about information in living systems. The difference is made when the information is differently treated. As Sabine Brauckmann says: «Therefore one concludes from a cellular or organismic viewpoint that genes are the signs for life processes which are in general dependent on the trick of construing causality through semiotic control»[98]. Genes are sequences that originally do not have any meaning for the organism. They would have some meaning when they will be transcribed and translated into proteins. These proteins, when properly folded and assembled, would be finally semiotically controlled by the cell as a whole in order to perform their activity to achieve the functions required for survival.

4. **Summing up**

Shannon's model of communication is an informational framework focused on the reducing of noise in the act of transmission. It is a model where the most important issue is to faithfully transmit a message from the source to the receiver, regardless the meaning of the message.

There have been many attempts to apply the concept of information to biology following Shannon's model. As it has been showed, they make some contribution to better understand how biology has to deal with the concept of information. Moreover, two approaches that dare some explanation in which semantic content is involved have been show. However, those models are not fully satisfactory.

They miss a more complete framework where information can be linked to its content and the use the cell makes of it. It has to be taken into account that the aim of the cell or the organism is not just to deal with information. In fact, this information has to be used in order to maintain the system alive. Dealing with information, with its content and its linkage to vital functions is not optional for the system, but a necessity to keep itself alive.

[97] Cf. N. SHEA, «Representation in the Genome», 320.

[98] S. BRAUCKMANN, «On Genes, Cells, and Memory», 157.

When we talk about information in a cell, we cannot use concepts as *semantics* or *meaning*. These concepts are used only when information is used by a system able to perform abstract representations, to have intentional thinking and symbolic capacity. The cell does not *understand* the information it deals with, neither does it *represents* it as we represent the concepts in our brain. The cell deals with information semiotically, linking signals with referents and integrating those couplings with the needs of the system. That is the reason why we need an approach to the concept of semiotics to understand properly how the cell deals with information.

Chapter IV

Semiotics

1. The role of semiotics in biology

The category of semiotics is paramount in linking the informational paradigm to the concept of biological function. Indeed, information is a concept that has to do with the description of a system and all its possible configurations. As has been shown, the information of a system can refer to its actual arrangement or to all the potential organizations the system may achieve.

From a biological point of view, it is clear that not all the configurations in a system are assayed. The cell does not adopt all the possible arrangements that the combinatorial permutation of its items can permit. If we think, for example, about the potential number of proteins that can be synthesized with the 20 amino acids, we would see that the number is extremely high. However, the cell uses only a little fraction of all the possible proteins and combinations that can be achieved.

From another perspective, there is also the question of how the cell deals with all the molecules that are in it at a given time. They have to interact in a certain order, and need to be regulated both in space and time. Not all the molecules in the cell are permanently reacting to each ohter. They have to be activated properly and at the right time. For achieving this, a kind of organization, an integration of all the elements for the sake of the whole is needed. It is important for a cell that a given protein can perform its work properly, but it is no less important that this protein has to be integrated into the whole net of the cell to be linked and controlled by it.

Semiotics is seen as a bridge that links information with function. Semiotics sees information as something significant for the sake of the system. That is how the "informational talk" can re-enter in biology without being just a metaphor but something real.

2. A short history of semiotics

The Greek word for sign, *semeion*, was first used in the Ancient Greece as a medical term. A sign was seen as the external manifestation of something that is hidden. A sign is always a sign about something that is not evident. There is a straight relation between the sign and what it stands for. Having a yellow face is sign of liver malfunction, and these two facts do always go together and tell something to everyone that can establish a link between them.

The relation between the sign and what it stands for could be seen from two different perspectives. From a certain point of view, the fact that the sign is always related to the object it points at it can be stressed. However, the same sign pointing at the object could be seen as the item that leads an observer to become aware of something that was not evident. Notice that in this latter approach the observation and interpretation of the sign is paramount, and the whole scheme entails three items rather than two[1].

It was Augustine of Hippo the first to take the second approach to the question of the sign seriously. He defined the sign as «anything *perceived*, which in so doing, causes something other than itself to come into *awareness* (*Signum est quod se ipsum sensui et praeter se aliquid animo ostendit*)»[2]. At the same time, Augustine stresses the ontological status of the relation between sign and object: «*a sign is something* which, offering itself to the senses, conveys something other to the intellect (*Signum…est res praeter speciem quam ingerit sensibus, aliud aliquid ex se faciens in cogitationem venire*)»[3]. This relation is not accidental but essential; it is something that affects the very nature of the sign and the object. Hence, an interesting question emerges: «Is it "perception" and "awareness" on the part of some agent that *gives* a sign its representational efficacy – or does the agent merely "apprehend" a relation in the world that is already there, regardless of its apprehension?»[4].

Aristotle also dealt with the question of signs, but the influence he had is tightly related with the availability of his works. Indeed, for six centuries there was only a Latin translation of Aristotle's *Organon* available. In this work he

[1] Cf. D. Favareau, «The Evolutionary History of Biosemiotics», 5.

[2] Quoted in D. Favareau, «The Evolutionary History of Biosemiotics», 6.

[3] Quoted in D. Favareau, «The Evolutionary History of Biosemiotics», 6.

[4] D. Favareau, «The Evolutionary History of Biosemiotics», 7.

only talks about signs regarding human language. The translation of Aristotle's biological works came later, but the doctrine of signs was already applied only to human experience, leaving out biology from that categorization[5].

We have to wait until the 17th century to see how a theory of signs embracing the whole reality begins to appear and to be taken seriously. John Poinsot defines the sign as a triadic relation: *x* as *y* to *z*. This formulation «resuscitates, the naturalistic Aristotelian understanding of a world of creatures whose internal organization give rise to their external interactions and vice-versa. In such a world, mind-dependant relations and mind-independent relations are tightly woven»[6]. Poinsot tried to recover a formulation that had been lost for centuries. However, his colleagues did not catch up this intuition, since most of them followed the path opened by Descartes.

Descartes broke away with the philosophical ideas that he received and tried to reconstruct the building of knowledge by himself with his reason as the only instrument for such work. The separation of the *res cogitans* from the *res extensa*, cut in some sense the correspondence between the inner and the outer world. The outer world is something different from the inner one, where reason deals only with clear and distinct ideas. Then, what reaches our senses are no more the signs referred to objects, but something that separates our minds from things[7].

The concept of information was coined in that epistemological framework, becoming just a product of material interactions that could be formalized by mathematical methods. Information is then linked to the *res extensa*, to the physical properties of objects. That is the paradigm that ruled when biology began to be an emerging science. Consequently, the informational paradigm applied to biology leaves out the meaningfulness and usefulness of informational exchanges, as Favareau remembers us:

---

[5] «Thus, the breaking apart of the subordinate study of human words and propositions in *De Interpretatione* form the superordinate study of animal organization and interaction in the world that Aristotle develops in *De Anima* – a more or less accidental bifurcation owing to the contingencies of history – became the starting point of a developmental pathway whose alternative trajectory would remain *terra incognita* long after the end of the Middle Ages and right up to the last half of the twentieth century». D. FAVAREAU, «The Evolutionary History of Biosemiotics», 10.

[6] D. FAVAREAU, «The Evolutionary History of Biosemiotics», 12.

[7] Cf. D. FAVAREAU, «The Evolutionary History of Biosemiotics», 16.

> For Francis Crick, articulating his "central dogma" of genetic inheritance, "information" was synonymous with "the sequence of amino acid residues" *per se* while for Claude Shannon and Warren Weaver, "information" was the diminution of uncertainty in a system absolutely without regard to cognitive or semantic considerations. Both Crick's notion of "information flow" from gene to protein and Shannon and Weaver's mathematical theory of "communicating across a channel" thus explicitly deny that the "information" that they are talking about "means" anything in the sense that we associate with the word "meaning"[8].

The point here is that "meaning" is strictly associated with language and, hence, with symbolic relations. That is why very often the "informational talk" in biology is seen as a metaphor, not as a truly explanation of what really happens in nature. The symbolic nature of our language is projected into the whole living world, trying to explain the structure of the living appealing to one of the most impressive traits of the most evolved creature: human language. Semiotics, on the other hand, try to do the other way round; just to find the common semiotic structure that pervades the whole living tree, the very structure out of which has emerged also human language[9].

After the discovery of DNA structure there was a group of philosophers and biologists that begun to think about how to match signs and living beings. People from different disciplines began to reflect about semiotic as an interdisciplinary category that could help to understand the basilar structure of life[10]. The semiotic model of Charles Sanders Peirce inspired most of them. I myself am inspired in the writings of this great philosopher, who has very acute intuitions when dealing with those issues.

---

[8] D. Favareau, «The Evolutionary History of Biosemiotics», 22.

[9] «A semiotic process is not a ghostly, mental, human thought process. Rather, it is, in the first instance, nothing more nor less mysterious than that natural interface by which an organism actively negotiates the present demands of its internal biological organization with the present demands of the organization of its external surround». D. Favareau, «The Evolutionary History of Biosemiotics», 24.

[10] For an historical account of the people involved in these first steps of what would be biosemiotics, see D. Favareau, «The Evolutionary History of Biosemiotics», 25-46.

3. The peircean paradigm

If we talk about semiotics, two big names must be taken into account: Ferdinand de Saussure and Charles Sanders Peirce. Roughly speaking it can be said that Saussure establishes a dyadic relation between sign and meaning, between signified and signifier. On the other hand, Peirce thinks about semiotics as an act of interpretation. It is not a static relation but a dynamical one. Peirce defines a semiotic system as a triad of sign, object and interpretant.

Saussure applied his dyadic relations mostly to language and human communication in general. The clearest example of a sign is speech. The relation between the sign and the meaning socially established is something static. Once the meaning is fixed, the relation with the sign is well established. Saussure studied the signs as human facts. Signs regard human communication and their sharing within a social community. This kind of study is named semiology, and because of its nature it is linked also to social psychology.

Peirce's model is known rather as semeiotics[11], not semiology. He started considering the dynamical act of human cognition, but he broadened his theory to the whole nature. Peirce's concept of sign is triadic, and is founded in the irreducible relation between sign, object and interpretant.

3.1 *One, Two, Three*

I cannot explain in detail here the whole conceptual universe of Peirce's semeiotics, but it is needed to understand some key categories in order to grasp the significance of the use he makes of signs. Peirce introduces three categories that help him to understand and develop semeiotics as a science of the signs that can be applied to everything. Those categories are irreducibly present in every sign.

Firstness makes reference to feeling, to the presence of something here and now with its immediate sense of being present. Chance is the paradigm of firstness, the novelty that arises before being associated with any other thing.

[11] «Semeiotic, as a branch of philosophy, is a formal, normative science that is specifically concerned with the question of truth as it can be expressed and known through the medium of signs, and serves to establish leading principles for any other science which is concerned with signs in some capacity». J. Jakób Liszka, *A General Introduction*, 14.

Secondness introduces the sense of reacting upon something different. There is another thing to which the first object has to enter in relation with, and this opens the loneliness of the Firstness to the dyadic interaction of the Secondness. *Thirdness* implies something between the former two. It has to do with the habit-taking of the relation of the Firstness with Secondness. Thirdness establishes some habits that link the relation between the first and the second to a third that is an end. Thirdness is a means to an end, a superior instance to which Firstness and Secondness are referred in order to achieve an end[12].

3.2 *What needs a sign to be a sign?*

> A sign, or *representamen*, is something which stands to somebody for something in some respect or capacity. It addresses somebody, that is, creates in the mind of that person an equivalent sign, or perhaps a more developed sign. That signs which it creates I call the *interpretant* of the first sign. The sign stands for something, its *object*. It stands for that object, not in all respects, but in reference to a sort of idea, which I have sometimes called the *ground* of the representamen[13].

Peirce states the formal conditions a sign has to have in order to be considered as such. Firstly, the sign has to have a representative condition that allows to establish a correlation between the object and the sign. Secondly, this relation is established according to some respect or grounding that is known as the presentative condition of the sign. Thirdly, the sign has to somehow determine an interpreter. Roughly speaking, the sign has to have some meaning for someone (or something). This third characteristic is the interpretative condition of the sign. Lastly, as has been said, the relation between the sign, the object, and the interpretant is an irreducible triadic relation. A sign has to represent something to someone in a certain respect[14] (Figure IV.1).

---

[12] Cf. N. Houser – C. Kloesel, ed., *The Essential Peirce*, 4-5.

[13] Ch.S. Peirce, *Collected Papers*, 2.228.

[14] Cf. J. Jakób Liszka, *A General Introduction*, 18.

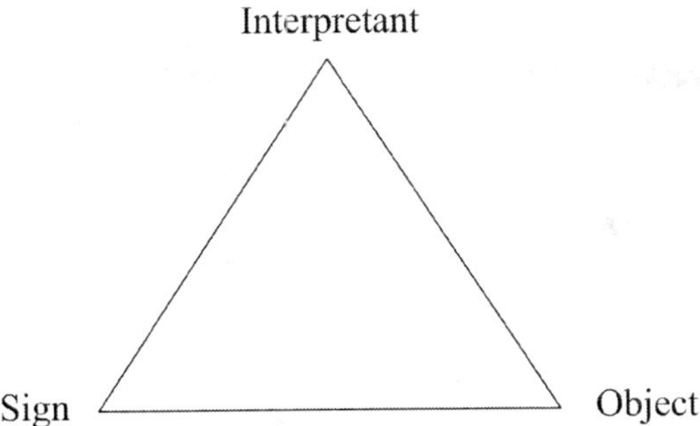

Figure IV.1: Semiotic diagram according to Peirce, where sign, object and interpretant have a triadic relation that cannot be decomposed into pair-parts.

Peirce makes a very detailed classification of signs, which we do not need to give a full account here. However, it is worth noting how the classification that Peirce makes has a correlation with the categories of Firstness, Secondness and Thirdness. Regarding the three main types of signs, Peirce says that

> the first is the diagrammatic sign or ***icon,*** which exhibits a similarity or analogy to the subject of discourse; the second is the ***index,*** which like a pronoun demonstrative or relative, forces the attention to the particular object intended without describing it; the third is the general name or description which signifies its object by means of an association of ideas or habitual connection between the name and the character signified[15].

The icon is a sign that is characterized by its likeliness to the object it refers to. But precisely because of this likeness, it is not necessary that the object to which it refers grasps the features of the icon. So we can say that the icon is a First that refers to its very characteristics, and it will retain its likeliness to the object even if the object disappears. The icon is possible due to its quality. Its properties can be used as a sign because they resemble some other objects. The icon can no carry any information on its own, since there could be many things that resemble it but they are not connected in any way. For doing that, there has to be an indexical relation.

[15] CH.S. PEIRCE, *Collected Papers*, 1.369.

The index, for its part, is a sign that is explicitly refers to an object. It is like a label that is tagged to an object, indicating the relation between them. The object to which it points modifies the index. There is a true modification of the index by the object, since now there is not only the iconic nature of the sign, but also an indexical relation to an object.

Finally, the symbol is a sign referred to an object through a law. Then, the symbol makes an association that establishes the habit through which it is referred to the object [16]. So we can see that there is a relation and a determination between the different kinds of signs. It would be difficult to find a pure icon, index or symbol. Often we will find entities or processes that would have iconic and indexical features interwoven.

Peirce also distinguishes an immediate and a dynamical object. The immediate object is the object seen from the sign's point of view. It is the feature of the object that is showed immediately by the sign. The dynamical object has to do with the determination that the sign would act upon the object. It refers to a process of determination of the object by the sign; it is like the invisible hand that guides the semeiotic process towards its end. This semeiotic process could not be already achieved, but is the end towards which the determination tends to [17]. The dynamics of semeiotics is progressive, and needs to make its way towards the representation of an object by a sign under some respect. That is why «the sign can only be said to represent its object if there is an interpretant which correlates the two, and that can be done only if there is a ground upon which to make that correlation. This grounding is the *effect* of the dynamic object, the determination of the dynamic object» [18].

The third element needed for a sign to be a sign is the interpretant; some instance that establishes the relation between the sign and the object. In this case there is also a division of interpretants. The immediate interpretant is the total unanalyzed and raw effect. Is the Firstness that feels the effect of the relation with the object and the sign. The dynamic interpretant is the effect produced on some interpretative agency that sees it as a Secondness upon which a reaction has to be performed. In the end, the final interpretant is the law or rule a sign

[16] Cf. R. FABBRICHESI LEO, *Introduzione a Peirce.*

[17] Cf. J. JAKÓB LISZKA, *A General Introduction*, 21.

[18] Cf. J. JAKÓB LISZKA, *A General Introduction*, 24.

has in an interpretant, compelling it to take the habit of relating the sign with the object in order to determine the interpretant instance or Thirdness[19].

As it has been shown, this model allows for a very rich approach to the notion of sign. However, my interest about the application of this framework to biological systems makes it difficult to adapt the Peircean nomenclature as it is sketched above (Figure IV.1). The concept of "object" is not adequate when we talk about biological systems. An object in biology could be an item or a process, something that has a spatial-temporal organization. For example, an object could be the genetic material, but can be also the whole process that leads to a selective activation of a certain gene in this genetic material. That is why I prefer to use the term "referent" instead of "object".

The notion of "interpretant" is also problematic, since this concept is frequently associated with a subject that interprets something. I claim that the cell does not interpret anything in the sense that we understand that for the human mind; and that is why I retain the term "icon" to name this instance that establishes a relation between the sign and its referent. The icon somehow represents the object in the inner cell that the sign refers to. It can be said that the cell *hardwires* somehow a link between the sign and its object in order to establish the continuity of this linkage. The structure (understood in general as a spatial-temporal process) that emerges from this linkage is the icon, which in turn is the higher instance to which both sign and object are referred once the linkage is established. Then it could be seen that the icon has a very important role in coupling the sign to its referent, but there is no need to postulate a mind to do such coupling, as if it were the linkage between the words and their concepts[20].

Let us see an example from biology to shed some light on this terminology. It is well known that cells have a very sophisticated mechanism of communication between them. The cell has proteins on its surface called receptors that interact with other chemicals from the environment, which are delivered by other cells or simply present in the surroundings of the cell. The fact is that some of these chemicals interact with the cell receptors and generate some sort

---

[19] Cf. J. JAKÓB LISZKA, *A General Introduction*, 26-27.

[20] For the changes in the peircean terminology I have been inspired by G. AULETTA, *Cognitive Biology*, 248-251.

of response inside the cell. Normally this response begins with the activation and/or inhibition of certain proteins, transmitting the alteration to the nucleus, where finally there is some activation or repression of genes in response to that alteration (Figure IV.2). The whole biochemical pathways that are implied in the transmission of that signal from the cellular membrane to the activation of some genes in the nucleus, can be consider as the icon that links the sign (the molecule that interacts with the receptor) to its referent (that for what the sign stands for). Then, the whole biochemical pathway is the way the cell *hardwires* the link between sign and referent.

In the hypothetical example shown in Figure IV.2, the molecule that interacts with the receptor is the sign, which stands for something that the cell does

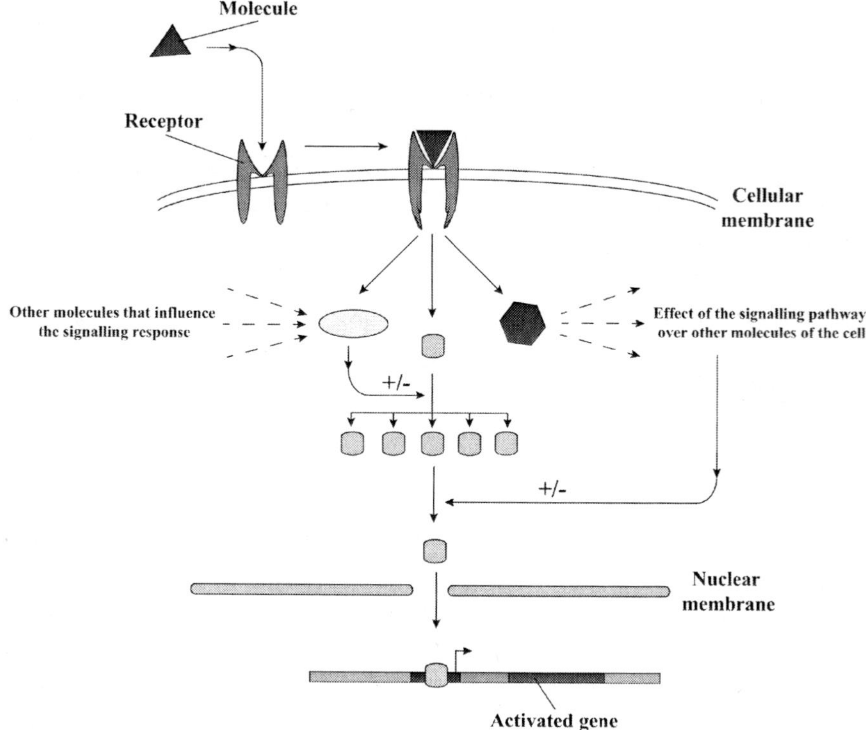

Figure IV.2: A hypothetical intracellular signal pathway. The signal begins with the interaction of a certain molecule with a specific receptor in the cellular membrane. Then the receptor changes its conformation, activating other molecules inside the cell. Those molecules are embedded in the metabolic web of the cell, and then there is a multiple influence of activations (+) and inhibitions (-) that modulate the signal. Finally, in this case, the signal reaches the nucleus in order to activate a gene to respond to the signal.

not experience directly (the referent, that could be an enemy, a source of energy or something else). The icon is the whole spatial-temporal network of chemicals that convey the signal from the cellular membrane to the nucleus. Since the gene activated is a response to the stimulus of the sign, we can see that the icon is what couples the sign with its referent in a certain respect. This coupling allows the cell to properly respond to a given stimulus. So, from a certain point of view, it can be seen that the icon is some sort of interpretative instance, but I prefer to call it just the icon to avoid the problems mentioned above.

This association between the sign and the referent through the icon is a semiotic relation. The importance of the chemical nature of the elements involved is relative, since «a sign does not represent physical or chemical properties but the *function* a certain item has, plays, or represents for the organism»[21]. What really matters here is that the linkage between the elements were done in such a way to allow the cell to properly respond to a certain situation, which does not control all its details, but just interacts with some signs that stand for something more. Then, a new scheme of the relations among the elements is needed (Figure IV.3).

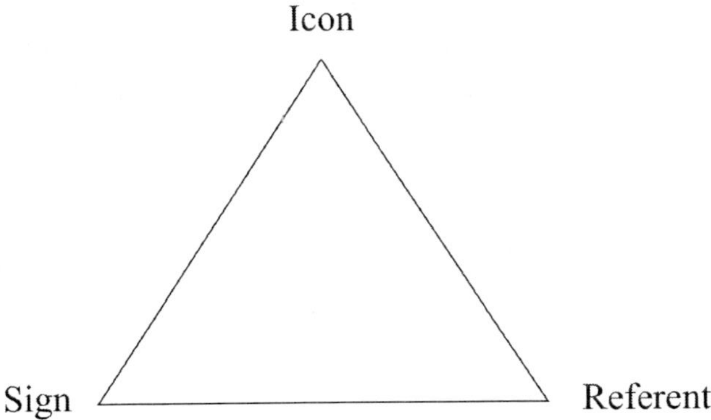

Figure IV.3: A new scheme for the application of the peircean theory of signs to biological systems. The sign stands for a referent, to which it is coupled through the icon, that is somehow an inner representation of the referent, since it establishes a link between it and the referent it stands for.

[21] G. AULETTA, *Cognitive Biology*, 249-250.

3.3 *Triadic relations and the process of semeiotics*

The relation between the sign, the referent and the icon cannot be reduced taking them separately or even trying to establish pair-relations (i.e. dyadic relations) among them. Peirce cannot think about something without referring to this triadic scheme. Reality is no longer something with which we establish a dyadic relation of object and subject and that is all. Peirce defends the fact that anything stands for another thing in a certain respect. The dyadic relation is open to a triadic scheme, but not as something added to the former, but as something that is essential to the system.

This triadic relation is irreducible, since its three components have no identity as sign, referent and icon taken separately. When a (first) thing or event is a sign that stands for another thing (second), which is the referent of the sign, there is the production of an icon (third) that establishes the relation among them[22].

As has been said previously, Peirce thinks in a semeiotic framework that could be applied to the entire universe, not only to human mind. Indeed, he himself used the term *quasi-mind* when referring to interpretant instances that are not human minds. This allows us to treat organisms, even unicellular ones, as icons of the dynamical process of semeiotics[23].

The concept of icon, then, can referr to the organism as a whole, pointing out the link of its very needs and the environmental cues that can bring some information about what is out there. The icon leads me to briefly consider the question of the self. When we talk about a self in cellular biology, we are not defending the fact that the cell has a mind or intentional behaviour properties. Neither does it mean that we have to suppose an external observer that interprets the linkage between signs and referents. The icon, at least in biology, could not be ontologically separated from the other elements of the triad. When we think of a human interpreter, the act of interpretation sees the sign-referent link from outside. On the contrary, the absence of a mind in a biological system like the cell, constrains us to think about the interpretant in a different way, as Barbieri remind us:

---

[22] Cf. T. VEHKAVAARA, «From the Logic of Science», 260-261. I do not use here the same terminology as Vehkavaara, but the relation among the elements is the same.

[23] Cf. CH.S. PEIRCE, *Collected Papers*, 4.550 and 4.551.

> The Morse code is built by a human being (the interpreter) who attaches a meaning to arbitrary combinations of dots and dashes (the signs) by setting them in correspondence with the letters of the alphabet (the objects). In protein synthesis, combinations of three nucleotides (the signs) are set in correspondence with amino acids (the objects) by special compounds known as *adaptors*, i.e. *by molecules that perform two independent recognition processes* (one in the nucleotide world and one in the amino acid world). In this case the agent implementing the code (the interpreter) is the entire set of adaptor molecules. Protein synthesis, in short, is a true semiotic act because the three defining protagonists of semiosis are all present, the only difference with cultural semiosis being that the interpreter is not outside the system but very much inside it[24].

The object Barbieri talks about (I may say the referent) is something potential, not embraced in all its richness. It is somehow linked with the notion of potential information (Chapter III, section 2.1). The sign, on its side, is something actual, that stands for a referent in some respect, revealing a part of the potentiality the referent has. The potentiality of the referent could be actualized through new signs or through old signs with new respects. The semeiotic process, then, is subject to evolution.[25].

The icon is between the sign and the referent, between the actual and the potential, between function and information. The evolution of the semeiotic act depends on its boundary conditions; that is, the previous history of semeiotic acts linking referents and signs in certain respects. Hence, a certain semeiotic act depends on the history of the semeiotic acts that have preceded it. The icon has some kind of memory of the past semeiotic acts; a memory that records the linkage between signs and referents through a certain semeiotic act. The system,

---

[24] M. Barbieri, «Has Semiotics Come of Age?», 110.

[25] I am not going to deal here with the implications of semeiosis for the theory of evolution, but is something that is also studied under the peircean paradigm. «The idea of semiosis, in which semiosis is realized by the representations of initial potential intentional objects of the primary logical level of Peirce (Firstness) makes the semiotic model deep and applies particularly well to evolutionary process. The whole story of evolution can be perceived as an interplay between the potential intentional objects, most often not visible at all but hidden and underlying, and the actual representations of them. The evolutionary travel of the potential objects into representamens may take eons to realize. Apparently, there are a lot of potentials that will never be materialized». T. Jämsä, «Semiosis in Evolution», 79.

then, has in stock the semiotic solutions it has found in the past and those that it actually performs.

The case of the *phyB* (Chapter III, section 3.2.4) gene is an example of this semiotic linkage. *phyB* gene is a photoreceptor that has been used for different functions depending on the organism and the surrounding conditions. The gene codes for a protein sensitive to light, and this feature can be used by the cell (the final interpretant) as a sign that can indicate the time of flowering, the day-length or any other referent to which the sign can be linked to.

This is what happens for example in exaptation, which is «the evolution of a function of a gene, tissue, or structure other than the one it was originally adapted for; can also refer to the adaptive use of a previously nonadaptive trait»[26]. Going down until molecular level, this means that a gene that is used in a certain way, can be also recruited with a different respect, to have a role in a different function.

Besides boundary conditions, initial conditions, which are those conditions out of which the semeiotic act begins, also exist. These conditions are set by the elements that take place in the process, and depend on the nature and interaction of those items with each other. The initial conditions of a chemical reaction within a cell could be the pH of the medium, the availability of substrate for this reaction, the state of activation or repression of the gene that codifies for the enzymes that perform the reaction, and many other circumstances that can influence in the final result.

Semeiosis is a dynamical process. It is not a static relation, but a process that has a particular dynamicity. Peirce says that the semiotic process takes habits. The system adopts certain semeiotic processes rather than others. Why the system does that? My guess is that the criterion used to select certain semiotic processes rather than others is its linkage to the functions the system has to deploy in order to maintain itself alive. Being alive is the final goal of any semeiotic process.

Besides the final interpretant, Peirce also distinguishes a dynamical one. The dynamical interpretant becomes a sign for another round of the semeiotic act. The referent is always the same, but in every semeiotic act it is related to the interpretant through different signs, enriching their semeiotic content and

[26] D.J. Futuyma, *Evolution*, G-3.

gradually showing all the semeiotic possibilities they could have. «The unique character of *semiosis* is that it tends to be a *progressive* process. New signs in the chain expose piecemeal the whole information content about the object [referent] that the original representamen [sign] contained more or less hidden»[27]. The process ends with the final interpretant.

I would rather say that the relation between the sign and the referent, linked by the icon, is something dynamical and, therefore, can evolve. The referent is always the same, but it could be differently linked through its sign by diverse icon patterns, giving many possibilities of transmitting the information the sign conveys about the referent.

> So where is regulation? What regulates what? Because such complex networks are not exclusively determined by mass-energy restrictions, a random event, such as the building up of a given extracellular signal's concentration in the periphery of a cell, will produce a non-random response to such an event. This non-random response is not deterministic in the physical sense because the system that reacts to the random event has a repertoire of responses of which it will select the optimal one based on a global interpretation of the context – this is why we need to think in terms of semiotic (triadic) logic instead of exclusively in terms of mechanical (dyadic) logic[28].

3.4 *An example from biology*

Let us try to analyse in these characteristics of the sign using an example of life sciences to see the adequacy of this paradigm to be used in molecular biology. We are going to explain the Peircean semiotic paradigm illustrating it with the glyscosilation reactions in some proteins. As it is known, once proteins have been synthesized by the ribosomes, they have to pass through some *editing processes*. Some of those editing processes could be the adding of methyl groups to some amino acids, the association with cofactors that would help to protein's function (as it happens with the haeme group in haemoglobin), the addition of some carbohydrates (glyscosilation) and many others. I think that the process of protein glycosilation can be a good example to show the relevance of the Peircean paradigm to molecular biology.

Approximately half of the proteins of an eukaryotic cell are glycosilated.

---

[27] T. VEHKAVAARA, «From the Logic of Science», 261.

[28] L.E. BRUNI, «Cellular Semiotics and Signal Transduction», 375.

The addition of those carbohydrates takes place in the endoplasmic reticulum (ER). The carbohydrates are added to the amide nitrogen atom in the side chain of amino acid asparagine (and this reaction is called N-linkage). It can be also added to the oxygen of the side chain of serine or threonine (in which case is named O-linkage)[29].

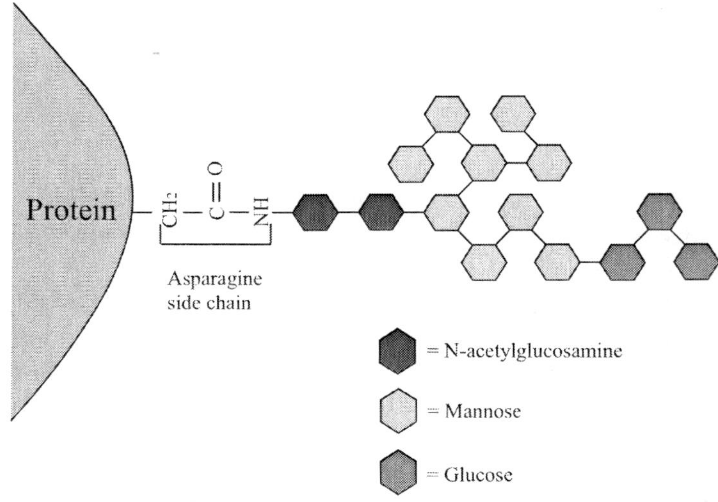

Figure IV.4: N-glycosilation of a protein. The addition of the carbohydrate structure is always done on the side chain of an asparagine residue. Two residues of N-acetylglucosamine, nine of mannose and three of glucose form the carbohydrate tree-like structure. Adapted from B. Alberts– al., *Molecular Biology of the Cell*, 737.

When a proper amino acid is exposed during protein synthesis to the glycosilation enzymes, a preformed oligosaccharide structure is linked to it. This common tree-structure is made of two N-acetylglucosamines, nine mannoses and three glucoses (Figure IV.4). All the diversity of glycoproteins comes from the modification of this basic structure. Normally, when they are still in the ER, the three glucoses and one mannose are removed. It is well known that those oligosaccharides attached to proteins are used to mark each stage of protein

[29] Cf. J.M. Berg – J.L. Tymoczko – L. Stryer, *Biochemistry*, 316.

folding. So we can see here that an oligosaccharide structure is used as a sign of the folding state of a protein (the referent).

In the ER there are two chaperone proteins, named calnexin and calreticulin, which are very important in helping other proteins to achieve their proper folding. These proteins can bind the oligosaccharide structure and their role is to retain incompletely folded proteins in the ER until they are properly folded. Indeed, calnexin and calreticulin bind the N-linked oligosaccharides that have just one of the three glucose residues, when they have lost the other two. Then, they bind those proteins that have noy yet removed the third glucose residue. Only when the third glucose is removed, calnexin and calreticulin dissociate from the protein and allow it to leave the ER (Figure IV.5).

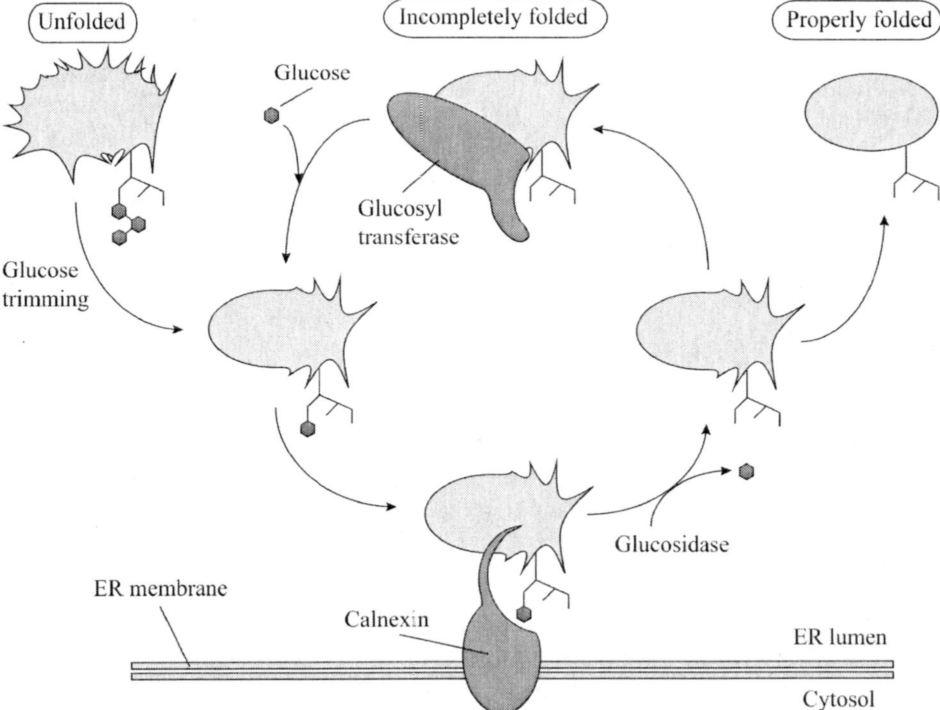

Figure IV.5: N-linked glycosylation is a control mechanism to make sure that proteins leaving ER are properly folded. Glucosil transferase is a protein that binds unfolded proteins and adds them a glucose residue. This residue is recognized by the ER membrane protein calnexin and the soluble protein calreticulin (not shown). On the other hand, there is a glucosidase that removes glucose residues from those folding proteins. Calnexin and calreticulin, with other chaperones, help proteins to reach their proper folding. When the proteins are correctly folded, glucosil transferase stops from adding glucose residues; then, its glucose residue is removed by glucosidase and consequently the protein cannot be recognized by calnexina and can leave the ER.

The question is: how do calnexin and calreticulin *know* that the lack of the third glucose residue is sign of the protein's proper folding? The answer is that there is a third partner at play that helps to make the selection. There is a glucosyl transferease that keeps on adding glucose every time a folding protein looses it. This transferase adds glucose only to those proteins that are unfolded, and then prevents their dissociation from calnexin and calreticulin. Only when the protein reaches its proper folding glycosyl transferase stops adding the last glucose residue, and then calnexin and calreticulin can dissociate from them when the third glucose residue is definitely lost, allowing the protein to continue its way through the ER to the Golgi apparatus. This is a dynamical system in which «an unfolded protein undergoes continuous cycles of glucose trimming (by glucosidase) and glucose addition (by glycosyl transferase), maintaining an affinity for calnexin and calreticulin until it has achieved its fully folded state»[30].

We can see that the carbohydrates that are linked to these proteins are treated, in this particular context, as signs. Indeed, they stand for another thing, which is the folded state of the protein. This is the respect, the ground of the sign; carbohydrates are used here as signs of the proper three-dimensional structure of a protein. There are other contexts in cell metabolism where carbohydrates are used in other different respects; but in this case they are signs for protein proper folding. Calnexin and calreticulin do not have to control the whole structure of the protein, but just to make sure if the last glucose residue is there or not. Then, the last glucose standing is taken as a sign of a not-already-achieved-proper-folding or, in its absence, an achieved-proper-folding.

In this example, if we think in the framework of the Peircean paradigm, the referent would be the protein that is being folded. Moreover, we can even distinguish here the immediate and the dynamical referent. The former is the referent seen from the point of view of the sign. If calnexin and calreticulin are bound to the oligosaccharide of a protein, the immediate referent it stands for is the unfolded protein, or the not-correctly folded protein. However, there is also a dynamical referent, to which the semeiotic process tends. Indeed, the aim of those interactions is not just to tag the unfolded proteins, but also to drive their folding to the correct three-dimensional structure they have to achieve for their proper functionality (the dynamical referent). The dynamical referent is

[30] B. Alberts– *al.*, *Molecular Biology of the Cell*, 739.

the final state to which the whole semeiotic process tends to in order to achieve the end-state that really contributes to the benefit of the system. Then, the dynamical referent is what constraints the sign to be such a sign; the dynamical referent determines the immediate one. The pattern of carbohydrates used as a sign of the folding state is a proper sign if it has the capacity to determine an icon to read it as a sign of the referent. The relation of the sign with the referent determines the icon, which somehow acknowledges and uses the relation between them for its own sake. In this case, the icon would be the whole network of spatial-temporal intereactions between the elements: calnexina and calreticulin, their spatial location in the ER membrae, the chemical differentiation of the folding proteins that have the last glucose from those that do not have yet, etc. All these elements constitute the icon that permits the linkage of the oligosaccharides' metabolism with the protein folding.

In this example, then, the icon would be the whole interactive network of elements that cooperate in order to make sure that the proteins that leave the ER are properly folded. It is very interesting to note that the icon does not have to be an item, could be just a process; as in this case the process of protein folding. Peirce distinguishes the immediate interpretant (I would rather name it immediate icon), the dynamical, and the final one[31]. The immediate icon is the total unanalyzed effect in its immediate and raw consequences. In this case, the immediate icon would be the process of folding.

The dynamic icon is the effect produced by the sign; in this case, it is an effect upon the folding process of the protein as the end of the whole protein synthesis process. Here we are in front of the particular effects, but we cannot talk about the meaning of this process. The final step of the process is the one that establishes and links the dynamics of the sign with the usefulness it could have for the system as a whole. The final icon establishes the habit through which the sign stands for the referent to which it is linked. I think that in molecular biology the final icon of every single semeiotic process within the cell is the cell itself. Jakób Liszka explains it as follows:

> Generally speaking, the final interpretant [icon] can be understood as the means by which a sign becomes connected or interrelated into a system of signs, that is,

[31] Cf. J. JAKÓB LISZKA, *A General Introduction*, 24-31.

translated "into another system of signs". "A system", according to Peirce, "is a set of objects comprising all that stand to one another in a group of connected relations". Signs become interrelated in such a way as to create certain generalizable effects, that is, the systematic translation of a sign[32].

Let us continue with the N-glycosilated proteins to see up to what extent the semiotic nature of these oligosaccharides is paramount. The fact is that, despite the work chaperones perform in the ER, there are many proteins that do not achieve their proper folding state. Those proteins have to be translocated to the cytosol and degraded. But the question is: how do the misfolded proteins are distinguished from those that are being folded? Both kinds of proteins are not properly folded, but the formers have to be degraded because there is no chance of achieving their folding, while the latter have to take their time to be properly folded by chaperones. The N-linked oligosaccharides are important also for this issue. The sign used to distinguish them is the velocity of the trimming of mannose residues by a family of enzymes called mannosidases. The slow trimming of the mannoses located in the oligosaccharide structure creates a new structural pattern that is recognized by the retrotranslocase (an enzyme that drives proteins from the ER to the cytosol), and then those proteins are delivered to the cytosol to be degraded. On the contrary, if the protein folds faster than the action of mannosidase, it is not translocated and escapes degradation. The misfolded protein is translocated to the cytosol and its remaining N-linked oligosaccharides are removed. Then it is tagged with another molecule called ubiquitin, which marks it in order to be recognized by a protein complex known as proteasome, which degrades proteins tagged with that molecule[33].

It can be seen here that the glycosilated proteins that are translocated to the cytosol, are in a context in which their N-linked oligosaccharides are sign of a protein that cannot achieve its proper folding state. The same oligosaccharide triggers a different response depending on whether it is in the ER lumen or in the cytosol. That fact triggers a response that removes this oligosaccharide and marks the protein with another tag to be degraded. The sign is the same, but just removing one mannose residue and locating the protein in the cytosol, it

---

[32] J. Jakób Liszka, *A General Introduction*, 27.

[33] Cf. B. Alberts– *al.*, *Molecular Biology of the Cell*, 739-740.

leads to a totally different effect. The protein is no longer trying to achieve its proper folding, and it has to be degraded by the proteasoma.

4. **Semiosis in living systems**

The cell has to have some structures, pathways, reactions, controls, etc, just to make sense of what happens outside and in its inner medium, to evaluate and see if what is happening affects its survival or not. The cell must to assess with what happens in the environment. It has to be sensitive to the proper signals and to integrate them for the sake of its survival. Not every piece of information is useful. The cell has to be able to distinguish mere impacts than useful information for its own sake[34].

Every sign is integrated according to the surrounding in which it is embedded. A given sign has many potential effects, depending on the surroundings and the linkage with its referent through the icon that somehow recognizes it. The structure and the chemical history of the cell act as the semiotic framework in which a given sign is integrated. From a certain point of view, the internal structure of the cell, especially all the pathways and reactions that are sensitive to environmental changes, are like an internal image of what happens outside. The cues received from the environment are integrated in the cell in a way that allows it to monitor what happens beyond the cellular membrane.

In this section I want to focus on the conditions that are needed for the cell in order to deal with information as a sign. It would be of great help to characterize the cell and its environment, not just from a chemical point of view, but also taking into account the semiotic content that those chemicals can carry when embedded in the proper context. This would lead us to consider that there is an epistemic cut, which differentiates the material stuff of items from their semiotic use. Then, the cell would no more be seen as a bag of chemicals, but as an organized system that can deal semiotically with everything it comes across.

4.1 *Umwelt*

A very important notion to understand how semiosis takes place in biological systems is Jakob von Uexküll's Umwelt. Roughly speaking, we can say that

[34] Cf. L.E. BRUNI, «Cellular Semiotics and Signal Transduction», 373.

an organism's Umwelt is the semiotic sphere in which it lives. Organisms live in a given environment, and there is a semiotic exchange between the formers and the latter.

> Current biosemiotics deals with biophysics and biochemistry, with molecular processes of life that are ultimately based on subatomic (quantum mechanical) and chemical substances and the biological factors maintaining life. We cannot draw a clear divide between the "dead nonsemiosis" and the semiosis of life. Life represents no doubt a new emergent level but it has not come into being from scratch and there is a certain link between the physicochemical and biological processes[35].

The environment, then, does not carry the same information for all organisms. Two different organisms do not equally experience a given environment. Physical and chemical cues do not have the same meaning for two different organisms. How an organism perceives its surroundings depends on its biological constitution. In fact, if an organism needs glucose for its survival, it must have some mechanism to perceive, internalize and use glucose. On the contrary, if an organism grows only with fructose, it must have mechanisms sensitive to that sugar. Then, as both organisms in the same environment, we can see that the presence of glucose would not have the same referent for them. The environment is physically and chemically the same, but organisms do not perceive it equally. The biological constitution of the organism constrains the semiotic referent that environmental cues could vehicle[36].

Umwelt has to do not with the elements of sensation but with the object of experience. An organism could receive many impacts from the environment; but coming across a certain item does not make it an object of experience for the organism. However, the difference between sensation and experience does not necessary depend on the stuff of the item encountered, neither with the type of environment in which that item is; the difference is not in physics but in semiotics. What matters is a relation between a certain item and the experience it can vehicle for a biological system, whose perception of that experience would depend on its very biological constitution[37].

---

[35] T. Jämsä, «Semiosis in Evolution», 73.

[36] Cf. J. Deely, «Umwelt», 126.

[37] Cf. J. Deely, «Umwelt», 127.

This relation links the environmental cue to the organism's needs. What is perceived is different from the sensed because of its linkage with the needs of the organism through the semiotic network. Indeed, the needs of the organism have modelled its constitution in order to interact with the environment paying special attention to those cues that need to be perceived, not only sensed. So objects of experience are actively received, not passively sensed. The sensory mechanisms of the organism are tuned to perceive what really matters for the survival; a tuning work that natural selection did during evolution.

Organisms do not deal with signs because of their physical or chemical properties, but because they are signs that point at another thing which has some importance for the organism. The relation is established according to the organism's needs. There is no need of special signs made of a different stuff or entelechies that would endorse matter with special features. The only thing needed for an item to become a referent for an organism is to have some relevance for its own survival and to be perceived by the organism through a sign. The key point is the relation, not the item *per se*. Deeley remember us how Uexküll understood this relations:

> Uexküll compared each Umwelt to an invisible bubble within which each species lives. The bubble is invisible precisely because it consists of relations, since all relations as such, in contrast to things that are related, are invisible. The objective meaning of each world and each part within each world depends less on physical being that it does on how the relations constituting the Umwelt intersect. The difference between objects and things makes mistakes possible, but it is also what makes for the possibility of meaning in life, and different meanings in different lives[38].

I agree with Deeley but, again, I would rather use the triad sign-referent-icon, instead of talking about objects and their meaning. The concept of meaning can be only used when the system is a human being, capable of abstract thinking and symbolic competence. I am interested in a more general framework, where the case of how humans deals with signs is just a part of a wider scheme where all life can be understood from an informational point of view.

[38] J. DEELY, «Umwelt», 130.

4.2 *Epistemic cut*

The relation between the matter of the sign and what the sign stands for is not a relation of necessity. There is a difference when we study a biological cue from a chemical standpoint than when we do that from the semiotic point of view. H. H. Pattee talks about this issue as the matter/symbol question. In fact, when we analyze a cell, it can be seen that everything within it is just matter. The cell is the basilar unit of life, it is true; but it is not made of a different stuff from the one we find in the non-living world. Neither do we find a special substance that is only present in living systems and which gives them their particular features. However, life has certain peculiarities, since life can deal with matter in a way that it is not reduced just to its material characteristics.

Pattee expresses this idea with the notion of epistemic cut. At a certain point, when matter acquires the degree of complexity that characterizes life, there is an epistemic cut that shows us that the cell does not use matter as a material source only, but assigns to it some semiotic content. I think that Pattee's epistemic cut takes place between the coding of a polypeptide chain and everything that comes after: its proper folding, its modification, its sorting and the control that is exerted over it by other molecules. Folding becomes a clear example of the semiotic use of information. Expressing a gene is just the activation of the coded information that is dormant in the DNA chain. But the polypeptide chain that comes out from the translation process is no more coded information and is modified by the cell as a whole in order to be more than just matter[39].

> The actual folding process is an entirely physical process of minimizing the energy under the semiotic constraints of the sequence. In other words, the strong-bonded sequence can be called informational because it is one of many physically equivalent alternative sequences, while the folding dynamics itself is not informational because

[39] «Folding transformations are the most fundamental semiotic processes in all living systems. Folding is fundamental because it is the process that transforms the passive symbolic gene sequences into the dynamic rate-control of enzymes. Folding transforms what are essentially rate-independent syntactically coded sequences into rate-dependent functional controls». H.H. Pattee, «The Necessity of Biosemiotics», 125. I would not use the term "symbol", as Pattee does. I rather prefer to use "sign" as a more global concept, since we will reserve the word "symbol" for the cultural signs of human beings.

no new information is added in the process of minimizing the energy. (There are special cases where folding information may be added from scaffolding molecules)[40].

The number of combinations of the amino acids of a single molecule is unfathomable. This poses the question of how the cell narrows the great amount of different possible combinations down to those that are really useful to it. Not every single possibility can be integrated in the cell's network. In fact,

> it becomes clear that only certain states in a very large number of possible ones could have significance for any finite system. This requires a standing reduction in the number of meaningful states by way of classifying recognized differences in only a few of the many degrees of freedom afforded by complexity, which few then carry the information used by, and meaningful to, the system. A functional complex system must, therefore, be able to classify its own states as well as those of its surroundings, and so must necessarily have information[41].

The dynamical constraints are the result of a history of contingencies. Then, the possible solutions that the cell has at hand depend on the past linkage of certain signs to certain referents. When the cell faces a new challenge, it essays if there is any already present possibility within it in order to modulate the perturbation and link it semiotically for the sake of its survival. If not, new combinations have to be tested.

4.3 *Semiotic closure*

The cell is a biological system that is clearly defined from a structural point of view. It is surrounded and limited by a cellular membrane, which individuates the cell and separates it from the environment. Although the membrane separates the environment from the inner space of the cell, it is also true that the cell has to exchange energy, matter and information with that environment to keep itself alive. In fact, the cellular membrane is not a closed frontier between the environment and the cell, but a semi-permeable boundary. The cellular membrane plays multiple roles in the cell; for example, it is involved in the transmembrane transport of small molecules, in the signal transduction of

[40] H.H. PATTEE, «The Necessity of Biosemiotics», 126.
[41] S.N. SALTHE, «What is the Scope of Biosemiotics?», 135.

environmental stimuli and the communication and adhesion to adjacent cells[42]. The cellular membrane is not the only membrane present in the cell. There are other types of membranes that further divide the inner cellular space, such as the endoplasmic reticulum, the Golgi apparatus, the whole network of vesicular traffic and the nucleus. These spaces are of great importance for other cellular functions such as genetic regulation, protein sorting, vesicular traffic and energy conversion[43].

However, there is still another characteristic of the membrane that is frequently forgotten or taken for granted. The cellular membrane makes the cell be a self. The notion of self is not reduced to the components of the system, but has to do with the system as whole. Auletta offers a definition of self that fits in with what I am talking about:

> By *self* I mean the systemic totality of the living organism, excluding everything that lies outside of it or that does not fully depend on this systemic organization. I wish to stress that the self is a *functional unity*, and for this reason it is only possible at a biological rather than physical or chemical level. The distinction between self and nonself is characteristic not only of the basic structures of life, but reaches all levels of organization and complexity, up to consciousness or the human language. So, an organism, by integrating in a new way systems that can be found separately in the physical world, somehow *duplicates* the world and is constituted as a universe apart[44].

The cellular membrane, then, not only closes the cell physically, but also allows everything that is in the cell to be part of the individual. The organelles, the cytoskeleton, ribosomes, etc, are no more autonomous parts and structures that play their tune regardless of the rest of the cell. The membrane is the first constraint that puts them together in a particular relation. This is what Cliff Joslyn calls the *structural approach* for the definition of a system:

> We call this view structural because it focuses on the specific given types of relations

[42] Cf. B. ALBERTS– *AL.*, *Molecular Biology of the Cell*, 615-650; 651-694; 879-964, respectively.

[43] Cf. B. ALBERTS– *AL.*, *Molecular Biology of the Cell*, 695-748; 749-812; 813-878, respectively.

[44] G. AULETTA, *Cognitive Biology*, 271.

> among specific types of entities. It entails entering into a relation among multiple entities, called parts, to form a new whole entity with new properties at a level hierarchically distinct from those parts. These new properties do not follow from considering the parts simply together as a collection, rather they must enter into a particular relation so that the whole is formed, resulting both from the parts as entities *and* from the particular way in which they are arranged, that is from their mutual interrelation and organization[45].

This structural approach allows us to understand the cell not as the sum of the physical and chemical properties of its parts, but as a system that, grounded in those physical and chemical properties, emerges from them due to their interaction as a whole. However, the structural approach is not the only way to describe the cell as a non-reductive system. Joslyn talks also of the *constructivist approach*, which «avoids concepts of existing entities with objective attributes, instead defining a system as a bounded region of some (perhaps abstract) space that functionally and uniquely distinguishes it»[46]. The cell, then, is not seen as a whole constituted by parts but by functional processes. The perspective adopted by the constructivist point of view is that of seeing the cell as a spatial-temporal coordination of interwoven functional processes. The important point is not the entities that are involved in one of those functions, but the processes that constitute that very function, and that of course, needs the material interaction of the entities described by the structural approach. «These systems are not *composed of* things, but are rather *defined on* things, and there is a clear distinction between their physical, "thinghood", and logical, "systemhood", properties»[47].

The notion of closure is essential to living organization, since it implies a configuration of elements that allow global properties to emerge. Pattee understands closure based on autonomous constraints, which controls the dynamical processes that take place inside the closed system. Constraints influence dynamics, but are decoupled from it.

> If the closure is observed from a hierarchical point of view then it is performed by an upper level of controls that, on the one hand limit the freedom of the underlying

[45] C. JOSLYN, «Levels of Control and Closure», 68.
[46] C. JOSLYN, «Levels of Control and Closure», 68.
[47] C. JOSLYN, «Levels of Control and Closure», 68.

> dynamic components and, on the other hand, give more freedom and creativity to the whole, precisely because autonomy from dynamics. Thus, whereas constraints of inanimate matter depend on the underlying components – in the *structural hierarchy* – the freedom of the constraints of the *control hierarchy* opens new possibilities for the system[48].

This point of view requires a new way of seeing biology. A cell is principally not a collection of molecules or cellular structures, but the result of a series of processes and relations, whose performances allow the cell to be alive. Of course the material structure furnished by the membrane, the organelles, the DNA, and the rest of cellular structures, is needed, but the cell, as a self, does not deal with this structures regarding its material composition alone, but its semiotic respect. The cell performs everything that is needed for its survival because it treats all its parts semiotically, linking the material aspect of its molecules to the processes that those very molecules have to be canalized to. Murphy says that this new perspective needs to change some of our mental habits when we want to fully understand the importance of processes over things.

> One has to give up the traditional Western philosophical bias in favor of *things*, with their intrinsic properties, for an appreciation of *processes* and *relations*; the components of systems are not things, but processes. Systems are different from both mechanisms and aggregates in that the *properties* of the components themselves are dependent on their being parts of the system in question[49].

4.4 *Semiotics and levels of organization*

Semiotic closure is a fundamental condition for a system that has to deal with information from a semiotic point of view. Some sort of closure is needed in order to establish the semiotic relations between all the elements present in the system. It is also true that a system can grow, learn, and establish new semiotic links between older elements and introduce also new elements in the already existing system. That means that the system has to be somehow re-opened to incorporate new elements and/or meanings to its items.

Jay Lemke proposes that the opening and closing of semiotic systems is

---

[48] A. Etxeberria, «Complementarity and Closure», 200.

[49] N. Murphy, «Reductionism», 32.

tightly related to emergent phenomena and the increasing of complexity. The phenomenon of emergence and complexity is linked to that of the hierarchy of organizational levels. As we have previously seen, the levels of organization in a system are hierarchically established. Higher levels emerge from lower ones, acquiring new properties that are not present in the items of the lower level when taken individually. From a semiotic point of view, the higher levels are constituted because of a rearrangement of the lower ones. But also the higher levels impose also some restrictions upon the lower ones. Formulating it more generally, Lemke establishes that «units on level *N* are constituted by interactions at level (*N-1*) among the units at that lower level, but that of all the possible configurations that such interactions might produce at level *N*, only those that are allowed by boundary conditions set at level (*N+1*) actually occur»[50].

The properties of items at (*N-1*) are the initial conditions of the system. However, those conditions do not determinate the dynamics of *N*, since the emergent dynamism of *N* out of (*N-1*) is constrained by the boundary conditions imposed by (*N+1*)[51]. Levels do not appear as a mere addition of the features of the basilar items. There is a dynamical interaction between them, regulated by their very structure. *N* filters the available configurations of (*N-1*), and (*N+1*) buffers the variations of (*N-1*) through the stabilization of *N*. The structure of the levels entails a restriction of the matter and information that passes through them. The system has gained in specificity and control over its components[52].

Reorganization of lower levels according to restrictions imposed from above enables the system to select certain information that is available from the lower level pool. Levels open and close in order to establish new degrees of emergence and complexity. The semantic closure opens and closes again in order to establish new semiotic links according to the functional needs of the system. The lower levels of organization have huge amounts of information; is precisely the encapsulation and differentiation of this information what allows higher levels to control it. Information is out there, but the system would only select the one

[50] J.L. Lemke, «Opening up Closure», 101.

[51] Cf. S.N. Salthe, *Development and Evolution*, 42.

[52] Cf. J.L. Lemke, «Opening up Closure».

that would be relevant for its survival and that can be controlled through the emergent structure and the dynamics of the system.

From a complementary point of view, we can say that the semiotic closure of the system opens and closes in order to acquire new habits, as Peirce would say. Each level of organization allows the re-organization of semiosis in the system.

It is worth noting that there is a difference between physical interaction and semiotic relation. When a photon hits an organism, there is a physical interaction between them. But there is something *more* than that physical interaction. This interaction could be seen as a sign of something else, as if the photon would be standing for something that the organism cannot experience by itself. We know, for example, that plant's tropisms and flowering depend on the availability of light. Depending on the hours of daylight and on their direction, plants grow towards a certain direction or flower at a certain time. What has there to be, then, for the plant to link the information offered by light with the pattern of flowering or the direction of growth? It has to be a semiotic relation that takes the light stimuli as a sign of the arrival of the suitable time for flowering. The stimulus of light is semiotically linked to the whole organism through the complex system of hierarchical levels that sense and process that stimulus. The inner Umwelt of the organism is structurally prepared to receive the light as a sign, as something that conveys information about another thing. The molecular and organular structure inside the plant's cells amplifies the stimulus and canalizes it to be fruitfully interpreted by the plant as a whole and properly responded. It can be said that the inner structure of the organism is tuned to translate the physical impacts of the stimuli into a signal that is useful for the plant's survival. The inner cellular structure is organized as a micro-cosmos that tries to internalize the outer macro-cosmos.

It is interesting to note that Lemke distinguishes two kinds of semiosis: topological and typological. Topological semiosis could be compared to analogical signalling, since the sign would vary in a constant and quantitative way. Typological semiosis, on the other hand, has to do with digital signaling. Typological semiosis then include those signs that are conveyed as discrete units[53]. Organisms adopt both types of semiosis, and many semiotic processes are a mixture of them. Signals can adopt a wave-like behaviour or a discrete one.

[53] Cf. J.L. Lemke, «Opening up Closure», 104.

Semiotic processes span through the levels of organization of the cell, amplifying, transmitting or dampening signals, and they can do that in a continuous or a discrete manner, depending on the case. Lemke proposes what he calls the principle of alternation, which says as follows:

> Each new, emergent intermediate level *N* in a complex, hierarchical, self-organizing system, functions semiotically to reorganize the continuous quantitative (topological) variety of units and interactions at level (*N-1*) as discrete, categorical (typological) meaning for levels (*N+1*), and/or to reorganize the discrete, categorical (typological) variety of level (*N-1*) as continuously variable (topological) meaning for level (*N+1*)[54].

It is worth noting that the emergence of a topological sign requires the emergence of a higher level that could deal with the sign in a typological way. On the other hand, for the interpretation of a typological sign, an emergent higher level is needed to interpret it topologically. Emergence is a necessary condition for a richer interpretation of the signs coming from the lower levels of organization. The scale of hierarchical organizations is the embodied structure that allows the sign to be linked to the referent to which it stands for, and according to the organism's needs. The more complex the organism, the more accurate would be the control that the system as a whole could exert in response to the sign; and then, the more usefulness of that sign would be for the organism.

> What the Principle of Alternation proposes is that the transformation of discrete to continuous and continuous to discrete *alternate* as we move from level to level of the dynamic hierarchy, and that in doing so they represent a *semiotic transformation* of the information content of lower levels as *signs* for higher levels, allowing *many-to-one* classifications and *one-to-many* context-dependent reinterpretations. A scale of dynamic processes in a system, in which such transformations occur, meets the logical condition for novelty that define for us a genuinely *emergent* level of organization[55].

[54] J.L. LEMKE, «Opening up Closure», 106.
[55] J.L. LEMKE, «Opening up Closure», 108.

5. **Summing up**

It has been shown that the category of sign according to Peirce helps us greately to understand how the cell deals informationally with chemicals. Molecules are not only considered from their chemical point of view, but the cell, which acts as an icon, also treats them semiotically.

The cell as a whole has the property of being not only a physical closure, but also a semiotic one. That means that the cell can be somehow aware of the semiotic aspect of its chemical stuff, and can control and constrain it according to its needs. Then the cell can take some habits that establish semiotic processes to properly respond to environmental cues and accommodate them to the semiotic world of the cell to better use them for its own sake.

Which are the criteria through which the cell takes the habit of linking a certain molecule in some respect to a given referent? My guess is that the cell uses its own survival as the inner criterion to drive semiotic processes. The cell keeps alive only if it is constantly actualizing the functions it has to deploy to maintain its physical integrity and its semantic closure. Let us pass to the next chapter to see how the concept of function fits in the peircean paradigm and can help us to better understand the semiotic aspect of life.

Chapter V

Functions

1. Has biology real functions?

The concept of function has always been the cornerstone of the biological agenda against reductionistic approaches. Notions such as "final cause", "teleonomy", "teleology" and "function" are the framework concepts where biology fights its autonomy before reductionism.

I think that one of the reasons that make the reflection on the concept of function so difficult is the searching for a functional paradigm that would embrace artefact and biological function. Many writings on this topic establish a strong analogy between the function of artefacts made by man and the function of organisms. The point is that artefacts receive their function from the designer's intention. The artefact does not establish what it is for by itself. It is another instance, different from the artefact that establishes its function. The subject that uses a given object uses it according to her intentionality. When men design objects, they put their intentions on them. It is very clear also when we use an object with an intention that is not that of its designer. We can use a chair to place a TV on, and if originally the chair was designed to sit down, it could also bear the intention of being a TV stand[1].

From my point of view, human intentionality on objects and non-human living beings cannot be treated the same way as the concept of function. The

[1] As Boorse states: «Cutting is the function of a knife or saw because that is the purpose for which humans use it. Objects identical to our knives and saws would have no function at all if produced by some random process on a planet devoid of life. Conversely, they can have many alternate functions here, if people use them differently: a knife can become a paperweight, a saw a musical instrument, an orange create a bookcase, a jelly jar a drinking glass», C. Boorse, «A Rebuttal on Functions», 68.

reason is that humans use symbolic thinking in order to deal with the purpose we have to achieve. We can think about our aims abstractly, *matter less.* We can separate in our mind the matter of the object we handle from the intention we have. Then we can "superimpose" our intention on the object and value if it would fit the function we are trying to accomplish. We can design that object according to the function we want it to perform[2]. But in living organisms, things go differently. Non-human living beings do not have abstract thinking; they cannot use symbols to represent their own self to themselves and the *Umwelt* they dwell in. They do not deal with information symbolically, as we do, but semiotically. That is why the notion of function cannot be applied straightforwardly from human artefacts to living beings or vice versa.

Michael Ruse thinks that teleological language is appropriate for talking about organic systems because they are artefact-like; that is, *as if* they were designed. Then, when teleological language is used, it is acknowledge that is only a way of speaking, that does not fits perfectly with reality, but that has heuristic value when dealing with living systems[3]. But in doing so we are doing the same than the biologists who think that "informational talk" applied to DNA is a metaphor, a way of speaking that, sooner or later, as we have more empirical data, would be left out.

I think that this position is misleading. There can be functions in biology without talking about them as a metaphor of how we humans deal with functions in our everyday life. After considering some historical landmarks, I shall introduce the concept of function and the different approaches with which it has been studied and applied to biology. Then, I will describe what a biological system is and how it can perform functions, achieve aims and control information.

---

[2] Michael Ruse explains the difference between the intentions and the design skills in humans: «Intentions apply to conscious beings whereas design is what conscious beings do to things». M. Ruse, «Evolutionary Biology and Teleological Thinking», 35.

[3] «Such language is thought appropriate because organisms or the parts of organisms (characteristics) are taken to be design-like: they are taken to be as if they were artefacts, or parts of artefacts, created by conscious intelligences in order to serve certain ends. Adaptations serve the end of survival and reproduction of their possessors, and, because they are artefact-like – as if they were designed – teleological language is appropriate». M. Ruse, «Evolutionary Biology and Teleological Thinking», 37.

2. Historical landmarks

The talk about functions is tightly bound to the concept of teleology. When we ask about the function of a given item, we are asking about the activity that item performs and the goal or end state this activity will reach. Another question would be whether this end state is reached consciously, intentionality programmed, fully determined or selected by the system where the item is embedded.

Plato and Aristotle reflect on teleology both in humans and in other living organisms. Aristotle distinguishes two types of teleological processes. On the one hand there are agency-centered processes, which could be of two kinds: behavioral or artefactual. The characteristic of behavioral teleology is that the goal is achieved through an agent's choice after an external evaluation. This evaluation is purposive, rational and intentional. In the case of artefact teleology the activity is for the sake of producing a certain object. The agent is also rational and intentional, but it projects its goal over an object that would help her to reach that goal.

On the other hand there is natural organisms teleology. In this case, the agent is neither rational nor intentional. Then, we talk of formal teleology to point out that the *telos* of the process is something inherent in the system, and helps it to develop properly. Internal teleology is used when the *telos* is inherent in the relation that a part of the organism has with the whole[4]. Peter MacLaughlin explains this internal teleology as follows:

> Internal teleology, on the other hand, presupposes neither an external agent nor an external source of evaluation for the system or process viewed teleologically; and the (internal) agency involved in such teleological processes is generally not intentional at all. Aristotle, of course, acknowledges the existence of external teleology, but he is also interested in an internal type. The goals or ends involved are those of the system itself, not those of its creator; and similarly, the evaluation of a goal as a "good" is made from the perspective of the entity whose good is involved, not from that of some external agent. The purposiveness is, so to speak, submental and in a sense naturalistic; there need be no ideal anticipation of the result[5].

[4] Cf. A. Ariew, «Platonic and Aristotelian Roots», 8-9.

[5] P. McLaughlin, *What Functions Explain*, 17.

Plato, for his side, stresses the fact that agency works for the best, for the search of excellence, represented in the forms. Any given system, either a human being or any other given organism, acts according to a model that is *external* to the system. For Plato the agency of any object is always for the best, from a cosmological point of view[6]. Summing up, «a Platonic *telos* is an agent that operates or creates purposively for the sake of the best. An Aristotelian *telos* is a property of an individual's functioning – its contribution to its own sustainability»[7].

Isaac Newton also reflects on the meaning of the universe and its finality. His reasoning uses an Aristotelian argument to reach a Platonic conclusion. The harmony of the universe, the fine-tuned balance of the solar system's planets, lead Newton to conclude that this arrangement suggests the existence of a designer (God). So Newton acknowledges with Aristotle that there is regularity and patterns in nature; but concludes with Plato that this is the consequence of an external designer, not of the immanent properties of matter[8].

The scientific revolution of the 17th century banished final causes as something that corresponds to mental representations in deity's mind. Immanent ends where accepted, but not the fact that these ends were God's intentions. Then, we can agree with MacLaughlin that «Descartes has excluded from science all discussion of God's intentions (final causes), but not forbid consideration of the mental blueprints God might have used (formal causes)»[9]. Then, formal causes where not excluded from science, but where assimilated to efficient ones.

Immanuel Kant, for his part, identifies two different kinds of *telos*: human purposive behaviour and the natural one, which we have to study *as if* it were rational and with a goal-directed activity as we humans have. What is important here is that our teleological and functional arguments are applied to non-human

---

[6] Cf. A. Ariew, «Platonic and Aristotelian Roots», 11-12.

[7] A. Ariew, «Platonic and Aristotelian Roots», 23.

[8] Cf. A. Ariew, «Platonic and Aristotelian Roots», 19-20.

[9] P. McLaughlin, *What Functions Explain*, 21. He also explains why final causes have been rejected and how formal causes have to be considered: «The divine artisan, when making the world or the things in it, always had two things in mind: a particular purpose or set of purposes that the created thing was to serve and a plan or blueprint of what the thing had to be like in order to serve this purpose. This second thing that God had in mind had not been banished from science by Cartesians; the plan of formal cause was assimilated to efficient causality». *Ibid.*, 23.

living beings as an analogy. Talking about living organisms' activity as if guided to a goal is something we cannot get rid of; it is just our way of thinking about nature[10].

William Paley, with his famous self-reproducing watch example, concludes that it has to be a designer to produce the complexity, order and arrangement we find in nature and living beings. Paley extrapolates his watch example to all nature, being a witness of God's designing activity. He applies the argument used in describing the organization and complexity of an artefact to the whole nature. Then, the designer is something external to the object; and, since that designer is God, has created the best of the possible worlds[11].

Darwin does not deny the existence of teleology in nature, but he replaces the Platonic demiurge and the God-designer by the blind forces of nature. «Darwin's argument against a creator and for a non-intentional force of nature is found in the awkwardness of developmental patterns, and the seemingly poor designs of nature»[12]. Darwin needs to show how a certain end can be achieved with non-intentional forces. «The point is that Darwin wanted to explain this "as if design" through natural selection (whether or not God stands ultimately behinds this selection), rather than through a directly intervening God of miraculous intention and design»[13].

3. **Systematic approaches**

To understand the different theoretical approaches to the problem of function, we have to begin with Hempel and Nagel's point of view. They study the category of function as an explanatory resource. Nonetheless, they perspectives are slightly different. Hempel holds that functional explanation «point out the causes of the existence and properties of the function bearers. Nagel says, they describe those effects of the function bearers that contribute to relevant properties of their containing systems»[14]. Hempel, then, focuses on the etiology of the

[10] Cf. A. Ariew, «Platonic and Aristotelian Roots», 21-22.

[11] Cf. A. Ariew, «Platonic and Aristotelian Roots», 23-25.

[12] A. Ariew, «Platonic and Aristotelian Roots», 27.

[13] M. Ruse, «Evolutionary Biology and Teleological Thinking», 39.

[14] P. McLaughlin, *What Functions Explain*, 66.

traits considered, while Nagel is interested in the causal contribution a certain trait makes to the goals of the system.

Both Hempel and Nagel think that the system has some goals that have to be achieved in order to keep it on going. For Hempel, the goal to achieve is the self-maintenance of the system, while for Nagel the crucial point is that the parts must causally contribute to the goals of the system. As McLaughlin says:

> The questions asked in Hempel's analysis of functional explanation deal with the causes (or, more generally, explanatory principles) of the function bearer *X*. Those involved in Nagel's analysis have to do with the effects of the function bearer. In this sense, Hempel's analysis is backward looking from *X* as an effect to the causes that give rise to it, and Nagel's analysis is forward looking from *X* to the effects it causes. Thus latter analyses in the tradition of Hempel tend to be called "backward looking", "etiological", or "teleological". Those in the tradition of Nagel tend to be called "forward looking", "dispositional", or "causal role"[15].

One of the interesting questions of Hempel and Nagel's approach, which is not be taken into account in the subsequent debate, is that they established a two-end relation between the function bearer and its end. Indeed, the function bearer is a mean to some end, which in turn is a means to a further end, and so on. The chain of means-to-an-end could be very long, but it has to stop somewhere. The last end in which the causal chain stops has to be something beneficial for the system. For Hempel, the chain stops where there is an explicit link with the system's welfare. On the other hand, Nagel considers that the means have to work for the proper activity of the system under consideration. Then, «both Hempel's and Nagel's notions of function attributions presuppose the existence of a special kind of system that is able to stop a functional regress»[16].

Then, we can see that the functional analysis could be made at different levels. We can ask for the function of a given item, and then Hempel would answer it appealing for its etiology, while Nagel would point out the causal role it plays in the contribution of the system's goals. The interesting point here is that we cannot talk about an item's function without referring somehow to a higher level of organization that helps us to fully understand it. In fact, the end

---

[15] P. McLaughlin, *What Functions Explain*, 70.

[16] P. McLaughlin, *What Functions Explain*, 77.

of a mean could be in turn a mean to another end, and then it is difficult to see an item's function until we reach the last end to which the chain of the means is linked. Considering an individual item, we can study what it does, but we would not be in disposition to say what its function is until we were not able to link what it does to the welfare of the system or to any of its goals.

> If the function bearer *X* is actually a means to the effect *Y*, but *Y* is neither an end in itself nor a means to anything that is such an end, then *Y* is not a function of *X*. That is, just because something actually works – in other words, has effects – doesn't mean it has a function. Function bearers are means to ends, not just causes of effects: They are causes of valuable or beneficial effects – whatever the source of the valuation[17].

A puzzling point about the functional analysis debate is that Hempel and Nagel do not talk about the role of evolution. On the contrary, as I shall show in the following sections, the continuators of the debate always have natural selection in mind when reflecting on biological functions. That is because Hempel and Nagel think about a common theoretical framework capable to all kind of functions, even those concerning biological systems. They do not explicitly talk about evolution or biological functions because they think that their theoretical framework is adequate for any kind of function, both artefactual or biological.

I think that this approach is not totally adequate because Hempel and Nagel are reflecting on a category that is not equally applicable both to artefacts and to living beings. It is not possible to talk about functions having a common framework for such different items as a steam engine and a cell. Furthermore, the concept of function in biology is not centred on natural selection, but how the cell as a whole can deal with information in order to keep it alive constantly satisfying all its needs. Of course, natural selection would play an important role selecting those linkages between causes and effects that are better for a given function, but what matters here is how the cell deals with all these elements in order to establish an habit that, afterwards, would be subject to natural selection.

[17] P. McLaughlin, *What Functions Explain*, 78.

3.1 *Causal-role analysis*

The "causal-role analysis" was firstly proposed by Robert Cummins[18], and characterizes an item's function as its causal contribution to the output capacity of the system. A functional analysis of a system implies the investigation into the causal role of each item in the final functional output of the system. Cummins' functional analysis explains a complex system ascribing functions to its parts. The function of a part is that of its contribution to the global capacity of the system. To say that an item has a causal role in doing *x* means that it has a disposition for *x*. «Thus, function ascribing statements imply disposition statements; to attribute a function to something is, in part, to attribute a disposition to it. If the function *x* in *s* is to ϕ, then *x* has a disposition to ϕ in *s*»[19].

Cummins makes an approach to the concept of function based on the causal role a given item has in a certain system. He rightly points out that we cannot explain the presence of a certain trait by appealing to what it does. In doing so, we are saying that a given trait is a necessary condition for a certain function. The point is not that the heart's function is to pump blood, but the fact that a vertebrate needs a blood circulator in order to survive[20]. The fact that a certain function is performed in a certain way, does not tell us nothing about alternative ways through which the function could have been performed. We cannot deduce the presence of the heart from the necessity of vertebrates to have a blood circulating system, as we cannot deduce chlorophyll from the fact that photosynthesis takes place[21]. Of course, in some sense we can answer the question "why *x* is there?" specifying *x*'s function as if *x* were part of an artefact, according to Cummins. It is an inference to our best explanation[22].

Hence, what helps us to individuate an item's function is its behaviour in a certain system. So we can see that the function-ascribing problem depends on

[18] Cf. R. CUMMINS, «Functional Analysis».

[19] R. CUMMINS, «Functional Analysis», 758.

[20] Ruth Garrett Millikan criticizes that the example of the heart as a blood pumping device is not adequate, since it tries to give «an account of biological function that presuposes evolution by natural selection on the grounds that Harvey didn't know about natural selection when he proclaimed the discovery of the heart's function», R.G. MILLIKAN, «In Defense of Proper Functions», 290.

[21] Cf. R. CUMMINS, «Functional Analysis», 745.

[22] Cf. R. CUMMINS, «Functional Analysis», 748.

the context the target item is placed. Cummins does not give importance to the etiological account that tries to explain the item's function appealing to its history of natural selection. The function of an item depends on its causal role in a given system. Then,

> functional analysis can properly be carried on in biology quite independently of evolutionary considerations: a complex capacity of an organism (or one of its parts or systems) may be explained by appeal to a functional analysis regardless of how it relates to the organism's capacity to maintain the species[23].

However, as Cummins himself admits, a certain function could be achieved if other item has the conditions to perform that function in a given system. Then, what really matters is to perform the function, not the way it is performed. Nonetheless, the way it is performed depends on the mechanisms the organism has at hand at a given time in order to perform that function. In that sense, the function does depend on the etiological history of that given trait that performs a certain function. So an etiological analysis could not been left out at all.

Moreover, as Valerie Gray Hardcastle stresses, it is difficult to ascribe any causal role without taking into account the context in which the item is. Functions of items, then, cannot be defined abstractly, but they depend on the context. The first thing to do when we want to know an item's function is to study the environment in which it is placed[24].

Then, there is not an explicit and unique function of a given item. That means that objects are not designed, as happens with human artefacts. This issue prevents functional analysis from having to accept some kind of designer that dictates what things are for. However, things are not as easy as this. The

---

[23] R. CUMMINS, «Functional Analysis», 756.

[24] «Hence, we can specify the function ofT only against a particular selective regime. The function of T is to do E for O because *in context C*, E increased the fitness of O's ancestors. However, if T increases the fitness of O's ancestors only in context C, then it would seem a mistake to claim that E is T's function *tout court*. Perhaps in context C*, having T do E does not help at all. If it turns out that all of O's ancestors lived only in C and never in C*, we can speak loosely and talk about *the* thing T is supposed to do, because in all relevant environments (namely, the one with O's ancestors), T doing E promoted O's ancestor's survival». V.G. HARDCASTLE, «On the Normativity of Functions», 147-148.

history of a given object has something to say regarding its actual function. More precisely, the history of trait's interactions with its environment in a certain organism constrains that trait to be part of the complex that performs a certain function. The etiological account has importance not because it is a way to reintroduce design, but to take into consideration what a given trait has been doing in order to see if it can be used to perform another function or to be modified to better improve its functionality.

However, this causal approach to the study of functions generates many problems. Indeed, thinking of functions as cause-effect relationships widens the concept of function excessively and yields a picture where almost everything has a function. That concept of function exceeds the biological field I am interested in. Christopher Boorse points out the fact that this notion of function «implies that the function of mists is to make rainbows, the function of rocks in a river is to widen the river delta, the function of clouds is to make rain which to fill the streams and rivers, and the function of a piece of dirt stuck in a pipe is to regulate water flow»[25].

Ruth Garret Millikan notices that Cummins' function is too wide a paradigm to be applied to biology. It does not differentiate traits that are passed to the next generation through reproduction from others that do not have any incidence during reproduction.

> Griffiths wishes to admit the snail shell that the hermit crab (soldier crab) carries on its back as having the function of protecting the crab. Since the systems that are hermit crabs do not participate in the reproduction of snail shells, this would be analogous to admitting the eggs that you eat as having the function of nourishing you or admitting the atmosphere as having the function of helping you breathe. Yes, of course one *could* consider the egg as part of the human Cummins system and also the hen that makes the egg, and one *could* consider the oxygen as part of the human Cummins system – and also the sun that helps photosynthesis hence the production of oxygen, and so forth. But a more reasonable place to draw a line around a Cummins biosystem excludes factors that are not reproduced by the organism. Probably non-reproduced factors are better considered just normal supporting conditions or normal input to the biological system[26].

[25] C. Boorse, «A Rebuttal on Functions», 65.

[26] R.G. Millikan, «Biofunctions: Two Paradigms», 141.

I think that the problem of Cummins is identifying teleology with a sort of intrinsic end or purposiveness. He criticises the conception that every thing has been selected according to its function. Cummins is against a selecting-for explanation of everything. I agree that there are many traits that have not been selected and, nonetheless, they are present in organisms and passed through the next generations. Cummins, however, does not pay attention to the evolutionary history of traits. He wants to focus just on the actual role a given trait has in the biological system. For him, appealing to its past history is a way of looking at it as if it were designed. But I think that the important question is not that. Looking at the evolutionary history of a given trait, or of an entire organism, shows us that functions are not something that have to be performed *here and now*, but that life, since its very beginning, has had to strive for survival, performing functions in order to grow and reproduce. Otherwise life would have had to start from scratch every time a new function is requested to respond to changing environmental situations. The actual function of a trait has a history, the very history of life striving for survival.

On the other hand, the function of a trait is not fully determined by the function the trait has deployed in the ancestors. The evolutionary material has to be tested in the actual conditions where the organism lives, and these conditions could constrain a trait to perform a different function. Saying it with Peirce's words, a trait can be used with a different respect, or can even be a sign of a different referent.

Cummins says that we cannot longer stand by the claim that single traits are there because of what they are for. Teleology says that there are hearts because they pump blood and there are eyes because they enable vision. But to my mind the point is that there are hearts because in vertebrates, there are some vital functions that, in order to be fulfilled, used the stuff that was available at a given time and found the solution to these challenges in the form of the circulatory system. The solutions it gave could have been solved differently, but at the moment that life was challenged, life responded with what was at hand, and that is the reason why the heart appeared.

3.2 *Etiological analysis and selected effects*

Larry Wright makes an interesting distinction at the very beginning of his famous paper *Functions*. He distinguishes between *the* function and *a* function of a given item. This is something paramount for the functional characterization of an object. An item can perform different roles within a system, and it is not the same to consider a function as the more important role the object plays as to consider it as a secondary effect. Indeed, in functional analysis this often leads to talk about function and accident; or to distinguish between proper functions and other side effects. We can articulate our discourse referring to proper functions saying that the function of *X* is *Z*; or referring to a side effect or accidental function saying that *X* functions as *Z*. For example, we can say that the nose functions as an eyeglass support or that heartbeat functions as a diagnostic aid. Finally, Wright distinguishes natural from conscious functions[27]. The latter have to do with designed artefacts and how they are used according to the plans and intentions of their designers. Natural functions, on the other hand, are related to how the heart pumps or the kidneys remove wastes and are closer to the concept of function I want to deal with.

Wright does not want to take any useful effect of an item as a function of it. Many things can have many beneficial effects, but not all those effects are functions of those things. But talking about functions, they cannot be reduced to any beneficial effect of the considering object.

> Livers are good for many things which are not their functions, just like anything else. Noses are good for supporting eyeglasses, fountain pens are good for cleaning your fingernails, and livers are good for dinner with onions. No, the *function* of the liver is that *particular* thing it is good for which explains why animals have them[28].

Then, here is the point where Wright introduces his etiological analysis of functions. To ascribe a certain function to a certain object means to trace the etiological causal background that object has had. Functional explanations are etiological explanations, because they explain why that object got that function. Then we can distinguish other "good" and "useful" functions that the object does, from that function that really explains why the objet is where it is and doing what it does. So the etiological approach tries to distinguish the proper function from its side effects, whatever their usefulness would be[29].

---

[27] Cf. L. WRIGHT, «Functions», 141-148.

[28] L. WRIGHT, «Functions», 156.

[29] «The buckle, the heart, the nose, the engine nut, and so forth were not there *because* they stop bullets, throb, support glasses, adjust the valve, and all the other things which were

Wright's etiological analysis says that we know an item's function when we explain why it is there. Saying that the function of *X* is *Z* is to say that:

> *X* is there *because* it does *Z*.
> or
> Doing *Z* is the *reason* *X* is there.
> or
> That *X* does *Z* is *why* *X* is there[30].

It is important to distinguish between the things that are made by accident from those that are a consequence of its proper function. It is not the same to say that an item's function is *X*, or to say that this item functions as *X*. Then, the usefulness of a given trait is not the ultimate criteria for individuating its function. The nose is useful to sustain eyeglasses, but is not its function.

However, I think that this example is not appropriate. Indeed, Wright is always jumping from an artefact model to a biological model, as if the notion of function in those examples would be the same. The fact is that they are not the same at all. One can see the usefulness of the nose as a glasses-holder because of abstract thinking, because there is a human mind that considers the usefulness of using the nose as a glasses-holder in order to put them at the appropriate distance of the nose and achieve better vision.

Wright says that functions are ascribed to objects to explain what the objects are for. Of course, things can be functional objects because of effects that are not their function, as in the case of noses and glasses; but the function of an object has to be «that *particular* thing it is good for which explain why animals have them»[31]. Then, Wright holds that the function ascription procedure has to take into account the historical background of the trait in question. To distinguish a function from an accidental effect, we have to see its etiology.

> Putting the matter in this way suggests that functional ascription-explanations are in some sense etiological, concern the causal background of the phenomenon under

---

falsely attributed as functions, respectively. Those pseudo functions could not be called upon to explain how those things *got* there». L. WRIGHT, «Functions», 156.

[30] L. WRIGHT, «Functions», 157.

[31] L. WRIGHT, «Functions», 156.

> consideration. And this is indeed what I wish to argue: functional explanations, although plainly not causal in the usual, restricted sense, do concern how the thing with the function *got there*. Hence they *are* etiological, which is to say "causal" in an extended sense[32].

Wright's functional explanation focuses only on the origin of a given trait. The explanation would be just a characterization of the effects it has had and still has. Another striking thing in Wright's analysis is that there is no reference to the benefit an item's function has to the system that contains that item. The relation between the part and the whole is lost. This leads to a functional view according to a reductive point of view. Probably there is no reference to the system because it is assumed that the function of the system would be just the sum of the item's functions.

McLaughlin criticizes Wright's point of view of rejecting any reference to the function's contribution to the welfare of the system. Wright does not want to introduce any element of evaluation, but only to trace the etiological history that leads a given trait to perform the function it is doing. However, this model cannot answer the question about what the traits are for or how they contribute to the global benefit of the system[33].

Even if he focuses only on the etiological point, Wright admits different kinds of etiologies; that is, different answers to the question: "Why that trait is there?" For example, we can say that the function of oxygen is to combine with haemoglobin, but this is not the main reason for the presence of oxygen there. The function of oxygen is to be a paramount element in energetic metabolism. This is a *stronger* "because" than that of its interaction with haemoglogin[34]. To differentiate between these etiologies, Wright says:

> So when we say that *Z* is the function of *X*, we are not only saying that *X* is there because it does *Z*, we are also saying that *Z* is (or happens as) a result or consequence of *X*'s being there. Not only is chlorophyll in plants *because* it allows them to perform photosynthesis, photosynthesis is a *consequence* of the chlorophyll's being there[35].

---

[32] L. WRIGHT, «Functions», 156.

[33] Cf. P. MCLAUGHLIN, *What Functions Explain*, 97.

[34] Cf. L. WRIGHT, «Functions», 159.

[35] L. WRIGHT, «Functions», 160.

To avoid such inconvenience, it is necessary to identify the type of etiology the object comes from. That selection comes from designer's intention for artefacts and from natural selection for organisms. The etiological analysis, then, is focused in individuating through which process of selection items have been selected for a given function. This approach stresses the previous history of the object, rather than the causal role it can have here and now. Colin Allen and Marc Bekoff summarize the main lines of the etiological account as follows:

> (1) Functional claims in biology are intended to explain the existence or maintenance of a trait in a given population;
> (2) Biological functions are causally relevant to the existence or maintenance of traits via the mechanism of natural selection;
> (3) Functional claims in biology are fully grounded in natural selection and are not derivative of psychological uses of notions such as design, intention, and purpose[36].

The problem is that, if we talk about living organisms, there are traits that «can persist for a long time for reasons other than selection – genetic drift, a lack of mutations, pleiotropy, genetic linkage, and so on»[37]. So natural selection does not always have the last word about which traits have to be preserved or which functions are they playing. Indeed, natural selection has to be understood not as something that looks for the best solution, but as something that eliminates the worse.

As it has been stressed, a given item in a certain system has a function, but there are more effects that derive from that item's activity. The example of the heart is very clear. Everybody would agree that heart's function is to pump blood, but it also produces noises when pumping, and also contributes to the total weight of the body. However, nobody would say that the function of the heart is to make noise or to contribute to body's weight. That is why Millikan sees the necessity to introduce the term "proper functions".

When Millikan talks about "proper functions" she underlines the fact that to determine an item's function it is necessary to look at its history rather than

[36] C. Allen – M. Bekoff, «Biological Function», 612.

[37] C. Boorse, «A Rebuttal on Functions», 66.

at its properties or dispositions. From that point of view, for an item *A* to have a "proper function" *F* one of the following conditions must be met:

> (1) *A* [is] originated as a "reproduction" (to give an example, as a copy, or a copy of a copy) of some prior item or items that, *due* in part to possession of the properties reproduced, have actually performed *F* in the past, and *A* exists because (causally historically because) of this or these performances.
> (2) *A* [is] originated as the product of some prior device that, given its circumstances, had performance of *F* as a proper function and that, under those circumstances, normally causes *F* to be performed by *means* of producing an item like *A*[38].

Griffiths agrees with Millikan's notion of proper functions, but notices that there is no distinction between currently functional traits and vestigial traits that worked in past adaptations[39]. For Millikan, an item has a proper function if it has had a proper history of selection. Cummins defends that what really matters is the role the item plays here and now in the system, not the historical path the item has walked to arrive where it is now. If we reproduce a given object exactly, an object that does not have a previous history, simply «it has just come into being through a cosmic accident resulting in the sudden spontaneous convergence of molecules which, until a moment ago, had been scattered about in random motion»[40], Millikan says that such an object would not have any function at all, since it does not have a previous history that explains why that object is there. Cummins, on the contrary, would defend that such a doubled object would have the same function than the original.

Millikan acknowledges that Cummins dispositional approach sheds some light on the question about function, but cannot explain the notion of purpose. «Moreover, being preceded by the right kind of history is *sufficient* to set the norms that determine purposiveness; the dispositions themselves are not necessary to purposiveness»[41].

Cummins functions and etiological functions are not as different as it may seem. Millikan says that it is just a matter of emphasis on one point or other.

[38] R.G. Millikan, «In Defense of Proper Functions», 288.
[39] Cf. P.E. Griffiths, «Functional Analysis and Proper Functions», 413.
[40] R.G. Millikan, «In Defense of Proper Functions», 292.
[41] R.G. Millikan, «In Defense of Proper Functions», 299.

«Cummins emphasized the project of finding out *how* the biological system works, not just finding out what it *does*. But, of course, finding out what it does in detail, what *all* its proper functions are and *all* the proper functions of *all* its parts, is finding out how it works»[42].

Hardcastle also thinks that there are no great differences between the etiological and the causal account. «The real difference is in attitude: the pragmatists [causal-role supporters] publicly embrace the relativity of functional attribution and then deny any sort of prejudice in assigning a goal state. We look to the various domains of science to set their own explanatory goals and we use those to make functional assignments»[43]. According to Hardcastle, dispositions and functions only differ in the goal searched for the latter; and that goal is established by the inquiry interest.

We get the "extra ingredient" that separates dispositions from functions by relying on the previously accepted explanatory goals and theories of a particular scientific discipline. The teleological goal for some trait or organism is neither arbritrary nor *ad hoc*; it depends upon the discipline generating the inquiry[44].

Hardcastle tries to overcome the distinction between proper functions and side effects saying that it depends on the epistemological interests of the investigator. To my mind this cannot be the way of dealing with this issue. Shifting the debate of functions on the investigator's interest is a way of moving the question from ontology to epistemology. I am not interested in what the scientific is looking for when she investigates certain biological phenomena; I want to know how the biological system itself deals with everything it comes across with and how it can be treated in order to contribute (or at least to not disturb) the self integrity of the organism. Hardcastle's proposal has already a point of view that sees function as something that has to be imposed from outside, as happens with human intentionality. On the contrary, my interest is in depicting a framework where the cell can distinguish and select what to do according to certain internal aims; rejecting any functional approach that could come from the intentions and interests of the investigator. Biology would be still being biology even if no human investigator were interested in it; and the cell would still be looking for

[42] R.G. Millikan, «Biofunctions: Two Paradigms», 140.

[43] V.G. Hardcastle, «On the Normativity of Functions», 150.

[44] V.G. Hardcastle, «On the Normativity of Functions», 153.

new strategies with which to respond to the challenges the environment imposes to it.

Finally, it has to be said that Millikan and Neander do not give account of those traits that have lost their proper function. It can be seen that in some biological systems some traits are maintained despite the fact that they apparently do not contribute anymore to the cell benefit. They are not being positively selected because they are not involved in any function. However, it is known that those traits could be recruited by natural selection when there are appropriate conditions where they can offer a positive contribution to a function. There are cases where old traits are rescued to collaborate in different functions than those they deployed in the past. This leads me to consider the cell not as a system where everything has its precise role and contribution to the whole. There are genes that are not contributing to the whole, but they could be seen as a reservoir pool of elements that the cell has at hand if it needs them to take different habits. Then, the functions of traits do not depend exclusively on their recent history of selection, but mainly on the respect the cell can use them in order to be a sign of a referent that can be interpreted by the cell for its own sake.

3.3 *"Modern History" approach*

Paul E. Griffiths says that it is important to distinguish causal effects from proper functions, as Wright did. He admits that the Cummins' causal-role analysis is important for the definition of a function, but this definition has to include an explanation of the fitness of a given trait in the ancestors that had it.

The presence of the liver can be partially explained by its capacity to store glycogen and secrete bile. These functions enter into an evolutionary explanation of the presence of the liver. But the presence of the liver cannot be explained by its capacity to accommodate liver flukes. This is the Cummins-function of the liver relative to the capacity to die of fluke infestation, but it is not a proper function of the liver[45].

Then, Griffiths stresses that Cummins' analysis and proper function approach respond to different questions. The biological fitness of a trait is explained by functional analysis according to Cummins' model. On the other hand, «the proper functions

[45] P.E. GRIFFITHS, «Functional Analysis and Proper Functions», 411.

of a biological trait are the functions it is ascribed in a functional analysis of the capacity to survive and reproduce (fitness) which has been displayed by animals with that feature. This means that a feature will have a proper function only if it is an *adaptation* for that function»[46]. Griffiths stresses that for Karen Neander the function of a trait is tightly linked to its recent evolutionary history. But things are more complex and traits are not always selected for an immediate function. Indeed, some traits can be selected as side effects of other traits, and only later on would they acquire or contribute to a function. There are also cases where traits lose their functions and take new ones[47].

Godfrey-Smith thinks that Cummins' and Wright's approaches are different paths to answer the question about function. In fact, both senses of function are right, but they do not answer to question under the same point of view.

The difference is in the type of explanation. So if it is claimed, for instance, that the function of the myelin sheaths round some brain cells is to make possible the efficient conduction of signals over long distances, it may not be obvious which explanatory project is involved. This may be intended as an explanation of why the myelin is there, or it could be part of an explanation of how the brain manages to perform certain complex tasks. Sometimes the same assignment of functions will be made from both perspectives, but this does not mean the questions are the same[48].

I think that this issue has to do with the role of the icon in Peirce's approach. From my point of view, we should reserve the term function for those processes that are vital for cell survival. Moreover, these processes are those that belong to the highest hierarchy of organization. Then, the point is that we cannot talk about the proper function of myelin unless we consider the higher structure in which it is inserted. Only when we understand the role that the higher structure plays can we infer the functional role of myelin in the whole organism. Then, it is important to see how the system treats myelin in order to be integrated in such a way, allowing it to have a causal role in functional processes that would not have taken place separately from the system where it is integrated.

---

[46] P.E. GRIFFITHS, «Functional Analysis and Proper Functions», 412.

[47] Cf. P.E. GRIFFITHS, «Functional Analysis and Proper Functions», 414.

[48] P. GODFREY-SMITH, «Functions: Consensus Without Unity», 201.

Then, to understand the function of a given item, it is not only necessary to have a causal-role analysis and an etiological account, but it is also necessary to see how the higher levels constrain and tune those items in order to perform a certain function. In other words, it has to be seen how the system exert some kind of top-down causation contributing to the function of the lower level items. Moreover, it has to be seen if the object under study is treated semiotically by the whole, and discover in what respect it is linked to its referent. That would give us valuable information about many details that would be very useful in understanding its function.

3.4 *Goal-contribution account*

This functional approach tries to define the functional role of an item in a system according to its contribution to the goal the system tends to. It is difficult to talk about goal-seeking without making reference to some kind of intentionality or human purposiveness. However, this account tries to analyze the concept of function and its teleological aspect as something common to all living beings, not only humans. Then, some way to talk about purposiveness in some non-mental, naturalistic way has to be found. From that point of view it can be said with Boorse that,

> a system S is "directively organized", or "goal-directed", toward a result G when, through some range of environmental variation, the system is disposed to vary its behavior in whatever way is required to maintain G as a result. Such a system, it is said, shows "plasticity" and "persistence" in reaching G: when one path to G is blocked, another is available and is employed[49].

This issue shows us that the organism is a self that strives for reaching its goals, and the important think is not the way these goals are reached, but that they are indeed achieved, whatever may be the path followed to reach that goal. In other words, biological systems display functional equivalence classes which allow them to perform the same function through different pathways to make sure that the blocking of one of them does not threaten the organism's survival.

[49] C. BOORSE, «A Rebuttal on Functions», 69.

This framework fits much better with my proposal. The important question that functional equivalence classes shows us is that the paramount point for a cell is not *how* to achieve its goals but to achieve them. Auletta *et al.* define functional equivalence classes as follows:

> When one speaks of equivalence in general, one understands a "possibility of substitution in any context". It is fundamental that functional equivalence classes are on the contrary context-sensitive. The concept of equivalence class [...] is therefore not a formal-logical construct but a pure functional-biological category, where different operations are considered functionally equivalent if they produce the same outcome for some functional purpose (the goal)[50].

3.5 *Function and design*

One of the more confusing questions when we talk about the notion of function is to take it for granted that we can use it in the same sense when talking about artefacts and natural systems. McLaughlin devotes one chapter of his book *What functions explain* to make it clear that this assimilation is misleading.

Artefacts are the paradigm of external teleology, as it has been shown. In fact, there are no artefacts without an agent that confers them a function. Then, an effort from the agent it is needed in order to confer a function to an artefact. This effort could be physical (if we strip twigs and smaller branches from the branch we want to be a walking stick) or just mental (we can find a branch that already fits perfectly for the function we want it for). Then, we can see that artefact functions depend on the agency of someone that ascribes it a certain function. It is not necessary for the artefact to always have this function. «While it is only on the basis of an intentional act that an entity acquires a function, it is not necessary that the entity itself come into being only after the intentional act. An existing (functionless) entity can later be subsumed under a functional kind»[51].

Let us take a step further in this conferring functional ascription to artefacts through external agents. The agent confers certain function to an artefact in order to achieve some good. A man strips twigs in order to have a walking stick

---

[50] G. AULETTA – G.F.R. ELLIS – L. JAEGER, «Top-down Causation by Information Control», 6.

[51] P. MCLAUGHLIN, *What Functions Explain*, 45.

that would help him in his walk. This implies a valuation component. The final state of the intentional ascription has to be realistically possible. Perhaps we would prefer that our walking stick were made of fire, but it is physically impossible.

I think that the evaluation component is not only about the feasibility of achieving the mental function to the artefact, but has also to do with the mental representation we have when thinking of that object with that function. We can see a branch as a wooden stick because we somehow see us using it as a wooden stick, and anticipating how useful it would be to us in helping our walking. This is something that the cell cannot do. It has to test it, there cannot be mental anticipation of the function a protein may have in the system. The cell cannot stop from living and try different possibilities to see which fits better to perform a certain function.

Artefacts have a very curious characteristic when considered from a functional point of view. As McLaughlin states, «the particular product of the effort (potentially) expended need not actually be particularly successful at producing something that fulfils the function»[52]. For example, the function of an anti-mosquito spray is to avoid them biting us. And this will still be its function even if one mosquito bites us[53].

Here we see a difference between the functional achievement that is in the agent's mind and the real achievement of the artefact. It is used as if it is going to fulfil the purpose of the agent; but even if it does not reach that purpose, it still has its function in agent's mind. This cannot happen in a cell. Functions are related to what is actually done, not to what it should be done but is not properly achieved. There is no extra instance, such as a mind, that can keep on ascribing a function to an artefact that does not achieve the purpose that is sought.

---

[52] P. McLaughlin, *What Functions Explain*, 46.

[53] Another example put by McLaughlin: «For years, the function (purpose) of plastic cutting boards in many butcher shops was to discourage bacterial growth even though it subsequently turned out that, compared to wooden boards, they were in fact conducive to the growth of bacteria. The function or purpose of an artifact is the end to which it is a means – whether successful or unsuccessful – for whoever made it, acquired it, used it, is expected to purchase it, or is supposed to be given it as a present». P. McLaughlin, *What Functions Explain*, 46-47.

Function ascription can change with regard to the same artefact, depending on the agent's intention. For example, a brick can be used as a door-stopper, even if originally it was designed to be part of a wall. Then, the function tells us little about the original design in the artefact's mind. The function of the artefact depends on how it is used here and now[54]. We can see that the difference between the designed intention and the actual use lies in the fact that, in the case of the designed object, the intention temporally precedes the existence of the object.

Another important question to understand an artefact's function is to take the context into consideration. «We have, nonetheless, a strong tendency to forget the context and treat design functions as indelible. If something comes into existence as an instantiation of a functional type, we tend to think it always retains this design function even in other contexts»[55].

The same object can be taken to fulfil different functions, depending on the agent's interest. I can use different functional names to refer to the same object, but which is used differently according to the agent's intentions. The functional ascription depends on the agent's intentions.

Another question to take into account is the consideration of complex artefacts and the parts that form it. It is more difficult to ascribe different functions to a part of a system that plays a very precise role in that system. Then,

> in more complex systems it is harder for a part of the system to acquire a new relative function without any change in its structure or in that of the system (its material context). For instance, the function of the pressure control valve on my steam engine cannot be changed as arbitrarily as the function of a stick found in the woods or as the function of the steam engine itself[56].

The system as a whole imposes physical constraints and relations that narrows down the possibilities of a given part. The contextual conditions of the whole make a given item very specialized to fulfil the role that is required of it, and this makes very difficult that this very item to play another function in other part of the system. At the same time, it is not easy to replace that item

---

[54] P. McLaughlin, *What Functions Explain*, 47.
[55] P. McLaughlin, *What Functions Explain*, 48.
[56] P. McLaughlin, *What Functions Explain*, 53.

with another, unless the new one will also fulfil the requirements imposed by the system's constraints.

3.6 *The notion of function in nature*

All these considerations I have made regarding artefact's functions cannot be straightforwardly applied to natural functions. Things go otherwise in nature: «The arbitrariness of function attributions and the non-dependence of function on actual success of performance are restricted to artefact functions. As opposed to artefacts, organs and (non-intentional) social institutions cannot have the function of doing something they do not in fact do»[57].

The confusion about the concepts of function and design has to do with our anthropomorphic view of design. It is clear that, when we talk about design when dealing with artefacts, we are saying that an external subject designs artefacts. The design does not come from the properties of the object, but has to be imposed from outside, by a human intentionality, to become the object man has in mind.

Problems arise when we try to apply this conception to biological systems. It seems as if, when we talk about design in nature, God has to be the designer, introducing confusion in the scientific debate. It is difficult to imagine how life has reached its complex structure and its fitness to the diverse ecological niche without the existence of a design. The question is not unimportant, and Kant tried to solve it saying that we have to study biological systems *as if* they were designed.

But the fact is that biological systems are alive; so they are constantly achieving the minimum conditions to stay alive. They function and all their items are interwoven, each of them playing its part, and cooperating as a whole in order to keep certain parameters within the narrow window of life.

Then, it is worth asking if function implies design in biological systems[58].

---

[57] P. McLaughlin, *What Functions Explain*, 61.

[58] The debate about design is not always welcomed. This «reluctance to discuss design may be due to a couple of factors. First, the notion of design may be considered metaphorical and offputting because it suggests a strong directional component in the evolution (or development) of a behavioral phenotype. Second, many authors seem to accept a principle that straightforwardly assimilates the notion of design to that of function such as: T is na-

Are function and design things that go together in the biological world? Can we talk about a natural design when considering the function of biological items?

Allen and Bekoff think that there is not a total correspondence between function and design in biology. When we talk about artefacts designed by humans, we produce an item according to a previous design. There is a goal-driven design that dictates the process of production. Then, the object has to be tested by trial and error and eventual modifications have to be made. In this case, design determines function. A plane is designed to fly, for example. Even if the plane would never take off, it has been designed to fly. However, if everything that is designed has a function, it is not true that everything that has a function has been designed. We can use seashells to make a necklace, but they were not designed for that function[59].

When referring to biological systems, Allen and Bekoff say:

> *Trait T is naturally designed to do X* if and only if
> (i) X is a biological function of T, and
> (ii) T is the result of a process of change of (anatomical or behavioral) structure due to natural selection that has resulted in T being more optimal (or better adapted) for X than ancestral versions of T[60].

The process of change is due to an optimization of individual traits. However, we can find traits that have biological function without having been naturally designed for that function; yet they could have been naturally designed for another function.

Then, it is more difficult to demonstrate that a trait has been naturally designed than that it has a given function. Indeed, «to show that a trait T is naturally designed for some effect X, in addition to showing that X is a function of T one must also show evidence of structural changes in the phylogeny of T so that T is better suited for X than ancestral versions of T»[61].

---

turally designed for X if and only if X is a biological function of T». C. ALLEN – M. BEKOFF, «Biological Function», 611.

[59] Cf. C. ALLEN – M. BEKOFF, «Biological Function», 613-614.

[60] C. ALLEN – M. BEKOFF, «Biological Function», 615.

[61] C. ALLEN – M. BEKOFF, «Biological Function», 619.

David Buller identifies some problems with this account, since its notion of design is very narrow. Firstly, Allen and Bekoff's view «counts a trait as an instance of natural design only if it has undergone directional selection, in which selection modified the trait towards greater optimality. But some traits are simply maintained by selection in a population, with active selection against any emerging variants»[62]. Perhaps a given trait does not need to be improved by natural selection, and the only pressure that natural selection has is to maintain that trait over the other possible variants.

Buller stresses that the important point is that not all etiological theories are equal. In fact, two main branches can be identified. The first is "strong etiological theory": «A current token of a trait *T* in an organism *O* has the function of producing an effect of type *E* just in case past tokens of *T* contributed to the fitness of *O*'s ancestors by producing *E* and were selected for (over alternative items) because of this contribution to the fitness of *O*'s ancestors»[63]. Then, there is also the "weak etiological theory": «A current token of a trait *T* in an organism *O* has the function of producing an effect of type *E* just in case past tokens of *T* contributed to the fitness of *O*'s ancestors by producing *E*, and thereby causally contributed to the reproduction of *T*s in *O*'s lineage»[64]. The difference between the theories is that the strong one establishes that the trait has to be positively selected. On the contrary, the weak one only claims that the trait has somehow contributed to the fitness of the ancestors, giving a wider framework in which traits can be passed on to following generations without being positively selected for a certain function.

Hence, we have here a notion of function that does not focuses on the fitness of individual traits, nor on the fact whether a trait has been selected or not; Buller proposes an idea of organism centred in the concept of function. What are selected are not individual traits, but organisms that are better fitted to their environment. These organisms, of course, have improved their traits, but it is not the trait itself that is selected, but the trait in the relational network of the organism that has contributed to its adaptation.

---

[62] D.J. BULLER, «Function and Design Revisited», 225.

[63] D.J. BULLER, «Function and Design Revisited», 230.

[64] D.J. BULLER, «Function and Design Revisited», 231.

> Here the function of a trait is not defined in terms of *selection for the trait*, but rather in terms of the trait's contribution to the *selection of organisms* with that trait, where the selection of organisms is a matter of overall fitness differences among organisms, which may ultimately be due to fitness differences in other traits. Consequently, this formulation does not require that a functional trait has a history of variation within a common selective environment, but only that it has a history of contributing to the fitness of its bearers[65].

These two approaches to the etiological account reveal two different perspectives in understanding natural selection. There is a trait-centred perspective that looks at the optimization of a given trait without taking into account its relation with the rest of traits. It is as if every trait would have an individual struggle with natural selection in order to improve its fitness. Natural selection would work to get the best version of each trait.

However, this is not the unique way to see it. The object of natural selection is not single traits but complex organisms. These organisms show complex structures and behaviours, and their parts are tightly interwoven. What really matters is not to have the best version of a given trait, but to have an organism capable of facing its environmental challenges. Then, functions are no longer a property of traits, but of wholes. Traits are not designed for a particular function. Rather, the organism as a whole can select which traits and the way to connect them in order to achieve the better adaptation to a given environment. If we want to keep on saying that traits are designed for something, we are only allowed to say that they are designed for contributing to whatever would be the needs of the organism in a certain moment[66].

Then, there is a strong difference between artefact design and natural design. One of the characteristics of biological traits is that they are rather complex and articulated. They can be divided into parts; each one makes its own causal contribution to the functional effect of the trait. There is, for example, a great difference between a stone that is used as a paper-weight and a protein. The rock is not internally articulated, while the protein is composed of domains, with different properties such as catalysis, recognition of other molecules, regulatory domains, and so on. The proper interaction and folding of all these domains and

[65] D.J. BULLER, «Function and Design Revisited», 232.

[66] Cf. D.J. BULLER, «Function and Design Revisited», 234-235.

the insertion of the protein in the whole metabolic network is what determines the contribution that the protein could make to the overall fitness of the organism. Then, besides the internal articulation of the trait, its external integration in the system as a whole is also needed.

> When a functional item is, like an organismic trait, both internally articulated and externally integrated, it is nested within a functional *hierarchy* of subsystems. There are thus subsystems "below" the functional item in the hierarchy whose effects are causally necessary for its producing its proper effect – in particular, those subsystems that constitute its internal articulation. In addition there are (sub)systems "above" the functional item in the hierarchy to whose proper effects its own proper effect is causally necessary –in particular, those (sub)systems with which it is externally integrated[67].

The fact that a given trait is nested in such a hierarchy of systems imposes certain constraints on an item's function. Is in that sense that we can talk of naturally designed. The trait does not only have to have a certain causal role *per se*; it has to be introduced in a network of relations with other traits, it has to influence (directly or indirectly) other traits, it has to let the regulatory influences of other traits act upon itself. In other words, it has to fit in the system. The designed item needed for a certain locus in the metabolic network has to fulfil many requirements, not just a causal-mechanical role.

4. **Biological Systems**

It has been shown that the concept of function is different when we talk about artefacts or about organisms. I think that the straightforward application of the artefact notion of function to living beings is misleading. In doing so, the same thing that biologists did when they used Shannon's informational model of communication to talk about biological information, has been done, especially when referring to that contained in DNA. The main difference that we can find between an artefact and a biological system has been already stressed: the function of the artefact comes from outside of it, from its human designer. In the case of a cell, for example, it is something that has its own aims and conditions

[67] D.J. Buller, «Function and Design Revisited», 236.

for survival, and is something that has to be reached by itself. From that point of view, it is the very cell that has to strive to achieve the proper functions that will yield as a result its own survival.

This different functional paradigm leads us to consider what a biological system is, how it is organized and how it can manage information for its own benefit. In the last chapters we have seen that a biological system is the result of a process that permits matter to be organized in such a way that allows the emergence of life's characteristics.

In this section I have shown the most elemental biological system, the cell, and seen which elements are present in such a system. Moreover, I want to see how these different elements are interconnected and regulated, since it has been stated from the very beginning of this work that the result of a living being could not be just the sum of the characteristics of its parts.

4.1 *Three levels of organization in nature*

In chapter II it has been seen how the concept of emergence can help to understand the increasing complexity that we see in the organization of matter in living systems. Here I want to briefly expose the different kinds of organization that are tightly interwoven in biological systems.

4.1.1 Self-organization systems

Biological systems are sustained by certain dynamical behaviours that are also found in non-living systems. In the non-organic realm some basic dissipative structures that are far from thermodynamic equilibrium can be found and can give rise to a coherent global behaviour or pattern. This behaviour could be maintained if certain parameters remain within certain critical limits. This kind of system tends to stabilize their components, allowing spatial and/or temporal patterns to emerge. This pattern shields the entire system against fluctuations and can buffer them, maintaining the global stability. These self-maintaining systems can appear only in far from equilibrium systems with a constant exchange of energy with the environment. The characteristic of that kind of patterns is just its recursive self-maintenance. This pattern would keep on going only under certain critical values of the parameters that maintains the energetic flux and the formation of the pattern. The system itself cannot modify or control those parameters. If they change, the system disappears. The system, then, can only be identified when the relations among the components of the system stabilize a certain macroscopic pattern[68].

An example of this self-organizing behaviour can be found in Bénard cells, which are a structural pattern that emerges in a fluid when certain variables are observed. Consider a fluid between two parallel plates. Initially the system is homogeneous, since the fluid is motionless and the temperature is the same both for the plates and the fluid. If we heat the upper plate slowly, the heat will pass to the fluid and from it to the lower plate through a process of thermal conduction. However, when the temperature reaches the critical point that overcomes the viscosity of the fluid, there will be movement of the fluid, changing the transmission of heat from conduction to convection. The movement of the fluid is not random, but it follows patterns that are structured in small hexagonal convection cells[69].

We can also find examples of self-organization in the metabolic pathways of biological systems. If we take any of the metabolic pathways of the cell, we shall see that in a given step of the pathway we can find an enzyme that reacts with the product of the previous step, and yields another chemical for the next step of the pathway performed by another enzyme. So when there is an input at any step of the pathway, we can see how all the downstream reactions are also performed. Then, a spatial-temporal pattern of reactions that are well organized appear, since every step is prepared to work with the chemical product of the previous one.

Auletta establishes the minimum conditions to have a self-organizing system. There has to be positive feedback that can amplify a given signal. There also has to be a negative feedback to dampen the signal when the amplification of the signal threatens the stability of the pattern. A self-organizing system does not have a single stable state, but can have a multiplicity of them. Finally, there can also be bifurcations in the system when some parameters are varied[70]. We can find these traits in many processes that take place within the cell. There is positive feedback, for example, in all the processes that deal with signal amplification, which multiply the stimulus that has arrived in the cell in order

---

[68] Cf. A. Moreno, «Closure, Identity and the Emergence», 114.

[69] Cf. G. Nicolis, «Physics of Far-from-equilibrium Systems», 318-319.

[70] Cf. G. Auletta, *Cognitive Biology*, 181.

to transmit it to the targets that have to respond to those stimuli. Negative feedback is found where final products accumulate at the end of a metabolic pathway. In order to avoid further synthesis, a final product inhibits a key step of the pathway to dampen the positive feedback. Finally, the multiplicity of stable states is linked to the possibility of bifurcation in metabolic pathways, since there are many cases in which the pathway can be feed by different metabolites and yield predominantly a different end product.

We can see that, as happens with Bénard cells, the metabolic pathways do not exist as such, but only when are fed with the proper metabolites. These pathways are patterns of interaction that allow the transformation of chemical stuff into molecules that can be used by the cell for a certain purpose. These biological self-organizing systems can be seen as patterns of complexity that are located between the physico-chemical and the biological world. There are some complex features that allow the cell to take advantage of an ordered interaction of the physico-chemical characters of certain molecules.

Complexity is a necessary condition for the cell to make good use of the characters of the physico-chemical level. The four main characteristics of complexity pointed out by Auletta are: the presence of hierarchical structures mutually connected, a relation of top-down causation from the higher levels upon the lower ones, the existence of modular recurrent structures and the capacity for plasticity and adaptiveness to certain variations of the system[71].

4.1.2 Self-production

This term regards the particularity of biological systems to be fed with matter and free energy and their capacity to transform them into structures for their own benefit. These kinds of systems are those based on self-replicating templates. In these systems there is a certain structure that acts as a blueprint that guides the production of a copy of it. A simple self-replicating system would be one where the «reordering of the components that are found in the environment and that will assemble to constitute the copy is caused by the material specificity and the spatial shape of the target structure»[72]. An example of simple self-replicating system could be the growing of a crystal structure, the

---

[71] Cf. G. AULETTA, *Cognitive Biology*, 182.

[72] A. MORENO, «Closure, Identity and the Emergence», 115.

dynamics of which depends on the structural pattern that is already formed and that acts as a template for the growing structure.

We can also find complex self-replication template systems, in which the process of self-replication depends not only on the template but also on the mediation of other structures. In this case, the identity and closure of the system does not depend on the extrinsic energetic flow, as happens in the dissipative systems. Here there are catalytic structures that mediate the replication process; entities that can regulate and perform the replication process shielded from the variations of the energetic flow. The replication depends on the availability of the catalytic entities necessary for the process[73]. This is what happens, for instance, in cellular protein synthesis.

We can talk at this level about autonomous systems that can control their own organization by controlling the external boundary conditions. A system is truly autonomous when it is able to manage and modify the external conditions in order to keep the goals of its actions.

The appearance of an autonomous agent is subjected to the existence of a selective membrane that separates it from the environment. The membrane acts as a constraint upon the components of the system. It closes the recurring autocatalytic networks and coordinates them in order to manage the energetic flows that pass through the membrane to keep the system alive.

> In causal terms, an autonomous system carries out an activity that functionally restructures its environment (formal causation), whereas the environment only interacts in a physicochemical sense with the system. Hence, its identity as an *agent*, that is, as a system that interacts with its environment in a different fashion than the environment acts on the system. It acts functionally on that environment and thanks to this, the system acquires autonomy with respect to external conditions[74].

Finally, it is worth distinguishing the difference between self-organization and self-production. Self-organization allows the emergence of new structures thanks to feedback mechanisms, as it has been shown in the example of Bénard cells. Self-production, on its side, is «the self-generation by the system of its own

[73] Cf. A. Moreno, «Closure, Identity and the Emergence», 115-116.

[74] A. Moreno, «Closure, Identity and the Emergence», 117.

structures»[75]. Self-production implies that the system has some sort of self-reference, allowing it to take care of its own maintenance. It is not just a matter of organization but of maintaining this organization through time[76].

4.1.3 Self-reproduction

One of the differences between living systems and other kinds of self-organization or self-producing systems is that the latter systems are based on an immanent mechanism, while living systems can pass those mechanisms to the next generation. «This problem can only be solved with the creation of a mechanism that allows the evolution of the constraining components to *go beyond* the limits of the existence of the individual systems in which they operate»[77]. It is necessary to have not only a self-replication system but also a self-reproductive one; a system that could produce another system, with all its organizational complexity. The production of a new system does not depend only on the self-organization activity of the already existing system. There has to be the transmission of heritable polymers that can act as templates in order to respecify the catalytic entities that would drive self-organization and self-production in the new system.

There is, then, an instructive domain and a dynamical domain. The information of the instructive domain has to be translated in order to give rise to its dynamical meaning. There is a relation between the metabolic system responsible for self-organization and self-production and the genetic system, whose aim is to transmit the codified information of that system to the next generation.

4.2 *Metabolic, genetic and selecting systems*

My guess is that it is not possible to think in a cell as a biological system without thinking about it as a complex ensemble of the metabolic, the genetic and the selecting systems that form it and that those very systems are related through an interchange of information between them. Those systems cannot be reduced to pairs, since they are mutually dependent and

[75] G. AULETTA, *Cognitive Biology*, 198.
[76] Cf. G. AULETTA, *Cognitive Biology*, 272.
[77] A. MORENO, «Closure, Identity and the Emergence», 118.

tightly interwoven. They cannot be reduced to pairs because the relation among them is a triad, not a dyad. As has been previously shown, the triad processor-regulator-decider fits in the biological system with the triad metabolic-genetic-selecting system.

The metabolic system constantly gives a variability of chemicals in the cell; it is the source of information. That variability is regulated by the genetic system which exerts control over the proteins that have to be synthesized and, then, over the chemicals that interact with those proteins. Not all the possible informational combinations are present in a cell, the regulator narrows down the possibilities. Finally, the selecting system is the decider that couples the processor with the regulator; it selects the information that is useful for the cell. In the cell, the selecting system is all the complex structure of the membranes that constitute not only the cellular membrane, but also the one of the endoplasmic reticulum, the Golgi apparatus, the nuclear envelop, the different vesicles that transport chemicals within the cell and from the environment, etc. Then, our triadic scheme, following that of Figure III.1, becomes the one showed in Figure V.1.

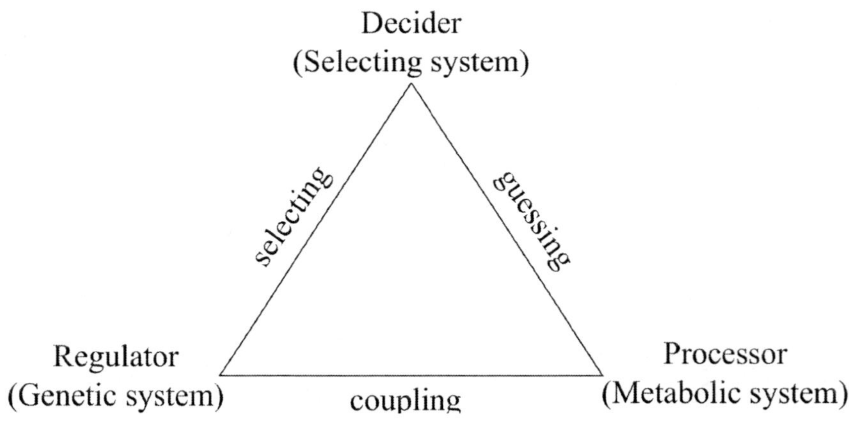

Figure V.1: The relation between the three biological systems. The metabolic system is the processor, the source of variability; the genetic system regulates this variability and the selecting system is the decider that selects the items that are relevant for the biological system.

This fits with the proposal of this work to see the function not as something ascribed to a given trait, but as something that needs the cooperation of many partners and the integration of the system as a whole. Let us see the systems that interact in the cell[78].

4.2.1 The genetic system

It has been shown that the genetic system should no longer be seen as an independent system that has the instructions for the whole cell to function. The genetic system has to be semiotically interwoven with the rest of elements of the cell to be expressed and regulated properly. One of the characteristics of life is its self-reproductiveness. Then, the metabolic system has to be somehow linked to the genetic one, since

> if metabolism did not become entangled with an informational machinery responsible for transmitting reliably self-referential records – and the changes these can suffer – to an offspring, it would not be possible for the system to get involved in a Darwinian evolutionary process. On the one hand, the appearance of genetic information and mechanisms is necessary; but on the other one, this information would have no internal meaning (or internal causal power) if it were not tightly connected to metabolism: functional dynamic units (like enzymes) allow its actual interpretation and expression in the system[79].

The storing of information in the genetic system is of great help for the cell. The elements coded are shielded against variations in the environment, and can be available according to the needs of the moment. The information is preserved for its use when needed. Having the information stored in the genes is of great help to have an organized access to it. It prevents all that information from being activated at the same time. The fact that DNA is a rather inert polymer permits proteins and other factors to have selective access to those genes that have to be expressed in a proper place and time.

On the other hand, genetic system allows the information to be preserved for the next generation. Then, the new individual is furnished with the genetic load in order to reconstruct the structures that would lead it to the adult form. However, as DST has pointed out, it is not only genes that are transmitted to

---

[78] A more detailed characterization of those systems can be found in G. Auletta, *Cognitive Biology*, 200-245.

[79] A. Moreno – K. Ruiz-Mirazo, «Metabolism and the Problem», 55.

the next generation. There are many other elements that pass trough the next generation because are also part of the developmental system. Moreover, it has to be said that the genetic information of the new individual would be expressed in the constant exchange of information with the environment. Transmitted genes are not just a blueprint that would be expressed anyway. They would also interact with the environment, whose informational cues would lead the developmental process along perhaps different paths from those of the previous generation.

4.2.2 The metabolic system

The metabolic system has been seen as the part of the cell that deals with energetic fluxes. It seems as if metabolism has to deal with the thermodynamic aspect of life, maintaining the cell far from equilibrium, supplying energy constantly and dissipating heat to the environment in order to keep entropy low. These characteristics of metabolism are absolutely true. It is impossible to talk about the cell as a semiotic system if the cell is not maintained far from thermodynamic equilibrium. However, metabolism is not only a notion that has to do with energy;

> rather, it comprises the biological system in its integrity, i.e. it relates to the energetic coherence of the system as a whole. Therefore, metabolism can be considered as the expression of the basic organizational gearing of life, which is realized through continuous structural, constructive and functional transformations. This means that metabolism is the energetic-material implementation of all the set of chemical reactions which self-maintain and self-produce the living being[80].

Moreno points out that the issue of metabolism has mostly been studied from the interest of artificial life. Investigators want to know what minimums are required to have a metabolic network that would reproduce the traits we find in the metabolic pathways of living systems. The paradigm used is that of seeing metabolism as a system that regenerates its own components. The problem is that very often it is seen very abstractly, as pieces that must be put together in order to have some recursive properties in the network. What matters for

[80] A. Moreno – K. Ruiz-Mirazo, «Metabolism and the Problem», 47.

them is just the relatedness of the items and what is required for the proper interactions among them. The question is that this approach does not take into account the physical and thermodynamic framework that is needed to the realization of the system[81].

When thermodynamic considerations are included, there are some modifications that have to be done to the model. It has to be seen how the organism manages the energy flow in order to ensure its self-maintenance. The dependence of the organism in those energetic flows is different from mere physico-chemical systems, since constraints and boundary conditions have to be taken into account in biological systems. In the case of self-organizing physico-chemical systems, variations in certain parameters could lead to great perturbations of the system and even to its complete dissolution. Biological systems, on the contrary, could store energy in order to face an external lack of it. Furthermore, they do not depend totally on the boundary conditions imposed by external variables, but they are able to construct their own constraints in order to couple reactions and steer the flows of matter and energy within them[82].

Metabolism, then, is a notion that is linked to that of autonomy, since metabolic reactions deal with matter and energy for the continuous construction and maintenance of the organism. So «metabolism should be universally defined as the recursive self-maintenance of controls upon the energy flows necessary for the physical realization of a component production system operationally closed»[83]. To achieve this, it is necessary to couple the reactions in space and time. That means that reactions have to be coupled in order to have energy, and these very reactions have to be tightly controlled to perform them at the proper time.

The system has to steer matter and energy for its own self-construction, to achieve the functional aims that have to be constantly performed in order to stay alive. The energetic flow and the relational structure of the organism are tightly interwoven, since «it is fundamental that the system becomes able to reinvest a big deal of the energy it receives in the generation and maintenance of its own constraints and their functional action»[84]. Neither the thermodynamics that

[81] Cf. A. MORENO – K. RUIZ-MIRAZO, «Metabolism and the Problem», 47-48.

[82] Cf. A. MORENO – K. RUIZ-MIRAZO, «Metabolism and the Problem», 49-50.

[83] A. MORENO – K. RUIZ-MIRAZO, «Metabolism and the Problem», 45.

[84] A. MORENO – K. RUIZ-MIRAZO, «Metabolism and the Problem», 52.

rules energetic fluxes, nor the relations established by the system elements can be taken separately.

4.2.3 The selecting system

In the last chapter the importance of the membranes in the cell has been underlined in order to establish a semiotic closure and to impose structural constraints over the elements that are in the cell. But there is also another feature of the cellular membrane that has to be stressed here. The cellular membrane is a selecting system that acts as a filter of the external stimuli, and decides which stimuli have to be transmitted and amplified and which ones are not worthy.

Then, the cellular membrane has to be linked to the needs of the entire cell, and has to be capable of evaluating which stimuli are worth transmitting and which are not. There are many proteins embedded in the cellular membrane, indicating the fact that the selecting and the metabolic systems are interwoven. Moreover, the proteins that are on the cellular surface have been synthesized through the genetic and the metabolic systems, and have been sorted out in the membrane thanks to the network of membranes of the endoplasmic reticulum and the Golgi apparatus. Then, the result of the selecting capacity of the membrane is due to the interaction of the genetic, the metabolic and the selecting systems.

The proteins that are in the outer membrane of the cell are as scanners searching signals that can bring some information of what is happening in the environment. But their role is not simply to interact with those chemicals that can be bound to those proteins, but to somehow transmit what those chemicals stand for and to trigger the necessary mechanisms inside the cell to properly respond to those chemicals. Then, the protein in the outer membrane is just a part of the story, since it has to be linked to the metabolic and genetic systems to adapt the cellular state to the new situation provoked by the stimulus.

As has been said, natural selection operates at this level, in the coupling of a given binding (of one environmental chemical to a given membrane protein) to a certain cellular response. It does not matter how the binding takes place in the membrane; the important thing is that the downstream chain of reactions and activations can lead to the proper cellular response according to the stimulus. The receptor could be different, some of the intermediate elements could also be organized in another way, but what is important for the cell in order for this pathway to be selected is that it can respond to a cellular need. All the chemical causal chain has to be coupled to a cellular function, no matter the peculiarities of the very items involved in such chain. To say it in other words: the important thing is to achieve the goal, no matter which items are used to get it.

4.3 *Teleonomy and teleology in biological systems*

The question about finality has been always of paramount importance when dealing with biological systems. It has been quite a problematic notion historically, since it has been linked to intentional behaviour (such as human behaviour) and to the deity. This has led to conceive everything out of humans and God as committed to the reductionist realm.

The notion of teleology is not very well received in science, since it seems as if one has to admit that future events have causal powers over present states. However, as Francisco Ayala points out, teleology has nothing to do with this issue. He acknowledges that teleological explanations are appropriate in biology and other life sciences. Probably the concept of teleology was coined when reflecting on human behaviour, since their actions are intentionally goal-directed. Nevertheless, there are other objects different from human beings that show also a goal-directed activity. «In this generic sense, teleological explanations are those explanations where the presence of an object or a process in a system is explained by exhibiting its connection with a specific state or property of the system to whose existence or maintenance the object or process contributes»[85].

Teleological explanations are appropriate when they show the contribution that an item makes to the whole system. For example, it is useless to think about the teleological aspect of the reaction of hydrolysis of ATP. It only makes sense when we place ATP as the most important energy exchanger of the cell and we see that it is linked to many metabolic reactions that work on the breakdown of energetic sources and in the synthesis of molecules for the cell.

Ayala's interest in the difference between teleological and non-teleological explanations is worth noting. In a teleological explanation the interest is focused on the consequences a given process or part has for the system as a whole. On the contrary, non-teleological explanations focus on the conditions of the possibility for the system to perform its activity. However, the teleological accounts give something more to the explanation of certain phenomena. They do not add

[85] Cf. F.J. AYALA, «Teleological Explanations in Evolutionary Biology», 8.

more data to the inner item or process under study, but helps to understand the structural complex organization that contributes to the benefit of the system as a whole. Teleological explanations show that the system is organized and structured in a certain way to allow items and processes of that system to be interconnected in order to perform whatever is necessary for the maintenance of the system[86].

Auletta goes further on the notion of teleology establishing a very clear difference between it and teleonomy. A teleonomic system is a system where different paths can lead to the same result. This can be done through feedback loops which control the process from the inside and integrate environmental cues in order to take one path or other. Teleology, for its part, requires a system to exercise an information control on another system. There is a semiotic relation among them in order to establish functional equivalence classes and select those that would be better for the benefit of the whole[87]. Then, teleology appears when there is an informational exchange between systems in order to fulfil a certain aim, selecting the way that has to be used to reach it.

Glucose metabolism in the muscle fibre can be a good example to understand the difference between teleonomy and teleology. The cell extracts the energy from glucose through a metabolic pathway known as glycolysis, which yields energy and a molecule called pyruvate. More energy can be obtained from pyruvate, but this would be done with more or less efficiency depending on the metabolic pathway followed. Indeed, pyruvate is metabolized into lactate in anaerobic conditions, but when there is an appropriate concentration of oxygen, pyruvate enters the mitochondrion and is fully oxidized to CO_2, generating much more ATP[88].

I think that the shuffle between anaerobic/aerobic glycolysis could be an example of teleonomy. The metabolic pathway has to yield ATP to supply the muscular fibres with energy. If there is enough oxygen, the way of the mitochondrion would be followed and the process would be more efficient. However, if oxygen is scarce, but energy is needed, a less efficient way has to be followed in order to maintain the necessary energy supply to the muscle anyway. Then,

---

[86] Cf. F.J. Ayala, «Teleological Explanations in Evolutionary Biology», 12.
[87] G. Auletta, «A Paradigm Shift in Biology?», 44.
[88] Cf. J.M. Berg – J.L. Tymoczko – L. Stryer, *Biochemistry*, 433.

teleonomy «is a mechanism based on the attraction exercised by a "final" or next stable state on a biological system»[89]. In this case, the stable state, the attractor, depends on the availability of oxygen, that steers the metabolic pathway, according to pure chemical rules, to the mitochondrion or to the production of lactate (Figure V.2).

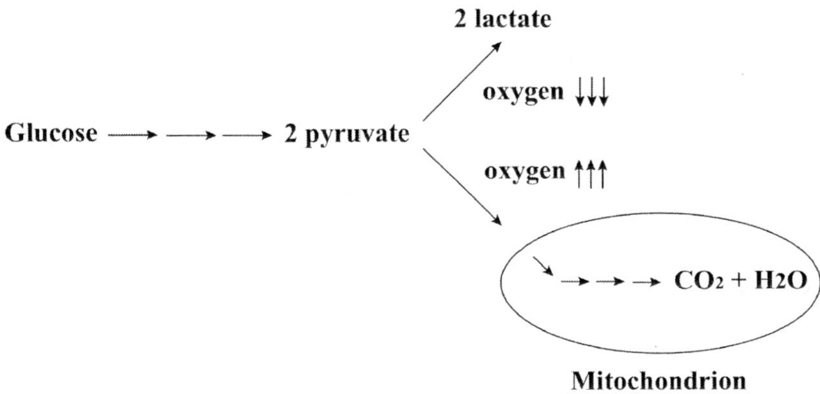

Figure V.2: The metabolism of glucose depending on the availability of oxygen. When oxygen is low, glucose is metabolized to pyruvate and then is converted to lactate. On the contrary, when oxygen is high, pyruvate can enter the mitochondrion to be fully oxidized to CO_2 and H_2O. This last option yields more energy for the cell than the former one. Both ways can be considered as chemical attractors that pyruvate enters depending on oxygen availability.

We can talk of teleology in the glucose metabolism when considering the informational exchange with other systems. In this case, the metabolic pathway of glucose would be the metabolic system, and we have to see how it is linked to the selecting and the genetic system. In mammals, glucose enters the cell through a family of transporters named GLUT and localized in the cellular membrane. Then, the selecting system decides when and to which extent glucose has to be left to enter the cell. For its part, the genetic system plays an important control role, since the transcriptional rate of the enzymes involved in the pathway is a way of controlling it (Figure V.3). When considering the interconnection between the three systems, we can talk of teleology, because

[89] G. AULETTA, *Cognitive Biology*, 255.

there is an informational control in order to better achieve the aim of obtaining energy, which is paramount for the cell[90].

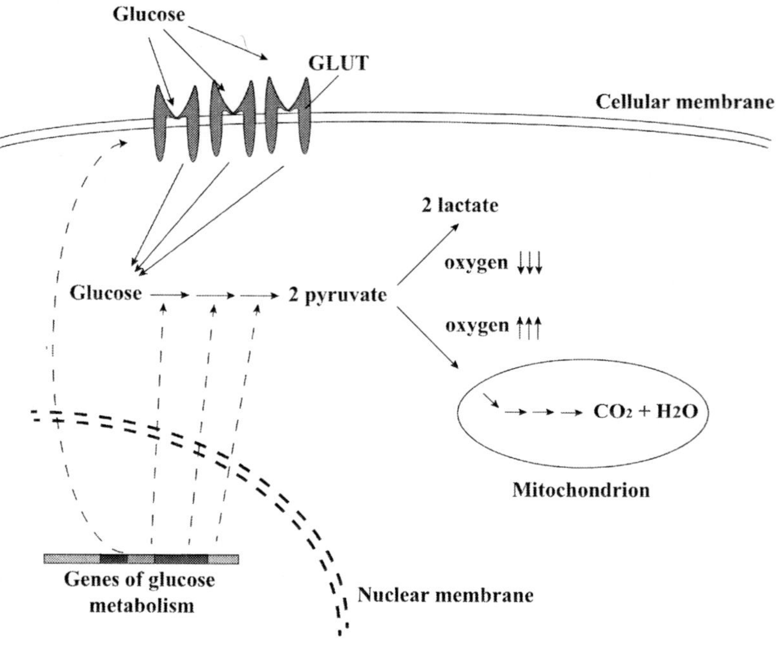

Figure V.3: The relation between the metabolic, the genetic and the selecting system in the metabolism of glucose. Here we can talk about teleology, since the chemical reactions are constrained by the biological system in order to better respond to the necessity of energy availability.

As I have previously said, Peirce conceived semiosis as the action of the sign. There is a signic action towards a final or intelligent causation. Semiosis, then, refers to the dynamical nature of signs. Peirce thinks that there are two kinds of actions in nature: a dyadic one, without purpose and unconcerned with the result, and a triadic one, which is governed by law. The dyadic action is called efficient causation and the triadic one final causation. As Peirce says,

> we must understand by final causation that mode of bringing facts about according to which a general description of result is made to come about, quite irrespective

[90] Cf. J.M. Berg – J.L. Tymoczko – L. Stryer, *Biochemistry*, 452-458.

> of any compulsion for it to come about in this or that particular way; although the means may be adapted to the end. The general result may be brought about at one time in one way, and at another time in another way. Final causation does not determine in what particular way it is to be brought about, but only that the result shall have a certain general character[91].

On the other hand, efficient causation is just determined by the particular condition of things, not by purposive behaviour. It is just a factual compulsion and a singular event that takes place in those circumstances. It is not subject to any kind of law or general rule, understood in a fully deterministic way[92]. While dyadic action has to do with the notion of force, the triadic one has to do with the notion of law. These two kinds of causes are always interwoven for Peirce. «Law, without force to carry it out, would be a court without a sheriff; and all its dicta would be vaporings»[93].

Of course, for Peirce the most obvious example of final causation is that of mind. However, he does not restrict the term "mind" to human mind. We can talk about a mind everywhere where we find a triadic relation between representamen, object and interpretant.

> Wherever there is a tendency toward learning, growing, or evolving, there is mind, no matter how rudimentary its action may be. Thus, final causation ranges from its most developed form, the intentional human act of reasoning under the guidance of self-control and self-criticism, to its most rudimentary form, wherever there is the working out of definite ends in the purely physical world[94].

According to what has already been said, we can make a distinction between final causality and purpose. In fact, purpose is the kind of final cause most familiar to human experience. It is «the conscious modification of final causation»[95]. Purpose is, then, just a type of final causation, but not the only one. Final causality is present in any goal-directed activity, regardless the fact of

---

[91] CH. S. PEIRCE, *Collected Papers*, 1.212.

[92] Cf. CH. S. PEIRCE, *Collected Papers*, 1.211.

[93] CH. S. PEIRCE, *Collected Papers*, 1.212.

[94] L. SANTAELLA BRAGA, «A New Causality», 503.

[95] Cf. CH. S. PEIRCE, *Collected Papers*, 7.366.

being physical, chemical, biological or mental. Any process that has a tendency to an end-state is characterized by a final causality.

To better understand Peirce's final causality it would be helpful to make the relation between law and chance clear. Peirce thinks that laws are present in nature, but they are not as absolute as the defenders of mechanicism may think. Peirce is against the deterministic model of causation, based on a closed mechanicism. He proposes a concept of law based on a habit that leaves enough room for chance, growing and evolution. Chance and law are not in an absolute contraposition, but they can coexist together[96]. Then, chance and law allow matter to take habits. Growing, learning and evolution is a triadic activity that takes habits in order to implement its dynamicity towards an end. Peirce thinks that to take habits is the background out of which all the laws of the universe have emerged[97]. Semiosis, then, is a process guided by final causality. The action of the sign is directed towards a goal. Santaella Braga expresses it as follows:

> This action can be intentional or not, it can be conscious or not, purposeful or not. What defines it is not the material or psychological matter through which the action is accomplished but its activity directed toward the production of another sign, the inerpretant, by means of which the sign signifies its object. Now, an action directed toward a goal is also what defines final causation. Consequently, Peirce's definition of the sign is a logical description of final causation[98].

Then, the action of the sign is only fulfilled when it determines the icon. But the process does not end here. In fact, the icon determined by the sign becomes itself a sign, continuing the semiotic process. This leads to

> a constant expansion of the process of semiosis since the interpretant [icon in the terminology I am using], in turn, determines a further sign, becoming thereby a sign to that further interpretant [icon]. Semiosis is, thus, an infinite process or an endless series in a process that operates in two directions, "back toward the object", and "forward toward the interpretant [icon]"[99].

---

[96] Cf. L. Santaella Braga, «A New Causality», 504-506.

[97] Cf. Ch. S. Peirce, *Collected Papers*, 7.515.

[98] L. Santaella Braga, «A New Causality», 514.

[99] L. Santaella Braga, «A New Causality», 515.

This does not mean that the object fully determines all the semiotic process. The sign acts through the mediation of a law; and we have seen that laws leave room for chance. «This law functions as a guiding principle for the generation of the interpretant of the sign. The interpretant actualizes this guiding principle by means of a habit. Whenever there is a habit change, there is a violation of the law, and the gradual constitution of a new law»[100].

5. Information control

In the last section the different systems that shape the basilar unit of life have been depicted. It has been shown that those systems have inbuilt goals and can manage different ways to fulfil those goals. The cell has its own survival as its main goal, and to constantly achieve this fundamental goal, there has to be tightly and fine-tuned interconnections between the genetic, the metabolic and the selecting system in order to keep the system far from equilibrium and to exchange matter, energy and information with the environment.

The cell is a very complex system that has a huge amount of information: the information contained in the genes, but also the information carried by the events that take place inside the cell and on its outer membrane. It is easy to imagine that it would be really a mess if all the interactions that can take place within a cell were not somehow controlled. I have talked about it (Chapter III) when the differences between actual and potential information were pointed out. The cell is a system where the actual information that is at play in a given moment is a small amount of the potential informational capacity of the cell. There has to be some control over all this dormant information, and a way to activate it at the proper time and place.

Information control is a concept introduced by Auletta, and can only be applied to systems that have a structure where there are feedback controls and a goal to be reached. When a system has a formal structure and a goal to achieve, information control can be exerted in order to make sure that the system will work to reach that goal. The existence of the goal has not to be thought of as something that is extrinsic to the system, but it matches the self-mainteinance of the very system.

---

[100] L. SANTAELLA BRAGA, «A New Causality», 516.

The aim of the information control is to make sure that the function that is needed for the cell to be alive will be performed. To do that, an organizational structure and feedback loops among them to link the different components are needed. A function needs to be instantiated in some operation; that is «any spatially and temporally coordinated pattern of physical-chemical interactions able to fulfil a function and therefore subjected to some top-down control»[101]. Then, a function is not reduced to the chemistry of the operation. Indeed, an operation is the instantiation of a function because it is connected with the needs of the whole organism and can be informationally controlled in order to perform what is needed for the self-maintenance of the organism[102].

We can see once more that the important issue is not to have a certain chemical items or a unique way to perform a certain function. Biological systems need to be integrated in such a way that allows them to have a formal structure suitable to be informationally controlled by the system as a whole. Functions need to be instantiated, no matter what the chemical details to achieve it are.

5.1 *Feedback control*

Cellular metabolism has numerous feedback loops that are at the core of the control that has to be exerted over the different metabolic pathways. The feedback loop allows a distant point of the pathway to return to some of its previous steps in order to control it. We can find many examples both of positive and negative feedback; where the former stimulates the controlled point activating it and the latter inhibits a key element in order to repress the pathway. The network of activations and inhibitions in the cellular pathways is extremely complex. There are some chemicals that not only control elements of their own pathway, but also regulate elements of other pathways of the cell. Thanks to that, the metabolism is a coherent network where all its elements are interconnected in order to achieve the same goals.

This feedback control that the cell exerts is linked to the consideration of the cell as a self. Indeed, the fact that an element of a given pathway can have an

[101] G. AULETTA, *Cognitive Biology*, 256.
[102] Cf. G. AULETTA, *Cognitive Biology*, 260.

influence on a previous element of the same pathway is a regulatory mechanism that allows the pathway to be treated as a whole. The cell does not need to control every single element of the pathway, but just the key points that are crucial for the activation, the inhibition or the bifurcation of the pathway. The cell selects some properties of certain chemicals, which fit with some of the enzymes, in order to select and couple them to the cellular needs. The feedback inhibition of a certain step is treated as a sign by the cell, since that inhibition does not concern only to the point where the inhibitor acts, but to the whole pathway.

Feedback loops, then, allow to that necessary semiotic closure that permits the pathway to be treated as an autonomous instance. The pathway is not only formed by the enzymes that interact among them, but also by the whole control process that makes the cell capable of semiotically controlling it.

Feedback loops make circular causality, the one that takes place among elements of the same-level, possible. It is at this same-level where bottom-up and top-down influences cross and act with their material and formal causes respectively, influencing in the dynamical causality of the level at stake.

5.2 *Modularity*

Modularity is a necessary condition for the cell to deal with complexity and to easily exert control over its different parts. The cell is made of soft-assemble parts that are used as basilar units and can construct complex structures.

This characteristic allows the cell to easily deal with its own molecules, since all of them are synthesized following few organizational patterns. For example, it is known that there are some basilar structures found in many proteins: alpha helix and beta-sheet are common structures out of which more complex structures can be built.

Let us see how modularity is a pattern of structure that can be found in the three systems of the cell.

5.2.1 Metabolic modularity

Modularity is also found in the very structure of metabolism. In fact, as Berg *et al.* state, even if the metabolic network seems rather unaffordable, «there are unifying themes that make the comprehension of this complexity more manageable. These unifying themes include common metabolites, reactions, and regulatory schemes that stem from a common evolutionary heritage»[103].

This modularity allows the cell to easily control the key points of the metabolic network, as well as to obtain a selective access to the encapsulated information hidden in those modules.

5.2.2 Genomic modularity

Frequently we have in mind the idea that genes are placed one next to the other all along the DNA chain. It would seem as if the only order that is observed in the chromosome is just the one given by the sequence of genes. However, the linear sequence of genes laid in the chromosome is not the only structure we can find in the genome. There are other elements that give another kind of structure and orderliness to it.

The first thing to be said is that only about 1-5% of the DNA sequences in eukaryotes are coding sequences that would be translated into protein. There is a great amount of non-coding DNA sequences that play an important regulatory role. Those sequences include short and long interspersed nuclear elements (SINE's and LINE's, respectively), micro-satellite sequences, pseudogenes, transposones and other unclassified DNA sequences. Here it is not important to consider the nature of each of those structures, but just to point out that they are non-coding structures that play an active role in the control of coding-sequences. Von Sternberg explains the regulatory role of those elements:

> Although individual nongenic elements have no direct phenotypic role, the locus-specific combination of sequences, the chromosomal arrangement of the various element families, and the interaction of these with chromatin proteins is a subtle determinant of genomic regulatory architecture, and numerous three-dimensional (conformational) codes appear to be present[104].

Here we have a clear example of the context-dependence of phenotypic expression. Depending on how those kinds of sequences regulate the coding ones, the phenotype could be expressed in one way or another. The message

[103] J.M. Berg – J.L. Tymoczko – L. Stryer, *Biochemistry*, 420.
[104] R. Von Sternberg, «Genomes and Form», 226.

coded in gene's sequence is context-independent, but its expression depends on the non-coding regions that in one way or another, interact with this gene, regulating its expression. Then we see here a picture quite different form that of the genes orchestrating the phenotype and organizing it out of the raw materials given by the rest of the organism. Here, however, we see the contrary. It seems as if genes are just the expressing units, while the other genomic structures of the chromosome are tuning and controlling this raw expression.

It has been seen that those non-coding regions are organized in isochores; that is, megabasepair-long structures that are maintained by the cell during generations. These long regions are actively maintained by the cell through its DNA repairing systems and through the distribution of chromatin proteins. Then, the cell can actively control the rules that establish the limits and functions of these isochores, modifying the chromosomal context and structure, as well as the point where recombination takes place[105].

«Transposable elements or transposons [...] are discrete sequences in the genome that are mobile – they are able to transport themselves to other locations within the genome»[106]. They can move through the whole genome, but their distribution is usually non-random. These elements can contain regulatory or even protein-coding sequences, origins of replication, splicing sites, enhancers and silencers of transcription, etc. Finally, their mobility could be enhanced in response to cellular perturbations and stress situations[107]. Then,

> as TEs [Transposone Elements] are (1) compact concatenations of regulatory sequences, (2) nonrandomly distributed along the chromosome, and (3) activated during periods of genomic shock, we can detect the lineaments whereby the cell can literally edit genome structures and codes in a bounded but not strictly determined manner. If the cell needs to break out of a semantic dead end it has the physical means to do so because TEs[108].

Then, as Sternberg says, this perspective gives us the possibility of seeing genomes as something written by the cell. If cellular elements are responsible

[105] Cf. R. VON STERNBERG, «Genomes and Form», 227.
[106] B. LEWIN, *Genes IX*, 523.
[107] Cf. R. VON STERNBERG, «Genomes and Form», 228.
[108] R. VON STERNBERG, «Genomes and Form», 229.

for the structure of those genetic elements that regulate the expression of certain genes, we can admit that the cell is exerting some control in the way the genetic system has to manage the information it has[109].

Sternberg names the process by which the cell tests new combinations of genetic elements in order to achieve a more robust morphological attractor that easily allows the cell to achieve its survival *teleomorphic recursivity*. To say it in Peirce's words, teleomorphic recursivity is the way the cell has to take habits to implement a given function.

We must not think of a change of paradigm that forces us to ignore genes and to focus on higher level DNA structures. What it is intended to be stressed is the fact that is difficult to dissociate the coding aspect of the genome from the other structural ones. «Given that DNA encodes enzymes, which in turn alter the same DNA in response to endogenous and exogenous factors, one is hard-pressed to determine where the boundaries between descriptor, constructor, and the constructed reside»[110].

5.2.3 Membrane modularity

The selecting system also has a modular structure of paramount importance for the distribution of the cellular molecules. First of all, it has been seen that the membrane limits the self-closing and differentiation from the environment. It is a physical differentiation, but also a semiotic one. The membrane maintains the cell's autonomy, but also allows the exchange of matter, energy and information with the environment. Indeed, the outer surface of the cellular membrane has many different receptor proteins that constantly scan the environment looking for cues that could be of interest for the cell functioning.

Often we may think of the cellular membrane as a lipidic structure where proteins are embedded in a random way, leaving them to freely move horizontally through the membrane. But things are more complex and there is also a control exerted over the distribution of proteins all along the membrane. Indeed, «the picture of a membrane as a lipid sea in which all proteins float

[109] Cf. R. Von Sternberg, «Genomes and Form», 232.

[110] R. Von Sternberg, «Genomes and Form», 233.

freely is greatly oversimplified. Many cells confine membrane proteins to specific regions in a continuous lipid bilayer»[111].

Another important selecting system found in eukaryotic cells is the nucleus. It separates the genetic material from the cytoplasm, permitting a closer control over genetic expression. The nucleus is a physical barrier that establishes a certain control over protein translation, since it controls the mRNA's passing to the cytoplasm in order to be translated by the ribosomes.

The endoplasmic reticulum and the Golgi apparatus are structures where protein folding takes place and, at the same time, they are sorted to different parts of the cell or to the cellular membrane to be secreted outside the cell. This sorting is done through a network of vesicles which, depending on their composition, deliver to the appropiate cellular locations.

We cannot forget other selecting systems that are present in mitochondria and chloroplasts. Those organelles have a membrane that allows them to be separated from the cytoplasm. The cell uses this in order to establish a chemical gradient between the cytoplasm and the inner space of those organelles. This chemical differential is very important to the energetic metabolism of the cell. In mitochondria the oxidation process that yields the most important production of ATP, the energetic currency of the cell, takes place. In the chloroplast, the energy of the light is transformed into energy suitable to be used by the cell. Then, in both cases, the modularity of the selecting system is the cornerstone of the energetic balance of the cell[112].

6. **Summing up**

The concept of function in biology is not as easy to define as it would seem. Many of the attempts to explain that concept have been focused on the functions of artefact, and they have tried to apply it to biological systems. However, biological functions are different from artefacts. Artefacts are designed by an external intentionality, and their function is linked to their external design. Life, for its part, is not designed externally, but has to seek new solutions to its

---

[111] B. Alberts– *al.*, *Molecular Biology of the Cell*, 645.

[112] Cf. B. Alberts– *al.*, *Molecular Biology of the Cell*, 813-855.

challenges arranging and organizing its inner structures. Life cannot stop from living to reshape its components according to an external design.

Two main approaches can be individuated in literature when dealing with the concept of function in biology. The causal-role approach stresses the importance of the role the items play in a given system here and now, without taking into account their previous history. The etiological analysis, on the contrary, says that the function of an item is known when the causes that have yielded its incorporation in the system are revealed. Both approaches show a part of the issue at stake: the former points at the causal role each item has, while the latter shows the history that has led a certain item to be there.

Causal-role and etiological analysis are complementary, but even if they are taken together, there is something that misses. They do not appreciate that the living system is not an artefact. Biological systems are the result of recurring processes of self-organization, self-production and self-reproduction. This makes organisms extremely complex systems, highly organized and tightly interwoven. Biological systems emerge from the connection among the genetic, the metabolic and the selecting systems. This allows the organism to regulate its own organization and have strategies to achieve its own goals, which are nothing more than its own survival. That means that the organism has the capacity to control and manage the different modules of its organization to achieve its goals; something that an artefact cannot do, unless it is handled by an external user.

The information control of the modules that constitute the different biological systems (metabolic, genetic and selecting) is done in a top-down fashion. The system as a whole maintains a control over its parts in order to constrain them to perform the necessary operations that will ensure the surviving of the organism. To do that, the system has established, over the whole history of evolution, links between signs and referents in order to deal with them informationally and couple their relation with the needs of the biological system. As it has been shown in the example of glucose metabolism, GLUT transporters have been selected because they are linked to glucose transport in a way that is useful for the cell in order to have energy for its own survival, not because of its chemical characteristics as such.

The function is something that happens, it is not linked to a single structure but to the goal that has to be achieved to stay alive. The system exerts information control to constraint the modules, the lower level items, in order to steer them to deploy what is needed for the organism as a whole. For doing that, the most important point is not the chemical nature of the items involved, but the semiotic relation that can be established between them. The paramount issue is that the system has to treat its items as signs of referents that have to be guessed by the system to keep its structure in a changing environment.

Evolution does not select individual items, but triadic relations among sign, referent and icon that integrate the metabolic, the genetic, and the selecting system in such a way that allows them to instantiate operations that perform a function needed by the biological system for its own survival.

THIRD PART

CASE STUDY

Chapter VI

Selenocystein insertion mechanism

1. Introduction to the case study

In this part of the work I am going to study an example of a biological mechanism. All throughout the previous chapters there have been some examples used to illustrate the concepts and notions explained. Here I want to make a very detailed study of one particular case of the cell's biochemistry. I think that showing a case study could help us to understand the theoretical and philosophical aspects I have considered in the previous part of this work. I shall point out the biological features that could led us to understand how the cell deals with information and selects the proper way to accomplish the functions it needs to survive.

Moreover, studying a single example in detail is the best way to see if the theoretical framework is really explanatory for biology or not. It will be shown that the notions of emergence, top-down causation, function, information control, etc, find their biological partners in this case study. I guess that this will show that philosophical work is paramount for the scientific research, since it provides a wider epistemological framework to better understand how life works.

The case study will be the biochemical mechanism through which the cell deals with the element selenium (Se). It will be seen that the cell cannot deal directly with Se as it does with, for example, calcium (Ca) or sodium (Na). In fact, the major role of Se in the cell consists in forming part of the modified amino acid selenocysteine (Sec) in order to act in the catalytic core of the so-called selenoproteins[1].

[1] «In mammals, 80% of the selenium accumulated in the body is present as selenocysteine, a cysteine analog in which a selenium atom is found in place of sulfur», N. Hubert

The important point of this process is the way the cell inserts Sec in the polypeptide chain. There is no triplet in the genetic code that codifies for Sec; and the Sec found in proteins is not the result of the modification of one of the twenty amino acids of the genetic code *after* its insertion in the polypeptide chain. Sec has its proper tRNA (tRNASec), and the way a tRNA without a codifying triplet can insert the amino acid it carries into the nascent polypeptide chain shall be studied. It will be seen that this includes an astonishing mechanism that allows us to talk about a new way of dealing with and selecting information.

2. The importance of selenium for the cell

When we take a look at Mendeleev's periodic table, we can see the wide variety of elements that are found in nature. On that table, elements are characterized according to their number of protons and electrons and to the distribution of the latter in the energetic shells. In fact, the number and distribution of electrons is what confers to a particular element its chemical properties. Moreover, it is the incompleteness of the outer shell that gives an element its chemical features and the possibility to react with other elements.

From that point of view, Se is characterized as a non-metal element with an atomic number of 34 (that is, it has 34 protons and so 34 electrons) and a weight number of 79 (what means that the sum of the number of protons and neutrons yields 79).

2.1 *The role of trace elements in living organisms*

In living organisms there is just a small fraction of the wide variety of elements we can find in nature. Carbon (C), hydrogen (H), nitrogen (N) and oxygen (O), make up 96,5% of an organism's weight[2]. Then, from the chemical point of view, we can consider living organisms as a specialized filter that accumulates only certain elements. Furthermore, these elements are found in a different proportion if compared to that of the non-living world (Figure VI.1).

– R. WALCZAK – C. STURCHLER – E. MYSLINSKI– *AL.*, «RNAs Mediating Cotranslational Insertion», 590.

[2] Cf. B. ALBERTS – *AL.*, *Molecular Biology of the Cell*, 46.

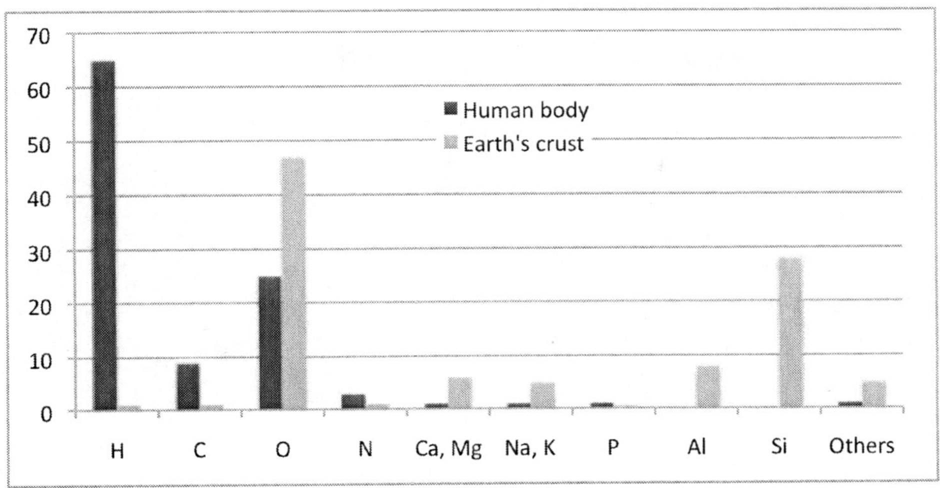

Figure VI.1: Relative abundance (%) of some elements of the non-living world (grey) compared with their presence in the human body (black). Adapted from B. ALBERTS –AL., *Molecular Biology of the Cell*, 47.

Living organisms do not use all the elements that exist in nature, but only those that have certain properties and can provide certain functions for life to success. The chemistry of the non-living world is broader than the chemistry of life. However, there are some elements that, despite being present in living organisms in very little quantities, are very important for the proper function of the cell. In the middle of the last century some of these trace elements were identified: iron (Fe), iodine (I), chromium (Cr), copper (Cu), zinc (Zn) and Se[3]. In recent years the list has grown with new elements that are necessary for life. In Figure VI.2 the elements of the periodic table that are known to have some function in living organisms are shown.

There are various functions that the trace elements are important for. Some of them are part of the enzymatic core of certain proteins as cofactors[4], allowing them to speed up the rate of conversion of substrates into products. Other trace

[3] Cf. W. MERTZ, «The Essential Trace Elements».

[4] A cofactor is a chemical required for the proper function of an enzyme.

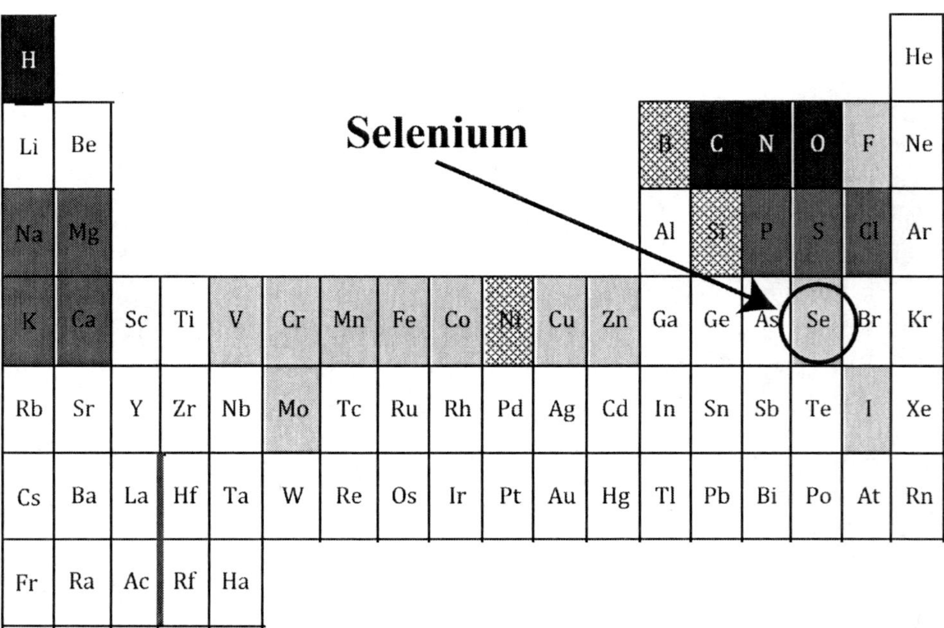

Figure VI.2: Mendeleev's periodic table of the chemical elements. The elements in black are those that constitute the vast majority of the weight of the human body (99%). The elements highlighted in dark grey are present in a 0,9% of the total. The light grey ones are really trace elements; that is, the organism requires them in quantities that are less than 100 micrograms/day. It is not clear if the dashed ones are essential for the human body or not. Adapted from B. Alberts – *al.*, *Molecular Biology of the Cell*, 49.

elements can be found in electron transport chains, which are essential for the energetic metabolism of the cell. They also have an important role in stabilizing the three-dimensional structure of some proteins. Finally, some trace elements contribute to facilitate the binding of molecules to the membrane receptors of the cell[5].

Se is present in all three kingdoms, and in all of them plays an important role for the proper function of the cell and/or the organism, but not all the species of each kingdom have Se incorporation in their metabolic pathways. Se has

[5] Cf. Y. Zhang – V.N. Gladyshev, «General Trends in Trace Element Utilization», 3393.

essential roles in oxidoreduction reactions, redox signaling, antioxidant defence, thyroid hormone metabolism, immune response and other cellular functions[6].

As it can be seen, the chemical analysis of the elements that are involved in biological systems confirms what has been said in the previous chapters. Life does not have any special element to which to ascribe the properties of organisms. Living beings have the same elements that can be found in the inorganic world. Moreover, life uses fewer elements than those that it has at hand, so there is yet a narrower window with which work. The novelty of life and the development of all its properties would depend not on the stuff used, but on the way it is used and organized.

2.2 *How does the cell uptake and use Se?*

When we talk about trace elements, we normally expect the entrance into the cell to be done through an active transport. In fact, there is a family of transporters called ATP-binding cassette, that actively uptakes certain trace elements into the cell, like Zn, Manganese (Mn), Molybdenum (Mo) or Nickel (Ni). There are also other types of active intake, thorough other transport systems that are responsible for the entrance of Fe, Cu, Ni and Cobalt (Co). Finally, there is also a general cation/ion influx, which is not as efficient as the transport mediated by a specific transporter, but also plays its part[7].

In the case of Se, how does it enter the cell? It seems that Se enters neither through an active system nor the general cation influx. Despite «Se-specific transporters have not been identified, [...] the uptake (in the form of selenate or selenite) is thought to be maintained up by the sulfate transport system»[8]. The cell can only use Se when it forms part of Selenocysteine (Sec), an amino acid that differs from cysteine (Cys) in the fact that Se has replaced its Sulphur (S) atom (Figure VI.3).

Only Sec is able to enter the biosynthetic pathway of the cell and then be inserted in the polypeptide chain, using a mechanism that I shall describe below (section 3.2). Furthermore, what counts is not just the presence of Sec,

[6] Cf. J. Lu – A. Holmgren, «Selenoproteins».

[7] Cf. Y. Zhang – V.N. Gladyshev, «General Trends in Trace Element Utilization», 3393.

[8] Y. Zhang – V.N. Gladyshev, «General Trends in Trace Element Utilization», 3393.

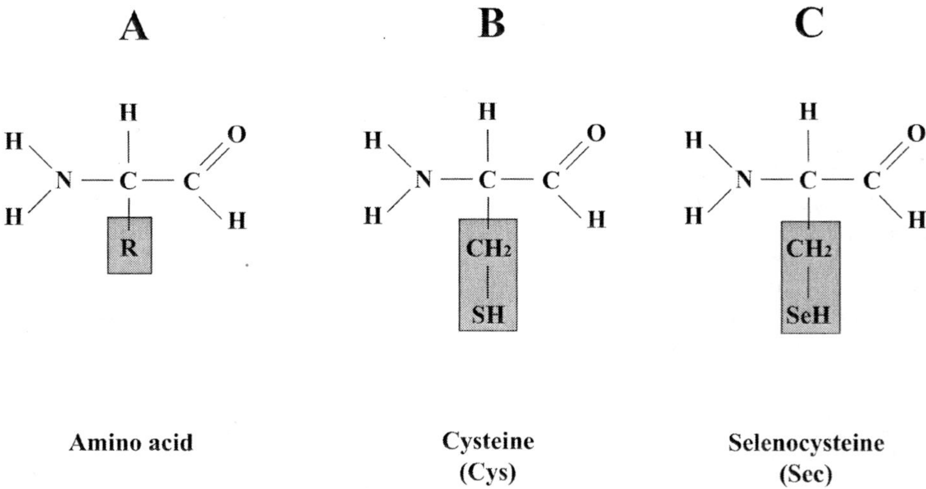

Figure VI.3: Chemical formulas of amino acids. A) The general chemical structure of all amino acids. R is the radical that is different in each single amino acid. B) Chemical formula of the amino acid Cysteine. C) Chemical formula of the amino acid Selenocysteine. Note that in Selenocysteine, the atom of sulfur of the Cysteine has been replaced by a selenium one.

but it's loading in the proper tRNA in order to be available for the translation machinery and be inserted in selenoproteins. So, if tRNASec is the special tRNA used for Sec insertion, Sec-tRNASec would be the same tRNA charged with Sec and ready to be used by the protein synthesis machinery (Figure VI.4).

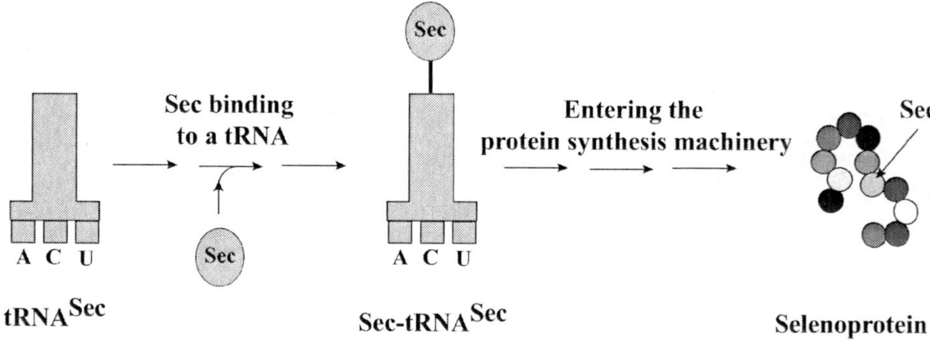

Figure VI.4: Insertion of Sec into a protein. Sec needs to be linked to a tRNA to be available for entering protein synthesis. Once Sec is bound to a tRNA, it can be used as the other amino acids for protein synthesis. Proteins that have one or more Sec in their structure are called selenoproteins.

2.3 *Selenoproteins*

As has been said, Se can only be managed by the cell in the form of Sec, an amino acid that can be co-translationally inserted in the polypeptide chain by a special mechanism. Then, the biological role of Se is developed in the catalytic core of a certain type of proteins that, due to this feature, are known as selenoproteins.

The numbers of genes involved in selenium metabolism (the selenoproteoma) varies from one organism to another. Indeed, there are twenty-five genes characterized in humans[9], eighteen in zebrafish[10], but only three in *Drosophila melanogaster*[11] and one in *Caenorhabditis elegans*. In prokaryotes a wide variety of selenoproteomes can be found, depending on the species we are considering[12]. Also in archaea selenoproteins are present[13]. However, although some selenocompounds are found to be essential in plants and yeast, no selenoproteins are reported in those organisms[14].

I guess that this differences in the distribution of selenoproteins among organisms are related with the medium those organisms dwell. Indeed, as it shall be seen (Chapter VIII), selenoproteins probably appeared in aquatic environments

---

[9] Cf. G.V. Kryukov – S. Castellano – S.V. Novoselov – A.V. Lobanov– *al.*, «Characterization of Mammalian Selenoproteomes». For a detailed review of the human selenoproteome see D. Behne – A. Kyriakopoulos, «Mammalian Selenium-containing Proteins» and S. Gromer – J.K. Eubel – B.L. Lee – J. Jacob– *al.*, «Human Selenoproteins at a Glance». For the relationship between selenium and human health see K.M. Brown – J.R. Arthur, «Selenium, Selenoproteins and Human Health» and B. Moghadaszadeh – A.H. Beggs, «Selenoproteins and their Impact».

[10] Cf. G.V. Kryukov – V.N. Gladyshev, «Selenium Metabolism in Zebrafish».

[11] Cf. F.J. Martin-Romero – G.V. Kryukov – A.V. Lobanov – B.A. Carlson– *al.*, «Selenium Metabolism in Drosophila».

[12] Cf. G.V. Kryukov – V.N. Gladyshev, «The Prokaryotic Selenoproteome».

[13] Cf. T. Stock – M. Rother, «Selenoproteins in Archaea». For a comparative study between Sec synthesis in Eukarya and Archaea, see X.-M. Xu – B.A. Carlson – Zhang,Y. – H. Mix– *al.*, «New Developments in Selenium Biochemistry».

[14] Cf. P.D. Whanger, «Selenocompounds in Plants and Animals». In 2002 an article was published about a selenoprotein found in plants, L.H. Fu – X.F. Wang – Y. Eyal – Y.M. She– *al.*, «A Selenoprotein in the Plant Kingdom». But the most recent reviews explicitly exclude plants of having selenoproteins; as it is shown in A.V. Lobanov – D.L. Hatfield – V.N. Gladyshev, «Eukaryotic Selenoproteins and Selenoproteomes».

with low oxygen concentration. The extreme sensitivity of some selenoproteins to neutralize oxygen species, which are very dangerous for the cell, makes them a very efficient instrument to keep those dangerous molecules low. On the contrary, the increasing of oxygen in the atmosphere and the colonization of the terrestrial ambient (where Se is more scarce), lead to use selenoproteins in other tasks and, in some cases, to replace Selenocysteine by Cysteine.

2.3.1 Selenoproteins' functions

One of the main functions of selenoproteins is their antioxidant activity, which preserves the cell against ROS[15] and other dangerous chemicals that can damage proteins, DNA or lipids. Glutathione peroxidases and Thioredoxin reducatses are selenoproteins that have an important role in controlling the redox status of the cell[16].

Selenoprotein P plays an important role in Se distribution in the organism, since is an extracellular glycoprotein that transports Se from the liver to the brain, testis, thyroid gland and kidneys[17]. Iodine metabolism, for its part, is

[15] The reactive oxygen species (ROS) are produced either by endogenous cellular oxygen metabolism or by exogenous sources. Notwithstanding which origin they have, the cell has to neutralize them in order to protect its structures from uncontrolled oxidation. However, it has also to be said that the scope of the cell is not to neutralize all the ROS, since the redox status of the cell is important in some types of signal transduction. In this sense, the definition of "oxidative stress" has been redefined as a "disruption of redox signaling and control", as is noted by H. Steinbrenner – H. Sies, «Protection Against Reactive Oxygen Species».

[16] Sec metabolism is linked with the cytoprotective response of the cell. Deletion of $tRNA^{Sec}$ gene induces a compensatory response of other antioxidant mechanisms of the cell, as it is shown in T. Suzuki – V.P. Kelly – H. Motohashi – O. Nakajima– *al.*, «Deletion of the Selenocysteine tRNA». In A. Sengupta – B.A. Carlson – J.A. Weaver – S.V. Novoselov– *al.*, «A Functional Link» it is shown that there is a functional link between housekeeping selenoproteins and phase II enzymes. When synthesis of selenoproteins involved in the stress response is inhibited, phase II enzymes are overproduced in order to cope with the stress situation.

[17] Many literature can be found about this topic: Cf. K.E. Hill – J. Zhou – W.J. McMahan – A.K. Motley– *al.*, «Deletion of Selenoprotein P», L. Schomburg – U. Schweizer – B. Holtmann – L. Flohé– *al.*, «Gene Disruption Discloses Role», B.A. Carlson – S.V. Novoselov – E. Kumaraswamy – B.J. Lee– *al.*, «Specific Excision», R.F. Burk – K.E. Hill, «Selenoprotein P: an Extracellular Protein», M. Scharpf – U. Schweizer – T. Arzberger

closely related to Se's, since some of the proteins that interact in that pathway have a Sec amino acid in their enzymatic core[18].

Immune response is also dependent on selenoprotein activity[19]. The role of selenoproteins in the inflammatory response is to regulate the redox status of immune cells. Moreover, selenoproteins are necessary for the proper expression of T-cell receptor[20] (TCR) and the receptor for interleukine 2[21] (IL-2R). Selenoproteins are also reported to be involved in cancer prevention[22], muscular activity[23], cardiovascular diseases[24], neurological function[25] and lipoprotein metabolism[26]. Nevertheless, there are selenoproteins whose function has not yet been discovered or has not been totally characterized, as Lobanov *et al.* notice it:

– W. Roggendorf– *al.*, «Neuronal and Ependymal Expression» and R.F. Burk – K.E. Hill, «Selenoprotein P-expression».

[18] Cf. M.J. Berry – J.D. Kieffer – J.W. Harney – P.R. Larsen– *al.*, «Selenocysteine Confers the Biochemical Properties». For an interesting review on that topic see J. Köhrle, «Selenium and the Control», and also J. Köhrle – F. Jakob – B. Contempré – J.E. Dumont– *al.*, «Selenium, the Thyroid».

[19] Cf. J.R. Arthur – R.C. McKenzie – G.J. Beckett, «Selenium in the Immune System» and B.A. Carlson – M.H. Yoo – Y. Sano – A. Sengupta– *al.*, «Selenoproteins Regulate Macrophage Invasiveness».

[20] Cf. R.K. Shrimali – R.D. Irons – B.A. Carlson – Y. Sano– *al.*, «Selenoproteins Mediate T Cell Immunity».

[21] Cf. M. Roy – L. Kiremidjian-Schumacher – H.I. Wishe – M.W. Cohen– *al.*, «Selenium Supplementation Enhances the Expression».

[22] Cf. D.L. Hatfield – M.H. Yoo – B.A. Carlson – V.N. Gladyshev– *al.*, «Selenoproteins that Function in Cancer». For a study of the role of GPXs in the different stages of cancer progression see R. Brigelius-Flohé – A. Kipp, «Glutathione Peroxidases in Different Stages».

[23] Cf. A. Lescure – M. Rederstorff – A. Krol – P. Guicheney– *al.*, «Selenoprotein Function and Muscle Disease».

[24] Cf. R.K. Shrimali – J.A. Weaver – G.F. Miller – M.F. Starost– *al.*, «Selenoprotein Expression is Essential».

[25] Cf. F.P. Bellinger – A.V. Raman – M.A. Reeves – M.J. Berry– *al.*, «Regulation and Function of Selenoproteins». An interesting review about Se metabolism and function in the brain can be found in J. Chen – M.J. Berry, «Selenium and Selenoproteins». For a map of the selenoproteins' expression in the brain see Zhang,Y. – Y. Zhou – U. Schweizer – N.E. Savaskan– *al.*, «Comparative Analysis of Selenocysteine Machinery».

[26] Cf. A. Sengupta – B.A. Carlson – V.J. Hoffmann – V.N. Gladyshev– *al.*, «Loss of Housekeeping Selenoprotein Expression».

> However, the functions of most other selenoprotein families remain unknown. These include Selenoprotein K, a short protein with a single transmembrane domain; Selenoprotein I, a recently evolved membrane selenoprotein; Selenoprotein O, the largest mammalian selenoprotein that has cysteine homologs in bacteria, yeast and plants; Selenoprotein T, a protein with a predicted redox motif; two related selenoproteins, Selenoproteins W and V (the C-terminal part of SelV is highly homologous to SelW); and Selenoprotein M, which is a distant homolog of the 15 kDa selenoprotein, Sep15[27].

It is difficult to say in a word what the function of Se in the cell is. As we have seen, it is present in many selenoproteins that are involved in many metabolic pathways. However, the conceptual framework proposed in this work could help to shed some light on this multitude of roles selenoproteins have.

First of all, it can be said that Se is used for many things in the cell. It is used in biochemical pathways that concern different processes such as antioxidant activity, iodine metabolism, immunity response, and many others. Then, Se is an element that is used in many *respects* by the cell. The different selenoproteins of which Se forms part provide the adequate environment for selecting the chemical properties of Se in that very environment that can be used for a given operation. Then, we have a one-to-many relation, where the surrounding conditions where Se is inserted are crucial for the role this element has to play in the cell's metabolism. From that point of view, we can see the different kinds of selenoproteins as different chemical contexts into which Se is inserted and allowed to participate in those different functions. Moreover, we can understand the folding process of selenoproteins as a top-down process where the three-dimensional structure of the protein constraints Se to be in the adequate chemical environment in order to deploy the operations that are linked to the functions of the cell.

Then, it is worth noting that the term "function" applied to single proteins does not have the same meaning as when it is applied to the cell as a whole. Indeed, a selenoprotein would have, for example, the function of Se transport to the brain; but this activity can be performed not only because it has the capacity to have up to ten Sec residues in its structure, but because there are also other

[27] A.V. Lobanov – D.L. Hatfield – V.N. Gladyshev, «Eukaryotic Selenoproteins and Selenoproteomes», 1425.

elements that allow its proper regulation and distribution. It is very important, for example, that the brain cells have a membrane receptor in order to recognize that protein. So the efficiency in performing a role is not enough to have a function, since that role have to be integrated and controlled by the whole biological system, which is the final criteria that would evaluate if that process is or is not for its own sake. There can be a protein that has a greater capacity to transport Se through the blood to the brain, but if this protein is not apt to interact with appropriate receptors, the brain could not uptake the needed Se.

3. **Co-translational incorporation of tRNASec in protein synthesis**

As mentioned above, Se is inserted into proteins forming part of Sec, an amino acid that does not have a specific triplet (or codon), as the others do. This particular feature constrains the cell to use the coding elements it has for inserting this amino acid, thus "forcing" the rules of the genetic code. It is during the process of translation of mRNA into protein that the Sec insertion in the polypeptide chain takes place. That is the reason why Sec has to be linked to a tRNA, because they are the molecules that establish the relation between the information codified in the RNA and the sequence of amino acids it codes for. The name of "translation" for this process comes from the fact that there is a true translation event that converts the linear string of the RNA nucleotides into an amino acid string, which forms the polypeptide chain. As I will show below, it involves more partners than those that are required for the "canonical" protein synthesis process, where each amino acid has one or more codons that codify for it.

This is the most interesting point of my study. The cell has to readapt an already working mechanism to incorporate a variant that will enable it to perform certain functions with greater efficiency. The metabolic network of reactions could be seen as a flexible web that reorganizes the already existing elements in order to achieve new functions.

I shall begin this section with a brief exposition of the main steps of protein synthesis. This will highlight the differences between the canonical process of protein synthesis and the particular steps needed to insert Sec into the growing polypeptide chain. After that, I will focus on describing the features of tRNA-Sec, the molecule responsible for Sec insertion. tRNA's are the molecules that read the genetic code and translate it into the proper amino acid. Then, some features of tRNASec are expected to be the same as the ones that are found in the canonical tRNA's. However, there has to be something different in order to deploy the special role tRNASec has.

Other elements that have to be taken into account are the secondary structures that messenger RNA (mRNA) adopts; they act as signal and instrument for the recoding event. From this point of view, tRNASec is not the only RNA molecule involved in Sec insertion. It will be shown that other RNA secondary structures that are formed in the very mRNA are essential for the mechanism we are about to study.

Finally, protein factors are considered. Special protein factors are needed to perform Sec insertion. As it has been said for the tRNASec, here, also, proteins that could perform the steps of protein synthesis are needed, but with some special features, that allow the Sec insertion event.

3.1 *General features of protein synthesis*

Transcription from DNA to RNA and translation of the latter into protein are the means through which the cell expresses the genetic instructions that are contained in its genes. That is the *central dogma* of molecular biology; that is, the genetic information flows from DNA to RNA to protein[28] (Nonetheless, it has to be said that the flowing of information can be also from RNA to DNA in the case of retroviruses). Here I will not explain the whole process of protein synthesis and its regulation, but just to sketch its main steps in order to see the differences between the incorporation of the 20 canonical amino acids and the incorporation of Sec.

Transcription is the process through which information contained in the DNA is complementary copied into RNA. There are two chemical differences between DNA and RNA. The first is that RNA uses the sugar ribose instead of desoxyribose. The other one is that in RNA we find the base Uracil (U) where DNA has Thymine (T). Despite this difference, the complementarity of the bases is preserved. Uracil (Thymine in the DNA) interacts through two hydrogen bonds with Adenine (A), and Guanine (G) establishes three hydrogen bonds with Cytosine (C) both in the DNA and RNA molecule[29]. Moreover, DNA and RNA differ also in their secondary structure. DNA has normally the form of a

---

[28] Cf. B. ALBERTS – *AL.*, *Molecular Biology of the Cell*, 331.

[29] Cf. B. ALBERTS – *AL.*, *Molecular Biology of the Cell*, 332.

double helix, while RNA folds adopting three-dimensional structures. In those RNA folded structures we can find conventional hydrogen bonds (Watson-Crick base pairs) and also non-conventional hydrogen bonds (non-Watson-Crick base pairs). The three-dimensional structure of RNA is a critical feature for the proper function of some types of RNA in the cell[30].

Once we have an mRNA as the product of the transcription of a gene, the translation machinery of the cell synthesizes a protein according to the instructions encoded in the mRNA sequence. The cell has to translate the information coded by the mRNA nucleotide sequence into the amino acid sequence of the protein. This translation process takes place in the ribosome. The tRNAs are the molecules that carry the amino acids to the ribosome in order to synthesize the new polypeptide according to the genetic code rules (Figure VI.5).

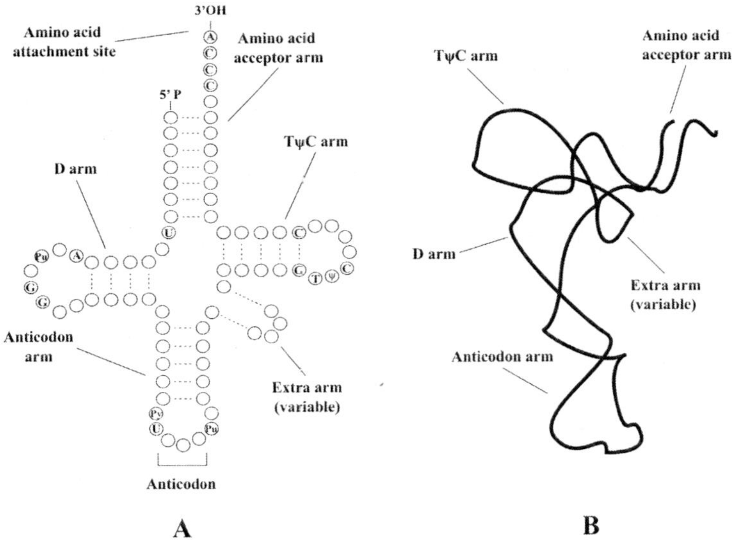

Figure VI.5: A) Secondary structure of a tRNA. This structure is also known as clover-leaf structure. Four arms can be individuated and are always present in all tRNAs. There is also an extra arm, whose length can vary between different tRNAs. The bases that are shown in the figure are always conserved in all tRNAs. Dashed lines indicate Watson-Crick base pairing between nucleotides. The anticodon is the part of the tRNA that matches with mRNA by basepairing. A: Adenine, C: Citosine; G: Guanine, U: Uracil, Ψ: Pseudouridine, Pu: Purine (A or G), Py: Pyrimidine (C or U). B) Tertiary structure of a tRNA, also known as L-shape.

[30] Cf. B. Lewin, *Genes IX*, 128-129.

Each triplet of mRNA (also known as "codon") codifies for an amino acid. The anticodon of the tRNA base-pairs through hydrogen bonds with the complementary three nucleotides of the mRNA. This complementation permits the translation of the nucleotidic units into amino acids[31]. The translation takes place in the ribosome (Figures VI.6 to VI.9), where not only the tRNA basepairs with its proper three nucleotides, but also the amino acid carried by each tRNA is bound to the previously amino acids that have been linked sequentially during translation. «This central reaction of protein synthesis is catalyzed by a peptidyl transferase contained in the large ribosomal subunit»[32].

The addition of an amino acid to the growing polypeptide chain requires a sequence of events that can be divided in four steps:

- In the first step we see the polypeptide chain attached to the tRNA that occupies the P-site of the ribosome. The tRNA that carries the next amino acid to be added, bounds to the A-site. That is possible because of the complementarities, according to the genetic code, between the three bases of the mRNA codon and the three bases of the tRNA anticodon. The name of P-site stands for peptidyl-tRNA site, since it is the place where the new amino acid that enters the ribosome is transferred to the polypeptide chain. The A-site is the aminoacylt-tRNA site, and is the point where newly bound tRNA's with their specific amino acid enter the ribosome. The E-site, for its part, is the exit site, where tRNAs leave the ribosome when they have already transferred the amino acid they carried.

---

[31] «The conversion of the information in RNA into protein represents a translation of the information into another language that uses quite different symbols», B. ALBERTS – *AL.*, *Molecular Biology of the Cell*, 367. Here Alberts *et al.* use the term "symbol" to indicate something that stands for another thing. It is just a way to denote that the nucleotides of mRNA are related with the amino acids of the proteins. I would rather say that the strings of nucleotides that form an mRNA molecule are codified information that is translated into amino acids in a polypeptide chain. It is a step of expressing codified information, not of interpreting symbols.

[32] B. ALBERTS – *AL.*, *Molecular Biology of the Cell*, 376.

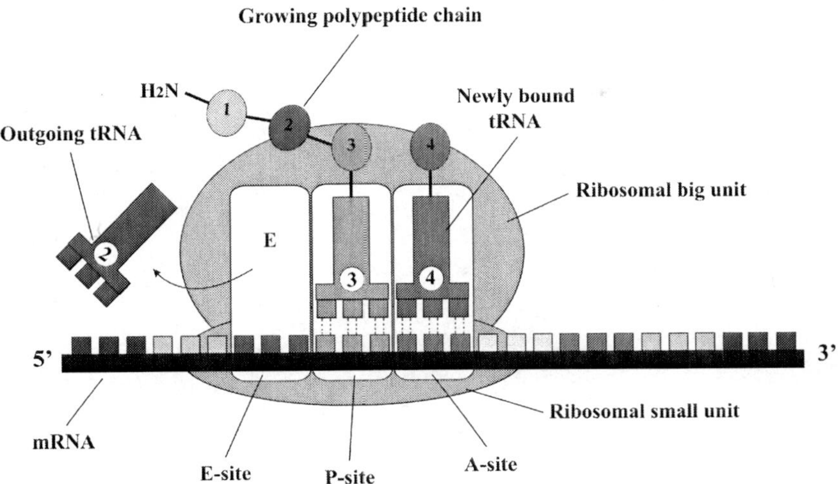

Figure VI.6: First step of protein synthesis. Here (and in the next three figures) only the mRNA molecule, the growing polypeptide chain, the tRNAs and the ribosome with the catalytic sites directly involved in base-pair recognition between the mRNA codon and the complementary tRNA anticodon are shown. Protein synthesis includes other factors that are not shown. Adapted from B. Alberts – *al.*, *Molecular Biology of the Cell.*

- In the second step, the carboxyl end of the polypeptide chain is released from the tRNA at the P-site, and then joined to the amino group of the new amino acid carried by the tRNA in the A-site. The polypeptide chain increases by one amino acid. The ribosome itself performs this peptidyl transferase reaction.

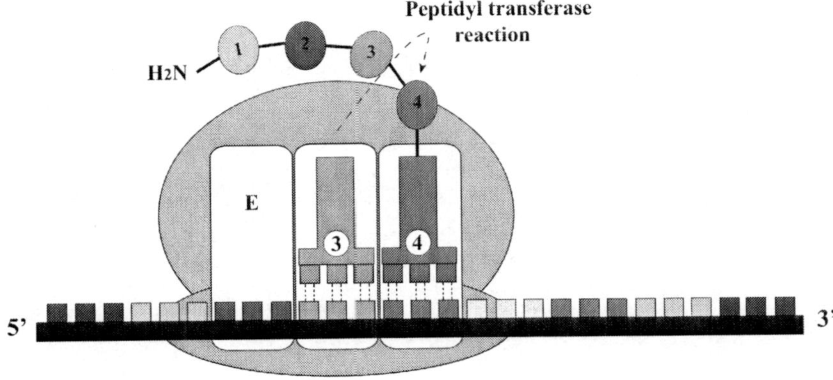

Figure VI.7: The peptidyl transferase activity of the ribosome attaches the already synthesized polypeptide to the newly bound tRNA.

- Then, the big unit of the ribosome goes forward through the mRNA molecule, always moving in a to 3' direction. This movement provokes a shift in the relative positions of the tRNAs. Then, the tRNA that occupied the P-site is now in the E-site, and the latest tRNA that has entered the ribosome in the A-site is now in the P-site, with the growing polypeptide chain joined to it.

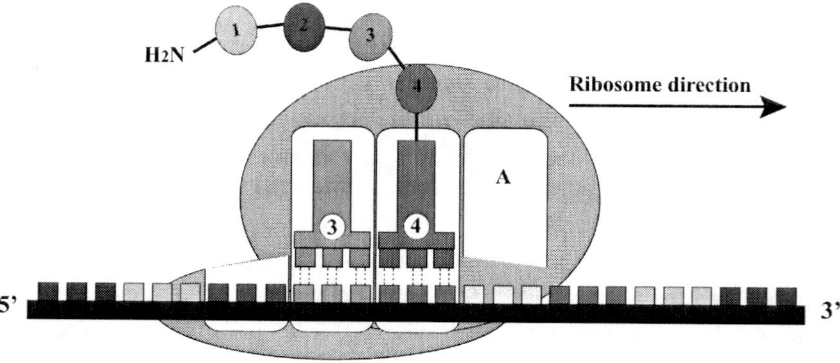

Figure VI.8: The ribosome's big unit goes downstream the mRNA to read the next three nucleotides.

- Finally, the small unit of the ribosome also moves forward, allowing the releasing of the tRNA located in the E-site and the cycle can start again[33].

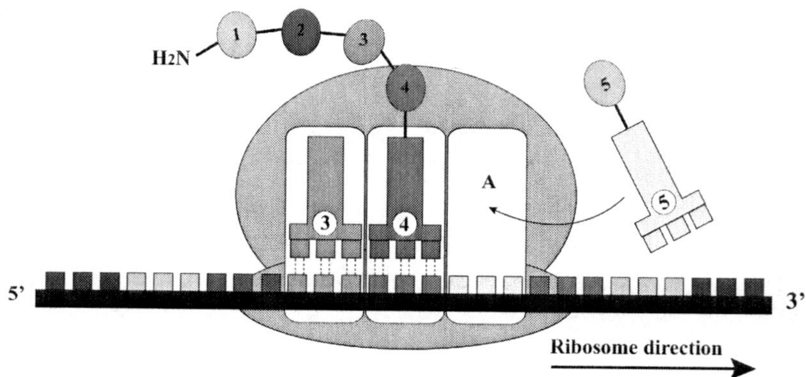

Figure VI.9: The ribosome's small unit moves forward. The whole ribosome has shifted three nucleotides and the next tRNA can enter the A-site and start the cycle again.

---

[33] Cf. B. ALBERTS – AL., *Molecular Biology of the Cell*, 376. Additional factors are involved

As can be seen, translation is the conversion of the information contained in the mRNA sequence into a polypeptide sequence. Since there are only four types of nucleotides in RNA and twenty different amino acids in proteins, this translation cannot be done in a one-to-one fashion. The nucleotides of the mRNA are read by the ribosome in groups of three, and they have to match with the anticodon of the tRNA that carries the proper amino acid. There are 64 different possible combinations of three nucleotides, but only 20 amino acids are found in nature. However, the fact is that all 64 possibilities are used (Figure VI.10), so we have to conclude that the genetic code is redundant: there are different codons that specify the same amino acid. Also three of these possibilities function as a stop signal just to end the synthesis process[34].

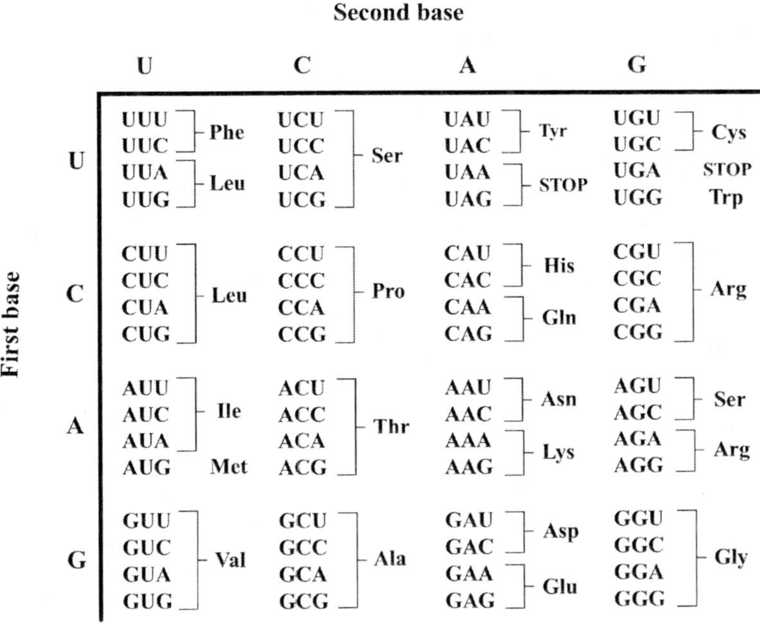

First base	Second base: U	C	A	G
U	UUU Phe	UCU Ser	UAU Tyr	UGU Cys
	UUC Phe	UCC Ser	UAC Tyr	UGC Cys
	UUA Leu	UCA Ser	UAA STOP	UGA STOP
	UUG Leu	UCG Ser	UAG STOP	UGG Trp
C	CUU Leu	CCU Pro	CAU His	CGU Arg
	CUC Leu	CCC Pro	CAC His	CGC Arg
	CUA Leu	CCA Pro	CAA Gln	CGA Arg
	CUG Leu	CCG Pro	CAG Gln	CGG Arg
A	AUU Ile	ACU Thr	AAU Asn	AGU Ser
	AUC Ile	ACC Thr	AAC Asn	AGC Ser
	AUA Ile	ACA Thr	AAA Lys	AGA Arg
	AUG Met	ACG Thr	AAG Lys	AGG Arg
G	GUU Val	GCU Ala	GAU Asp	GGU Gly
	GUC Val	GCC Ala	GAC Asp	GGC Gly
	GUA Val	GCA Ala	GAA Glu	GGA Gly
	GUG Val	GCG Ala	GAG Glu	GGG Gly

Figure VI.10: The genetic code has 61 triplets that codify for 20 amino acids and three triplets that serve as a stop-translation signal.

in protein synthesis, such as elongation factors and other proteins that are necessary to assure the accuracy of the translation process.

[34] Cf. B. ALBERTS – AL., *Molecular Biology of the Cell*, 367.

Having shown the mechanism of protein synthesis and the basic elements of the genetic code, we can now pose the following question: how does Selenocysteine (Sec) enter into all this process? No triplet that codifies for Sec can be found in Figure VI.10, so it is difficult to imagine how Sec would be inserted in a growing polypeptide chain. One possibility would be that Sec is present in proteins due to a modification of Cys once that amino acid has been inserted into the polypeptide chain[35]. But then the question that would arise would be: how does the cell recognize which Cys has to be converted into Sec?

The process by which Sec is inserted in selenoproteins is not a Cys modification into Sec after the incorporation of Cys in the polypeptide chain. The insertion of Sec takes place at the translational level, through an amazing mechanism that has been named as a natural expansion of the genetic code[36].

3.2 *The molecular mechanism of tRNASec insertion*

The cell uses a recoding strategy to insert Sec into the growing polypeptide chain. This strategy consists in reading a stop codon (UGA) not as a termination signal, but as a codon where a special tRNA (tRNASec) loaded with Sec (Sec-tRNASec) can insert this amino acid during the translation process. There is a reprogramming of the UGA codon in order to insert a new amino acid instead of finishing the translation. So we are looking at a novel way to insert an amino acid into a protein. There is not a specific codon to direct Sec insertion, but the reprogramming of certain UGA codons[37].

Facing this particular mechanism, some questions arise: how does the ribosome know where to insert Sec? What kind of tRNA is responsible for Sec insertion? How does the ribosome know when to really stop? What happens when no Se is available for the cell? Translation process takes place normally,

---

[35] Over 140 amino acid modifications after its insertion in the polypeptide chain have been reported, as it is showed in R. Uy – F. Wold, «Posttranslational Covalent Modification of Proteins».

[36] Cf. A. Ambrogelly – S. Palioura – D. Söll, «Natural Expansion of the Genetic Code».

[37] Cf. C. Allmang – L. Wurth – A. Krol, «The Selenium to Selenoprotein Pathway», 1415. One of the firsts characterizations of the Sec insertion process is found in F. Zinoni – A. Birkmann – W. Leinfelder – A. Böck– *al.*, «Cotranslational Insertion of Selenocysteine».

as has been described above. When the ribosome arrives at an UGA codon, something has to happen to redirect the "stop" signal of the UGA codon into a new informational role; that is, into "Sec".

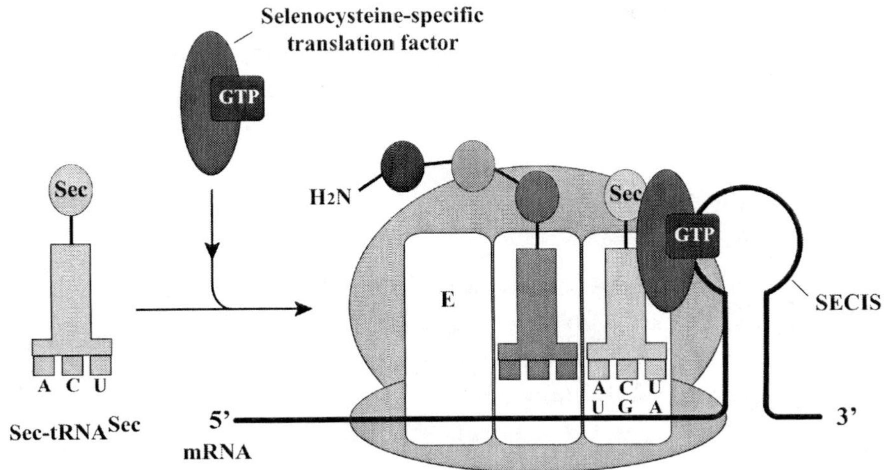

Figure VI.11: Schematic overview of the co-translational insertion of Sec into a growing polypeptide. Note that a specific translation factor (or more than one), a special tRNA and a stem loop structure (SECIS element) located downstream the UGA codon are needed. Depending on the organism, different factors are required. Adapted from B. ALBERTS – AL., *Molecular Biology of the Cell.*

As it is showed in figure VI.11, the recoding of the UGA codon to be read as Sec is determined by the secondary structure that mRNA adopts downstream[38]. This structure is the SECIS element (SElenoCysteine Inserting Sequence), which has the form of a stem-loop. A specific protein recognizes this stem loop. This protein, which has a GTPase activity, also binds the Sec-tRNASec in order to be inserted into the UGA codon. So we can say that the protein acts as a bridge between the SECIS element and the Sec-tRNASec.[39]

These are the general features of this intriguing process that allows the inser-

[38] Downstream and upstream are relational terms that indicate the relative position of a given item respect to the 5' end (upstream) or the 3' end (downstream) of a DNA or RNA molecule.

[39] Cf. B. LEWIN, *Genes IX*, 199.

tion of an amino acid that is not codified by DNA. The process is more complex, and requires more partners than those that have been showed in figure VI.6. In the next section all the elements shall be described. As it happens in many processes of the cell, the partners that are involved in Sec insertion have specific features when they are studied in the different kingdoms: Prokarya, Eukarya or Archaea. The machinery that synthesizes Sec and inserts it co-translationally is different in prokaryotes, compared with archaea and eukaryotes[40].

Let us see how this recoding event takes place in prokaryotes (bacteria and other organisms that do not have a nuclear membrane for separating the genetic material from the rest of chemicals of the cell). We can divide the selenocystein insertion process in six steps:

- Step 1. When the ribosome approaches the UGA codon, the complex Sec-tRNASec-SelB is already bound to the SECIS element. SelB protein is the Selenocystein-specific translation factor of prokaryotes, and we have introduced it generally in Figure VI.11. SelB is a protein with GTPase activity. That means that it can use the energy stored in the GTP (Guanosine triphosphate) to perform some work. GTPase proteins use the energy that is released when the bound of the third phosphate group of the molecule is broken, yielding a Pi (phosphate) and a molecule of GDP (Guanosine diphoshate).

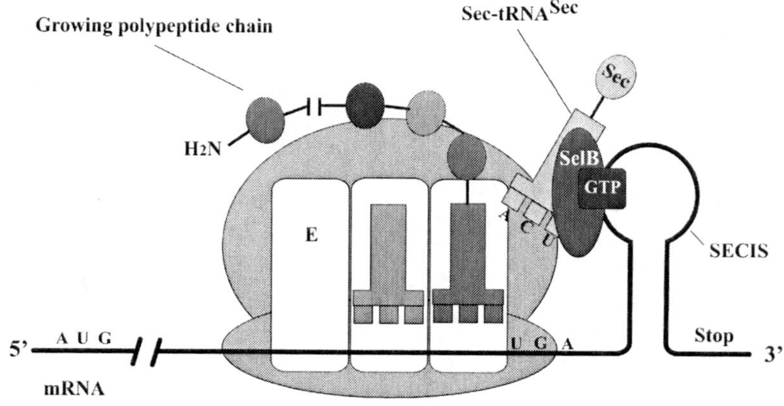

Figure VI.12: The ribosome arrives near the UGA codon.

[40] Cf. C. Allmang – A. Krol, «Selenoprotein Synthesis», 1562, and J. Yuan – S. Palioura – J.C. Salazar – D. Su– al., «RNA-dependent Conversion of Phosphoserine».

- Step 2. A signal is required to change the informational value of this triplet from "stop" into "Sec". This signal is provided by the SECIS element bound to SelB, the special translation factor with GTPase activity. This complex binds also the Sec-tRNASec.

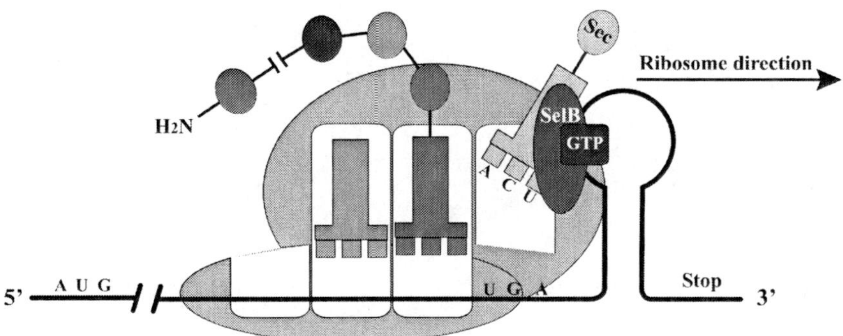

Figure VI.13: The insertion of Sec-tRNASec has to be done when the ribosome reaches the UGA codon, otherwise it would be read as a stop signal and translation would end there.

- Step 3. This complex inserts Sec-tRNASec in the A-site of the ribosome when it reaches the UGA codon.

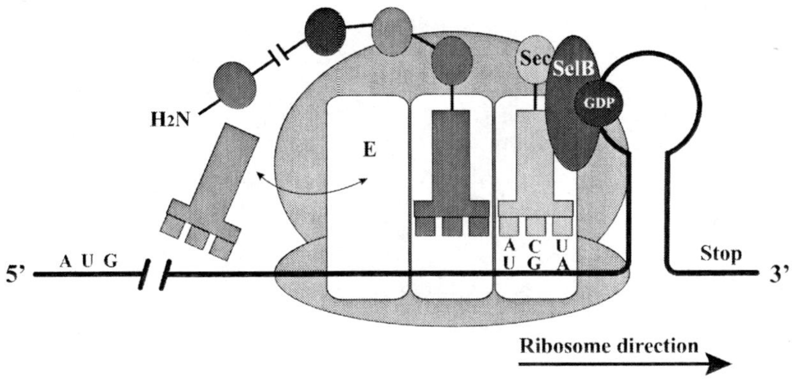

Figure VI.14: Sec-tRNASec is inserted into the A-site and prevents it from having a stop event.

- Step 4. Then, as it happens in canonical protein synthesis, the growing polypeptide is attached to the newly bound tRNA due to the peptidyl transferase activity of the big ribosome unit.

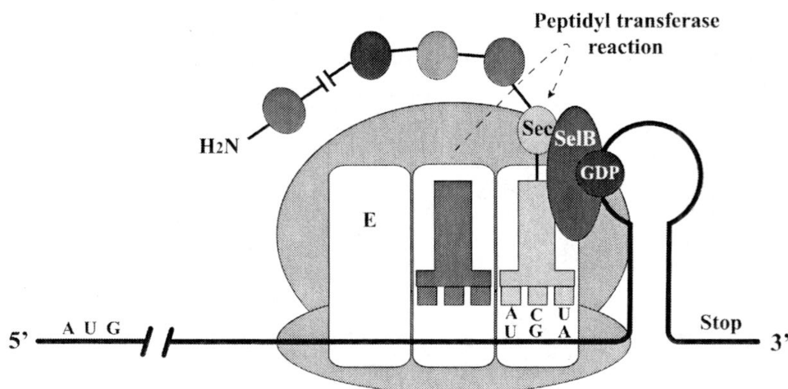

Figure VI.15: The growing polypeptide chain is transferred to the amino acid attached to the last tRNA that entered the A-site.

- Step 5. Once the Sec-tRNASec has been inserted, the remaining elements of the complex have to be disassembled in order to allow the ribosome to continue the translation of the mRNA. The ribosome exerts a physical force that breaks the hydrogen bonds that maintain the loop of the SECIS element.

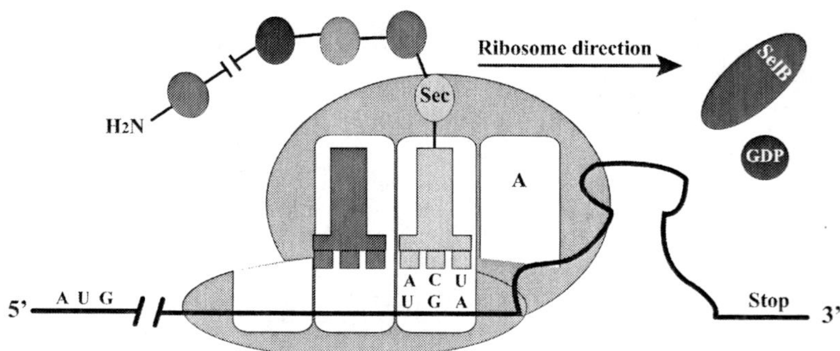

Figure VI.16: The ribosome moves forwards and the SECIS element is melted in order to allow the translation to continue.

- Step 6. Finally, the next amino acid enters the A-site of the ribosome and the translation continues normally.

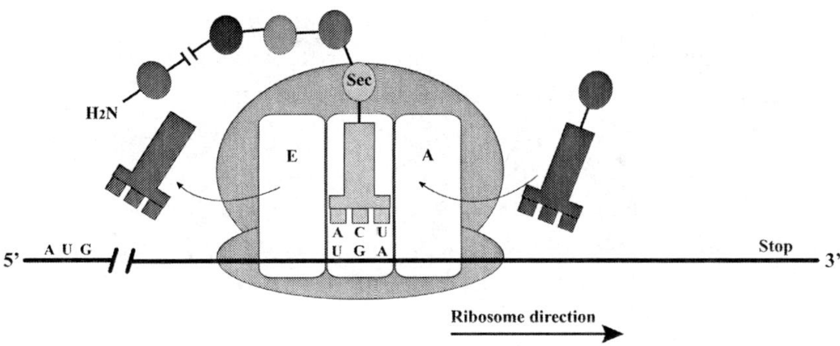

Figure VI.17: The translation continues until the "true" stop codon.

The selenocystein insertion mechanism in eukaryotes and archaea has two main differences from the prokaryotic one. The first one is that the SECIS element is not located in the open reading frame[41] (ORF) but in the 3'-untranslated region[42] (3'UTR), the end segment of mRNA after the stop codon that is not translated into protein. This allows separation of the codification sequence from the sequence that forms the stem loop of the SECIS element. In prokaryotes the SECIS element is located in the ORF, so it has two roles: being part of the recoding machinery and also carrying part of the sequential information that has to be translated into protein. This implies that, in prokaryotes, certain constraints have to be taken into account. Not any sequence could be in the SECIS element, but only a sequence that can be compatible with the secondary structure that allows the interaction with the other partners of the recoding machinery on one hand, and a sequence that codifies for a piece of the polypeptide chain that fits with the function the protein has to perform, on the other. Eukaryotes and archaea do not have this problem, since the nucleotide sequence that is in the SECIS element is located after the stop signal.

---

[41] «An ORF can be considered a stretch of DNA that begins with a start codon (ATG) and ends with a stop codon, with no stop codons in between, and thus could in principle encode a polypeptide.» B. ALBERTS – *AL.*, *Molecular Biology of the Cell*, 490.

[42] «The 3' untranslated region (UTR), [is] the region of RNA that extends from the stop codon that terminates protein synthesis to the start of the poly-A tail.» B. ALBERTS – *AL.*, *Molecular Biology of the Cell*, 487.

The other difference is the number of partners involved in the process. As it is shown in Figure VI.18 to VI.23, more partners are required in eukaryotes and archaea than in prokaryotes. This is a very important issue, since the separation of the roles and the increasing number of molecules that are needed, permits a greater control over the whole process. As it has been shown (chapter V, section 5.2), modularity is a necessary condition for the proper control of biological processes. One of the reasons why eukaryotes have more partners than prokaryotes, could be precisely because an increasing of modularity leads to a better information control and makes the organism more suited to better respond to challenges.

Let us see the different steps of the selenocystein insertion process in eukaryotes and archaea and the role each molecule plays:

- Step 1. The ribosome arrives near the UGA codon. In this case we can see that the special translation factor is not SelB but EFSec (Elongation Factor for Sec), which also has a GTPase activity. Furthermore, another protein is also needed, SBP2 (SECIS Binding Protein 2), that binds the SECIS element, because EFSec is not able to bind it. There is also the L30 protein, which is a very important partner for the binding between the ribosome and the SECIS element.

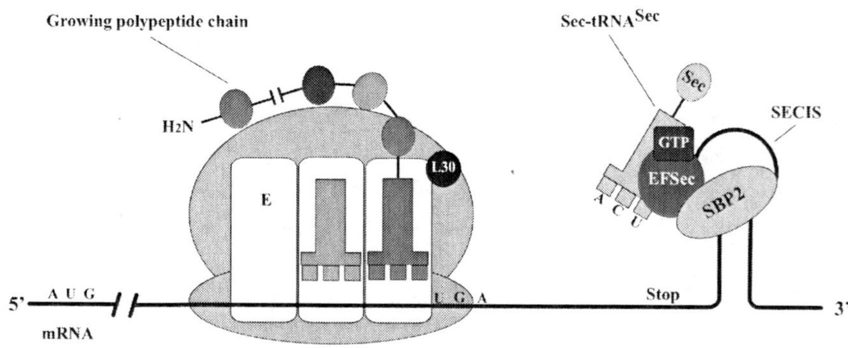

Figure VI.18: The ribosome arrives near the UGA codon. The SECIS element is not near the UGA codon, as happens in prokaryotes, but is located in the 3'-untranslated region, after the "true" stop codon.

- Step 2. The big unit of the ribosome goes forward. The ribosomal protein L30 is about to interact with the whole complex attached to the SECIS element.

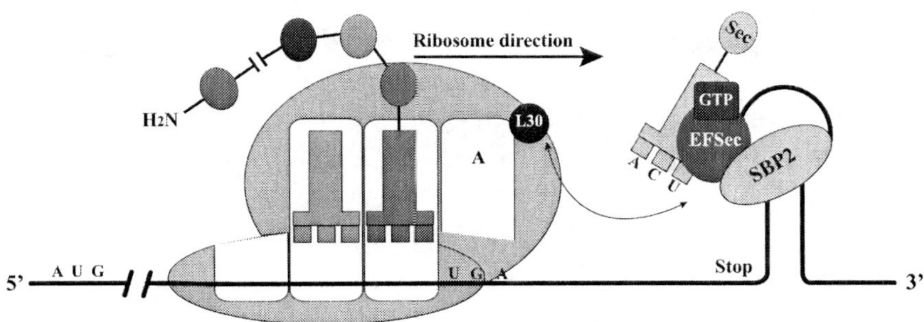

Figure VI.19: The ribosome and the SECIS element have to get close each other in order to interact.

- Step 3. The interaction between L30 and the complex provokes the bending of the mRNA, in order to come closer to the A-site of the ribosome.

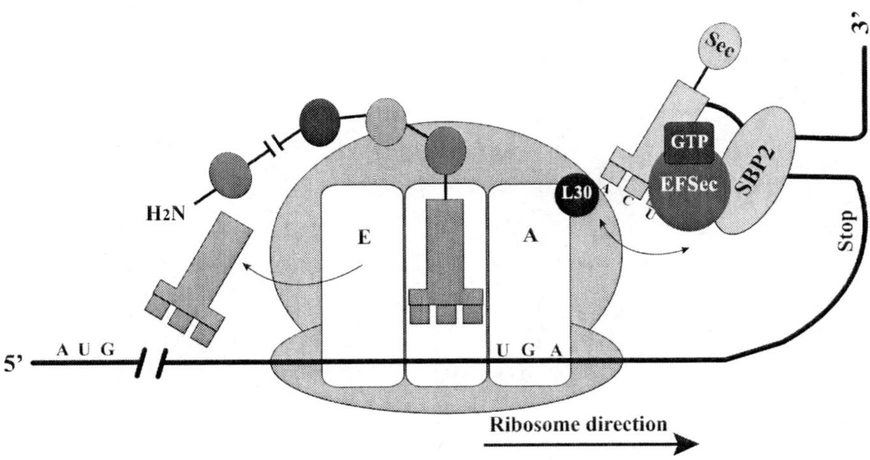

Figure VI.20: The SECIS element and the entire complex attached to it bend over to reach the A-site of the ribosome.

- Step 4. L30 displaces SBP2, provoking the change of conformation of the SECIS element. This change activates the GTPase activity of EFSec, performing the insertion of Sec-tRNASec into the UGA codon.

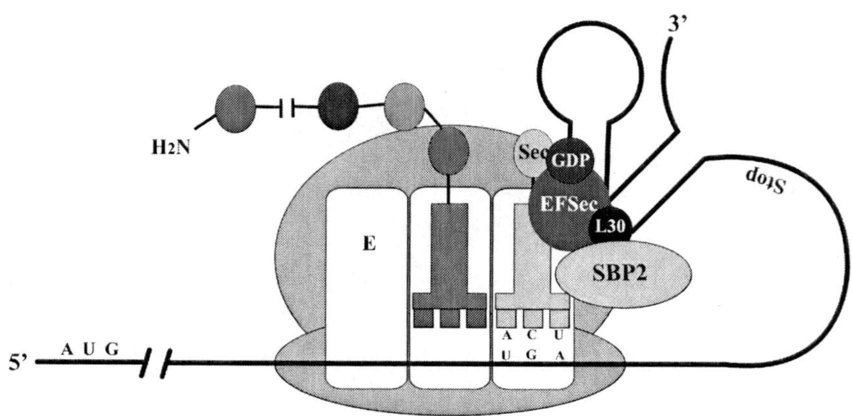

Figure VI.21: The conformational change triggers the GTPase activity of EFSec and the insertion of Sec-tRNASec.

- Step 5. Then the complex is disassembled and the SECIS element recovers its original conformation.

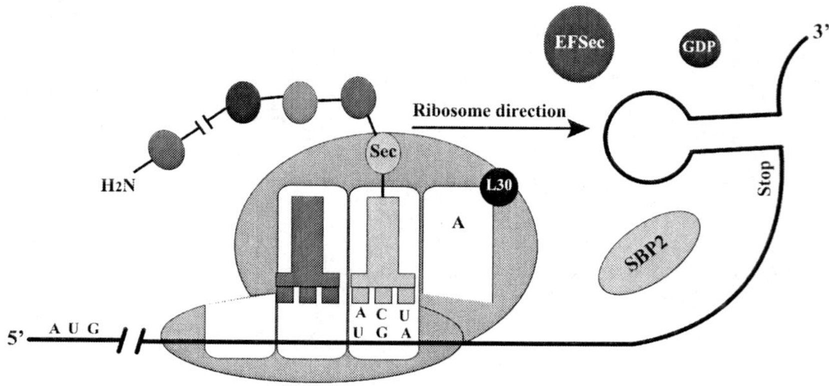

Figure VI.22: After the transfer of the growing polypeptide to the Sec-tRNASec, the ribosome continues its way forward disassembling the complex that has allowed Sec insertion.

- Step 6. The protein synthesis continues normally and the SECIS element maintains its secondary structure, so it can be loaded again with SBP2, EFSec and Sec-tRNASec in order to perform another round of selenocysteine insertion.

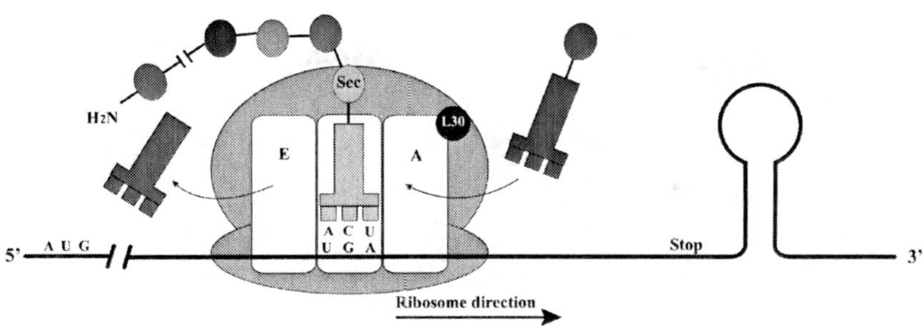

Figure VI.23: Translation continues until the stop signal.

Summing up, to recode UGA as Sec, the cell has to introduce four novel characteristics into the canonical protein synthesis: to synthesize the Sec-tRNASec, to distinguish between UGA Sec and UGA stop codons, to drive Sec-tRNASec into the A-site of the ribosome and to recycle the recoding machinery for another round of Sec insertion[43].

All these elements, which we are about to study in detail, can be considered as items of a higher level of organization (the three dimensional properties of the proteins and the RNA structures involved) that exerts a top-down causation upon the UGA codon in order to be read as "Sec" instead of "stop". All these elements use its chemical properties not to violate the rules of the genetic code, but to take advantage of them. There is no violation of the chemical closure, but the activity of some constraints that can ascribe to the UGA codon (the sign) a different event (the referent) which in this case is the insertion of Sec instead of stopping translation.

3.2.1 The tRNASec

A special tRNA (tRNASec)[44] that is loaded with Sec amino acid carries out its insertion during the translation process. tRNASec has a different structure in

[43] Cf. L. Chavatte – B.A. Brown II – D.M. Driscoll, «Ribosomal Protein L30».

[44] Cf. W. Leinfelder – E. Zehelein – M.-A. Mandrand-Berthelot – A. Böck– *al.*, «Gene for a Novel tRNA» and B.J. Lee – P.J. Worland – J.N. Davis – T.C. Stadtman– *al.*, «Identification of a Selenocysteyl-tRNA(Ser)». This tRNA^{Sec} has been characterized firstly in bacteria, but soon it has been sought that is widespread within the animal kindgdom, as

comparison with a canonical tRNA[45]. tRNASec acts both as a docking for Sec synthesis and as a donor of Sec for selenoprotein synthesis[46]. To have an idea of the importance of this type of tRNA and of the selenoproteins in general, it is worth saying that mice whose tRNASec gene has been knocked out, died in the first stages of embryonic development[47].

As it has been remarked in the previous section, there are some differences in the recoding machinery between prokaryotes, eukaryotes and archaea. Let us describe briefly the different metabolic pathways that we can find in prokaryotes and eukaryotes to synthesize the Sec-tRNASec.

a) *Bacterial tRNASec*

In bacteria[48], the gene that codifies for tRNASec is called selC. The Sec charging process into tRNASec begins with the charging of Ser in the tRNASec (Figure VI.24), yielding the Ser-tRNASec; a reaction catalyzed by seryl-tRNA synthetase, which is the same enzyme that charges Ser into tRNASer.[49] So we can see that the first step for the synthesis of Sec-tRNASec begins with the loading of serine into tRNASec. The same enzyme that charges Ser in its canonical tRNA, recognizes tRNASec as a putative substrate and charges Ser on it. We can talk here of a phenomenon of mimicry, since the tRNASec resembles tRNASer in order to be recognized by seryl-tRNA synthtase.

In this first step of Sec-tRNASec synthesis we can clearly see that Shannon's

---

it is reported in B.J. LEE – M. RAJAGOPALAN – Y.S. KIM – K.H. YOU– *AL.*, «Selenocysteine tRNA[Ser]Sec Gene».

[45] A structural comparison between both tRNAs is found in C. BARON – E. WESTHOF – A. BÖCK – R. GIEGÉ– *AL.*, «Solution Structure of Selenocysteine-inserting tRNA(Sec)».

[46] Cf. P.A. YOUNG – I.I. KAISER, «Aminoacylation of Escherichia Coli».

[47] Cf. M.R. BÖSL – K. TAKAKU – M. OSHIMA – S. NISHIMURA– *AL.*, «Early Embryonic Lethality» and E. KUMARASWAMY – B.A. CARLSON – F. MORGAN – K. MIYOSHI– *AL.*, «Selective Removal of the Selenocysteine».

[48] The first characterization of the bacterial genes that interact in the Sec insertion pathway can be found in W. LEINFELDER – K. FORCHHAMMER – F. ZINONI – G. SAWERS– *AL.*, «Escherichia Coli Genes» and W. LEINFELDER – T.C. STADTMAN – A. BÖCK, «Occurrence in vivo of selenocysteyl-tRNA(SERUCA)».

[49] Cf. T. OHAMA – D.C. YANG – D.L. HATFIELD, «Selenocysteine tRNA and Serine tRNA».

model of communication does not rule here. Shannon's model is based on the faithfully transmission of information, and we see here that an error of recognition is the first step used for the synthesis of Sec-tRNASec, and that can lead to two different results. Indeed, Seryl-tRNA synthetase is the enzyme that charges Ser in its canonical tRNA, which is tRNASer. The point is that the same enzyme is used to charge Ser into tRNASec. Then, the reaction of seryl-tRNA synthetase does not always have the same result, depending on which tRNA is recognized by it (tRNASer or tRNASec). Far from being problematic, this initial reaction is the first step of what further on would lead to Sec insertion.

Then, once Ser has been charged in the tRNASec, it has to be converted into Sec through selenocysteine synthase, the protein encoded by selA gene[50]. This conversion only takes place when Ser is attached to tRNASec. Free Ser cannot be substrate for selenocysteine synthase. This reaction needs a selenium donor, used by the cell in the form of selneophosphate, which is the product of the reaction deployed by selenophosphate synthetase, encoded by selD gene[51]. This enzyme prepares the selenium in its suitable chemical form in order to become a substrate for selenocysteine synthase (Figure VI.24).

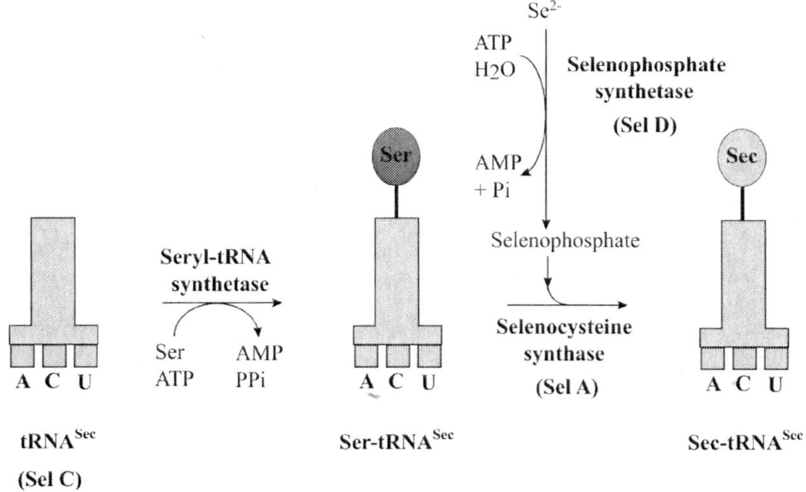

Figure VI.24: The selenocystein biosynthesis pathway in prokaryotes.

[50] This reaction has been studied in detail in K. FORCHHAMMER – A. BÖCK, «Selenocysteine synthase from Escherichia coli», and K. FORCHHAMMER – W. LEINFELDER – K. BOESMILLER – B. VEPREK– *AL.*, «Selenocysteine Synthase from Escherichia Coli (2)».

[51] The *in vitro* characterization of SelD protein and its role in Sec-tRNA^{Sec} synthesis has

b) *The tRNASec of eukaryotes and archaea*

In eukaryotes and archaea the process is more complex, since another intermediate is required to have a mature tRNASec. We can divide the pathway in three steps, as is shown in Figure VI.25.

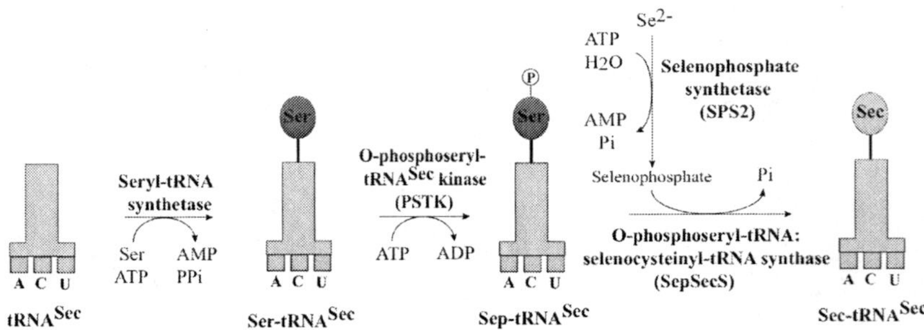

Figure VI.25: Biosynthetic pathway of Sec-tRNASec in eukaryotes and archaebacteria. Adapted from K. Sheppard – J. Yuan – M.J. Hohn – B. Jester– *al.*, «From One Amino Acid to Another».

- The first step is the same that we found in prokaryotes. tRNASec is loaded with a Ser, yielding Ser-tRNASec. This reaction is also catalyzed by seryl-tRNA synthetase.

- In the second step, Ser-tRNASec is phosphorilated by O-phosphoseryl-tRNASec kinase (PSTK). The product of this reaction is Sep-tRNASec (O-phosphoseryl-tRNASec; that is, the phosphorilated form of Ser-tRNASec)[52]. «From a chemical standpoint a Sep to Sec conversion is desirable, as Sep would provide a better leaving group (phosphate) than Ser

been studied in W. Leinfelder – K. Forchhammer – B. Veprek – E. Zehelein– *al.*, «In Vitro Synthesis of Selenocysteinyl-tRNA(UCA)» and in A. Ehrenreich – K. Forchhammer – P. Tormay – B. Veprek– *al.*, «Selenoprotein Synthesis in E. Coli.». Another example of an *in vitro* reconstruction of the whole tRNA^{Sec} byosinthetic pathway can be found in I.S. Choi – A.M. Diamond – P.F. Crain – J.D. Kolker– *al.*, «Reconstitution of the Biosynthetic Pathway».

[52] Cf. B.A. Carlson – X.-M. Xu – G.V. Kryukov – M. Rao– *al.*, «Identification and Characterization».

(water) for replacement with selenium»[53]. The activity of PSTK is strictly tRNASec dependent; it cannot phosophorilate Ser-tRNASer and, on the other hand, neither can hydrolyze ATP (another energetic molecule, as GTP) in the absence of tRNA. This high specificity is accompanied by an also high affinity of PSTK for its substrate. PSTK has a 20-fold higher affinity for Ser-tRNASec than any other seryl-tRNA synthetase has for its own substrate. This helps to compensate the low abundance of tRNASec in vivo[54].

- Finally, SepSecS (O-phosphseryl-tRNA:selenocysteinyl-tRNA synthase), transforms Sep-tRNASec into Sec-tRNASec, getting the selenium from the final product of selenophosphate synthetase (SPS2), the enzyme that gives selenium in the proper chemical form. SepSecS is a tetrameric protein, which means that is composed of four polypeptide chains that are assembled to become a single protein. Each of those polypeptide chains is a monomer. In the case of SepSecS, the tetrameric complex is formed by two groups of two polypeptide chains (two homodimers that form the tetramer). This tetrameric protein interacts with Sep-tRNASec exerting two points of control. In fact, «one SepSecS homodimer interacts with the sugar-phosphate backbone of both the acceptor-TWC and the variable arms of tRNASec, while the other homodimer interacts specifically with the tip of the acceptor arm through interaction between the conserved Arg398 [of SepSecS] and the discriminator base G73 of tRNASec»[55]. The

[53] J. YUAN – S. PALIOURA – J.C. SALAZAR – D. SU– *AL.*, «RNA-dependent Conversion of Phosphoserine», 18926.

[54] A detailed description of the molecular mechanism of this reaction can be found in Y. ARAISO – S. PALIOURA – R. ISHITANI – R.L. SHERRER– *AL.*, «Structural Insights». For a molecular characterization of the enzyme see B.A. CARLSON – X.-M. XU – G.V. KRYUKOV – M. RAO– *AL.*, «Identification and Characterization». Archaeal PSTK is characterized in J.T. KAISER – K. GROMADSKI – M. ROTHER – H. ENGELHARDT– *AL.*, «Structural and Functional Investigation», R.L. SHERRER – P. O'DONOGHUE – D. SÖLL, «Characterization and Evolutionary History» and Y. ARAISO – R.L. SHERRER – R. ISHITANI – J.M. HO– *AL.*, «Structure of a tRNA-dependent Kinase».

[55] J. YUAN – P. O'DONOGHUE – A. AMBROGELLY – S. GUNDLLAPALLI– *AL.*, «Distinct Genetic Code Expansion», 345. For a detailed study about the enzymatic reaction of SepSecS, see O.M. GANICHKIN – X.-M. XU – B.A. CARLSON – H. MIX– *AL.*, «Structure and Catalytic

phosphate group is a very important element in order to bring Sep into the active site of the enzyme (Figure VI.26).

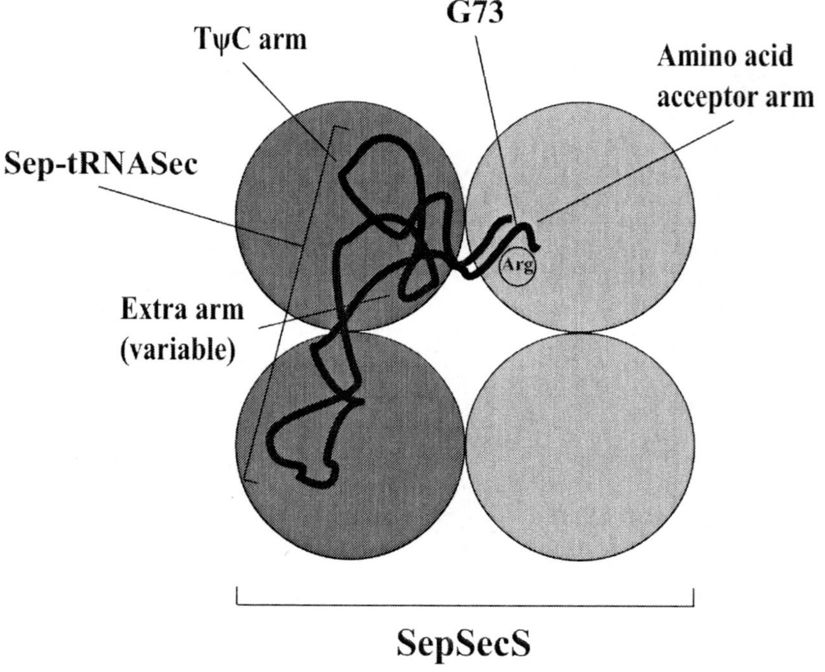

Figure VI.26: Tetrameric structure of SepSecS protein. The homodimer of the left (in dark grey) recognizes the extra arm and the TΨC arm of Sep-tRNASec. On its side, the homodimer of the right recognizes the amino acid acceptor arm, interacting with a specific residue of the amino acid Arginine of SepSecS. This homodimer also recognizes the nucleotide G73 of Sep-tRNASec.

Comparing this pathway with that of prokaryotes, we can see that two another enzymes are present in the reaction: PSTK and SPS2[56]. The fact that

Mechanism» and S. PALIOURA – R.L. SHERRER – T.A. STEITZ – D. SÖLL– *AL.*, «The Human SepSecS-tRNASec Complex».

[56] An extended study about the structure and the chemical reaction mechanism of SPS2 can be found in Y. ITOH – S. SEKINE – E. MATSUMOTO – R. AKASAKA– *AL.*, «Structure of Selenophosphate Synthetase».

SPS2 is a selenoprotien suggests the possibility of a feedback regulation[57]. An additional factor (SEC43p) is needed for the proper translocation of Sec-tRNASec into the nucleus; a feature that will be discussed below. It is important to stress that eukaryotes and archaea have more steps and more partners than prokaryotes. The growing complexity of the pathway permits the cell to control it more efficiently. The more steps there are, the more points where control can be exerted.

c) *Discriminating features between tRNASer and tRNASec*

The question that emerges is: how does the cell distinguish between the Ser residues that have to be modified into Sec and those that do not have to? Since both Ser-tRNASer and Ser-tRNASec are loaded, with Ser, is obvious that the discrimination should be done not through the amino acid that is charged but by another difference. The tRNA structure is crucial for those molecules to be processed differently[58]. In fact, as it is shown in Figure VI.27, the two-dimensio-

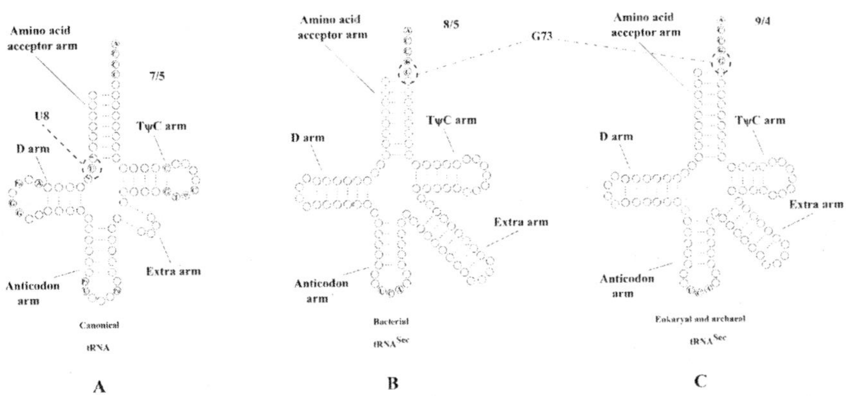

Figure VI.27: Secondary structure comparisons of canonical tRNA (A) versus bacterial tRNASec (B) and eukaryotic and archaea tRNASec (C). Note that the coaxial TΨC-amino acid acceptor arm has different base proportion depending on the tRNA type. In the canonical tRNA the positions that are invariable in all canonical tRNAs are indicated. Uridine located at position 8 is paramount for preserving the structure of canonical tRNAs. G73, for its part, is a very important element for bacterial, eukaryotic and archaeal tRNASec. Adapted from C. ALLMANG – A. KROL, «Selenoprotein Synthesis».

---

57 Cf. S.C. LOW – J.W. HARNEY – M.J. BERRY, «Cloning and Functional Characterization», M.J. GUIMARÃES – D. PETERSON – A. VICARI – B.G. COCKS– *AL.*, «Identification of a Novel selD», and P.R. HOFFMANN – M.J. BERRY, «Selenoprotein Synthesis».

58 In R. AMBERG – T. MIZUTANI – X.-Q. WU – H.J. GROSS– *AL.*, «Selenocysteine Syn-

nal structures of different tRNAs show typical features that allow the molecules to be discriminated each other.

The first differential feature is that tRNASec is 90 nucleotides long, being longer than the canonical ones[59]. This is because tRNASec has a longer extra arm. It is interesting to note that the recognition of the tRNA by the seryl-tRNA synthetase depends not on the extra arm sequence but on its orientation[60]. The stacking of the amino acid acceptor arm and the TΨC arm when tRNA is folded involves 13 base pairs instead of the 12 involved in the canonical tRNAs. This way of folding these two arms is also known as coaxial stacking (Figure VI.28). This discriminative feature allows SepSecS to bind only tRNASec and not canonical tRNAs[61]. Furthermore, the 6 base pair D stem is also an essential element for the tRNASec to become a substrate for SepSecS[62].

As we can see in Figure VI.27, other differences in their secondary structures distinguish canonical tRNAs from tRNASec. The D arm of tRNASec has six base pairs instead of four. This difference is the discriminating feature that allows tRNASec to be phosphorylated by PSTK[63]. Interestingly, in the case of archaea, the discriminating feature is not the D arm but the amino acid acceptor arm[64].

---

thesis in Mammalia», is shown a very interesting experiment in which the structural features of a $tRNA^{Sec}$ are transplanted to a $tRNA^{Ser}$, allowing then it's selenylation. Here are reported the discriminating features of $tRNA^{Sec}$ and the role they play. For a study about the crystal structure of human $tRNA^{Sec}$ and the differential traits with $tRNA^{Ser}$, see T. Itoh – S. Chiba – S. Sekine – S. Yokoyama– *al.*, «Crystal Structure».

[59] Cf. R. Amberg – C. Urban – B. Reuner – P. Scharff– *al.*, «Editing Does Not Exist».

[60] Cf. X.Q. Wu – H.J. Gross, «The Long Extra Arms» and M. Heckl – K. Busch – H.J. Gross, «Minimal tRNA(Ser) and tRNA(Sec) Substrates».

[61] Cf. S. Palioura – R.L. Sherrer – T.A. Steitz – D. Söll– *al.*, «The Human SepSecS-tRNASec Complex».

[62] Cf. T. Mizutani – K. Kanaya – S. Ikeda – T. Fujiwara– *al.*, «The Dual Identities of Mammalian».

[63] It has been demonstrated that, if two additional base pairs are joined to the D stem of a canonical tRNA, its Ser is also phosphorylated by PTSK. X.Q. Wu – H.J. Gross, «The Long Extra Arms», X.Q. Wu – H.J. Gross, «The Length and the Secondary» and S. Chiba – Y. Itoh – S. Sekine – S. Yokoyama– *al.*, «Structural Basis».

[64] Cf. R.L. Sherrer – P. O'Donoghue – D. Söll, «Characterization and Evolutio-nary History».

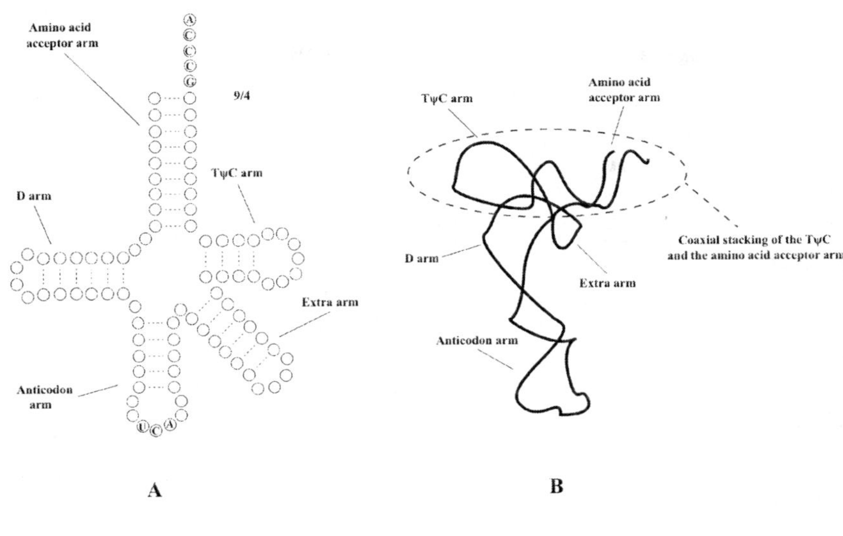

Figure VI.28: Structure of eukaryal and archaeal tRNASec. A) Secondary structure of tRNASec, also known as "clover leaf". B) Three-dimensional structure of tRNASec, where the coaxial stacking of the TΨC and the amino acid acceptor arms can be seen. The TΨC arm contributes to the stacking with 4 base pairs, while the amino acid acceptor arm contributes with 9. The three-dimensional structure of tRNASec is also known as "L-shaped".

Here we are in front of a very interesting issue. Indeed, depending on the organism (eukaryotes or archaea), the feature used in order to be the discriminating character to be recognized by PSTK is different. Eukaryotes use the D arm as the discriminating feature, but archaea use the acceptor arm. We are in front of a functional equivalence classe, since the same result is obtained through different ways. The organism only needs a discriminative sign that allows it to differentiate the canonical tRNA from tRNASec. It is not important what is the chemical feature that will establish such difference. What matters here is to have a difference that would be treated as a sign that stands for a different referent (a different tRNA), and then would be used in a different way by the protein synthesis machinery. The chemical differences are not important *per se*, but only because they can be treated semiotically. They can be used differently because are embedded in a context that can use them in a different way to achieve a different result. Then, it can be said that the cell deals with these chemical differences semiotically, since they stand for different referents, which can be used by the cell for its own sake.

Furthermore, the coaxial stacking of the TΨC and the amino acid acceptor arms adopt a diverse secondary structure. While in the canonical tRNA the ratio of base pairs is 7/5, in bacterial tRNASec it is 8/5 and in eukaryal and archaea it is 9/4[65]. The first number of the ratio indicates the number of base pairs that are present in the amino acid acceptor arm; while the second number reflects those of the TΨC arm. So the coaxial stacking of those arms in the case of tRNASec has an extra base pair if compared to the canonical ones. However, the difference between bacteria and eukaryal/archaea is not due to the number of base pairs involved in this structure but to the secondary structure distribution of these base pairs.

The three-dimensional structure of tRNA is L-shaped, and this is due to the stacking of the TC and the amino acid acceptor arms in one hand, and the anticodon and D arms on the other[66]. In bacteria, the extra length of the amino acid acceptor arm (is a base pair longer than the canonical) is needed for the interactions with SelB protein[67], the specific elongation factor used in Sec insertion, as has been said. In eukaryotes, the extra length of the amino acid arm structure serves, however, for the conversion of Ser to Sec through Selenocysteine synthase[68]. Hence, here we have another example of how the same structure can be used for different aims depending on the organism considered. Again, the extra length of the amino acid acceptor arm is used differently by prokaryotes and eukaryotes. In both cases the extra length of the amino acid arm can be used as a sign to differenciate tRNASec from canonical tRNA's. However, in the case of bacteria, this extra length is use for the interaction with SelB protein, while in the case of eukaryotes and archaea, is used to interact with

[65] Cf. N. Hubert – C. Sturchler – E. Westhof – P. Carbon– *al.*, «The 9/4 secondary structure» and T. Mizutani – C. Goto, «Eukaryotic Selenocysteine tRNA». Evidence of a 7/5 structure has been reported also in A. Ioudovitch – S.V. Steinberg, «Modeling the Tertiary Interactions» and S.V. Steinberg – A. Ioudovitch – R. Cedergren, «The Secondary Structure», but the 9/4 model has finally accumulated more experimental evidence.

[66] Cf. C. Allmang – A. Krol, «Selenoprotein Synthesis», 1562.

[67] Cf. C. Baron – A. Böck, «The Length of the Aminoacyl-acceptor».

[68] Cf. C. Sturchler-Pierrat – N. Hubert – T. Totsuka – T. Mizutani– *al.*, «Selenocysteylation in Eukaryotes».

Selenocysteine synthase for the transformation of Ser into Sec. In both cases the extra length is used as a sign, but for a different referent. The important issue is that a difference is used as a sign that stands for a different referent. The cell establishes the linkage between the sign and the referent depending on the best way to put together all the elements to perform the operation that is useful for the cell's sake.

One of the structural differences already mentioned between canonical tRNAs and tRNASec is the amino acid acceptor stem length. In tRNASec, the length of the amino acid acceptor is one amino acid longer (in position G73) than it is in the canonical tRNAs. This difference is due to an abnormal cleavage performed by RNase P, a nuclease involved in the maturation of tRNA molecules[69]. If the length of the amino acid acceptor stem is modified to be as long as the canonical one, no insertion of selenocysteine is observed[70]. This is due to the interaction between G73 and Arg398 of the catalytic core of SepSecS, which is essential for the performance of the last step of Sec-tRNASec synthesis[71].

One of the features that characterize canonical tRNAs is the presence of an invariant base at position 8. An Uridine (U) is always found at that position (Figure VI.27), and it plays an important role in some of the interactions that guide the proper folding of the molecule. It has been reported that no U8 is found in tRNASec. A purine occupies that position: an adenine (A) in eukaryotes or a guanine (G) in *E. coli*. This change has important consequences for the three-dimensional structure of tRNA. In fact, it is known that U8 interacts with A14 in canonical tRNAs, forming the elbow region of the tRNA[72]. How does this interaction shall be preserved when a purine replaces U at position 8? The molecule solves the problem involving another base in the interaction. So the tertiary interaction between U8 and A14 in canonical tRNAs is replaced by the interaction A8-A14-U21 in eukaryotic tRNASec. Prokaryotic tRNASec also involves the three partners that occupy theses positions: G8-U14-A21[73]. These

[69] Cf. U. Burkard – D. Söll, «The Unusually Long Amino Acid».

[70] Cf. C. Baron – J. Heider – A. Böck, «Mutagenesis of selC».

[71] Cf. S. Palioura – R.L. Sherrer – T.A. Steitz – D. Söll– *al.*, «The Human SepSecS-tRNASec Complex».

[72] Cf. E. Westhof – P. Dumas – D. Moras, «Crystallographic Refinement of Yeast».

[73] Cf. C. Baron – J. Heider – A. Böck, «Mutagenesis of selC».

interactions, with other ones that are also exclusive for tRNASec, allow the extra arm to be uncoupled from the rest of the molecule[74].

It has been also reported that the secondary structure of an archaeal tRNASec has its proper compensation mechanisms. As we have seen, the D stem of tRNASec is two base pair longer than canonical tRNA, and T loop is one base pair shorter. It was expected that this differences would disrupt the D-T interactions, but it does not happens at all. There is a mutual compensation of these two stems that allows the whole molecule to preserve its three-dimensional structure[75].

In both cases we can see that there is a balance to re-establish the overall three-dimensional structure of the tRNASec. If there is not the important base U8, other alternatives are searched in order to establish an internal relation among the rest of the items to preserve the structure that will keep the essential features of the tRNASec to be used by the protein synthesis machinery. The important issue is to retain certain structures that allow the item to be embedded in a relation of operations in a given biochemical pathway, not the chemical nature of each of its elements.

Another interesting feature of the tRNASec structure is that the non-Watson-Crick interaction between G5a-U67b is essential for the interaction with the specific elongation factor[76]. As we shall see below, the translation machinery needs a different elongation factor from the canonical one to insert Sec-tRNASec into the growing polypeptide. A Watson-Crick interaction is the one we found between the DNA and RNA bases. Guanine (G) and Citosine (C) establish three hydrogen bonds between them; while Adenine (A) and Timine (T) establish two hydrogen bonds (it is worth remembering that in the case of RNA, Timine is substituted by Uracil (U)). Any other interaction among nucleotides through hydrogen bonds that do not follow this rules, is considered a non-Watson-Crick

---

[74] Cf. C. Sturchler – E. Westhof – P. Carbon – A. Krol– *al.*, «Unique Secondary and Tertiary Structural». A more detailed description of the tertiary interactions in tRNA^{Sec} is reviewed in S. Commans – A. Böck, «Selenocysteine Inserting tRNAs: an Overview». The loss of invariable positions is compensated by new interactions between bases that are not related in the canonical tRNA structure.

[75] Cf. A. Ioudovitch – S.V. Steinberg, «Structural Compensation in an Archaeal».

[76] Cf. T. Mizutani – K. Tanabe – K. Yamada, «A G.U Base Pair».

interaction. Then, as it has been said, if there is an interaction between G5a and U67b, it cannot be a Watson-Crick interaction.

Those structural features are very interesting. As we can see, what matters here is not to have a certain sequence of nucleotides, but a three-dimensional pattern that would allow tRNASec to be recognized by the proper enzymes involved in the pathway. So it is not the sequence that is important, but the final shape that would let tRNASec interact with the rest of the elements that form the network involved in selenium metabolism. Moreover, it is not important that the same features would vehicle the different discrimination traits and reactions that have to be performed. We have seen that in different organisms those reactions and interactions are done according to different structures and recognitions. Then, what is important is that all the elements involved can achieve the reaction needed *as a whole*, no matter what is playing one role or other.

d) *The role of the base modification pattern*

Unusual bases are found in all tRNA molecules (as happens in many RNA molecules). tRNA-modifying enzymes produce those modifications[77]. tRNA-Sec contains only four base modifications[78]: pseudo-U55, m1A58, 6-isopentenyl-A37 (i6A37) and another modification at position 34[79]. The last one takes place in the third position of the anticodon. The modification at position 34 could be of two types: 5-methylcarboxymethylmethyluridine (mcm5U34) or 5-methylcarboxymethylmethyluridine-2'-O-methyl ribose (mcm5Um34), as it is shown in Figure VI.29. So we can find two different isoacceptors[80] of tRNA-Sec, depending on the modification they have at position 34. Those isoacceptors

---

[77] «All classes of RNA display some degree of modification, but in all cases except tRNA this is confined to rather simple events, such as the addition of methyl groups. In tRNA, there is a vast range of modifications, ranging from simple methylation to wholesale restructuring of the purine ring. Modifications occur in all parts of the tRNA molecule. There are >50 different types of modified bases in tRNA», B. LEWIN, *Genes IX*, 194-195.

[78] They are rather few, if we consider that a canonical tRNA has, on average, 15 to 17 modified nucleotides. D.L. HATFIELD – V.N. GLADYSHEV, «How Selenium has Altered», 3566.

[79] Cf. C. ALLMANG – A. KROL, «Selenoprotein Synthesis».

[80] Two tRNAs are isoacceptors if they carry the same amino acid; that is, if they have been recognized by the same tRNA synthetase. Cf. B. LEWIN, *Genes IX*, 200.

are known as the unmethylated (mcm5U) and methylated (mcm5Um) forms. Each of these isoacceptors may play a different role in selenoprotein synthesis. However the role they play is not absolutely clear and two models are proposed.

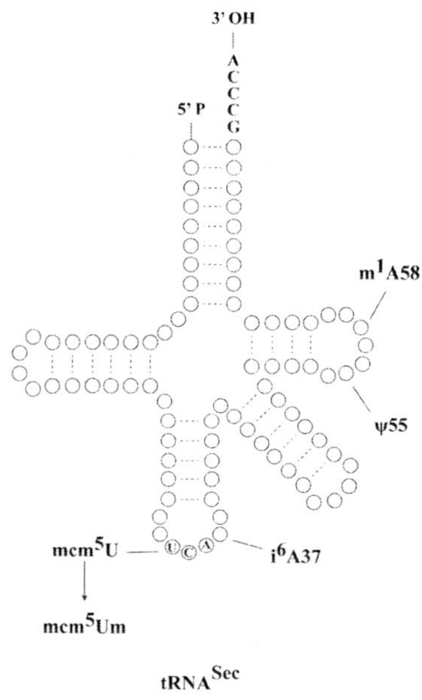

Figure VI.29: tRNASec secondary structure with the four bases modified. At position 34 it could be either mcm5U (unmethylated) or mcm5Um (methylated). Adapted from A.M. Diamond – I.S. Choi – P.F. Crain – T. Hashizume– *al.*, «Dietary Selenium Affects Methylation».

It has been suggested that methylation of mcm5U34 requires both the prior formation of the other modified bases and the intact tertiary structure[81]. So the position 34 cannot be methylated if the other three positions have not been modified previously and if the tRNASec has not reached its proper folding[82]. Indeed, the modification enzyme responsible for the methylation is present not

[81] Cf. L.K. Kim – T. Matsufuji – S. Matsufuji – B.A. Carlson– *al.*, «Methylation of the Ribosyl Moiety». The modification at position 58 has to be done first than the methylation in position 34, as it is shown in M.E. Moustafa – B.A. Carlson – M.A. El-Saadani – G.V. Kryukov– *al.*, «Selective Inhibition of Selenocysteine tRNA».

[82] If the 6-isopentenyl-A37 modification does not take place, the Sec incorporation

in the nucleus but in the cytoplasm[83]; and that means that the last modification cannot be performed in the cytoplasm until the previous ones have been fulfilled in the nucleus. Moreover, that single methylation provokes a conformational change in the whole tRNASec[84], indicating the importance it has for the proper functioning of the molecule. On the other hand, it's known that the level of Se influences the relative amount of each isoacceptor[85]. The methylated form (mcm5U34m) is increased in presence of Se; so the last step of the tRNASec maturation pathway depends on the availability of Se.

It can be seen here how a space-time coordinated process guides the whole pattern of tRNASec maturation. In fact, the proper folding of the molecule depends on a sequence of base modification reactions (with a precise order in time) and on a final step that takes place through an enzyme that is in the cytoplasm (separating spatially the last step of the maturation pathway). This model is very interesting, since it links the last modification of tRNASec to the availability of Se. Then, this last modification can be seen as a sign of the availability of Se (which would be the referent according to the Peircean paradigm). The icon would be in the middle; it is the very process of the dependence of the mature form on Se availability. This linkage is not random, since the cell has to somehow monitor if Se is available in order to synthesize all the necessary elements of the Sec insertion machinery.

Nonetheless, it cannot be concluded that the unmethylated form is not

---

into the growing polypeptide is impaired, as it is shown in G.J. Warner – M.J. Berry – M.E. Moustafa – B.A. Carlson– *al.*, «Inhibition of Selenoprotein Synthesis».

[83] Cf. C. Sturchler – A. Lescure – G. Keith – P. Carbon– *al.*, «Base Modification Pattern».

[84] «This was shown by three independent procedures: 1) elution properties on the ion-exchange resin, RPC-5; 2) differential patterns of formamide cleavage; and 3) mobility on non-denaturing gels. In all three approaches, the data indicate a shift from a compact to an open structure upon methylation at position 34. It is not surprising that 2'-O-methylation would alter the conformation of tRNA$^{[Ser]Sec}$. It has recently been shown that 2'-0-methylation of pyrimidine residues in the anticodon loop of tRNAs induces conformational rigidity of that nucleotide by stabilizing the C3'-endo form of the pyrimidine and thus stabilizes the conformation of that loop». A.M. Diamond – I.S. Choi – P.F. Crain – T. Hashizume– *al.*, «Dietary Selenium Affects Methylation», 14222.

[85] Cf. D. Hatfield – B.J. Lee – L. Hampton – A.M. Diamond– *al.*, «Selenium Induces Changes», and R.R. Jameson – A.M. Diamond, «A Regulatory Role for Sec tRNA[Ser]Sec».

functional. Indeed, as it shall be seen below, the relative amount of the two isoacceptors has an important role in the hierarchical regulation of selenoprotein synthesis. In fact, an alternative model defends the fact that the two isoacceptors (methylated and unmethylated form) are both functional, but they have different tissue distribution and different participation in the recoding process depending on the selenoprotein that has to be translated. Indeed, it has been shown that the methylated isoacceptor plays an important role in the expression of stress-related selenoproteins, while the unmethylated form of tRNASec is linked to the expression of housekeeping selenoproteins[86].

3.2.2 The SECIS and the SRE elements

The UGA codon and the Sec-tRNASec are not enough to direct Sec insertion co-translationally. Notwithstanding the importance of the UGA codon, another RNA structure must be contained in the mRNA. The SECIS element (SElenoCysteine Inserting Sequence) is a RNA stem loop structure located downstream the UGA codon, and it was first characterized in the early 90's[87] (Figure VI.30). As will be shown, this structure drives the interaction between the ribosome and the Sec-tRNASec. Other proteins are also required for the proper interaction of the different partners.

a) *Types of SECIS elements*

A main division can be made between the SECIS elements of eukaryotes and archaea in one hand, and those of prokaryotes on the other (Figure VI.31). As it

---

[86] This has been demonstrated in an experiment that analyzes the activity of mutants lacking some of the tRNA^{Sec} modifications. B.A. Carlson – X.-M. Xu – V.N. Gladyshev – D.L. Hatfield– *al.*, «Selective Rescue of Selenoprotein Expression» and B.A. Carlson – M.E. Moustafa – A. Sengupta – U. Schweizer– *al.*, «Selective Restoration». In general, housekeeping genes are those genes that are activated in all cellular types and, then, are always needed for the proper function of the cell.

[87] Cf. M.J. Berry – L. Banu – Y.Y. Chen – S.J. Mandel– *al.*, «Recognition of UGA», M.J. Berry – L. Banu – J.W. Harney – P.R. Larsen– *al.*, «Functional Characterization of the Eukaryotic» and Q. Shen – F.F. Chu – P.E. Newburger, «Sequences in the 3'-untranslated Region».

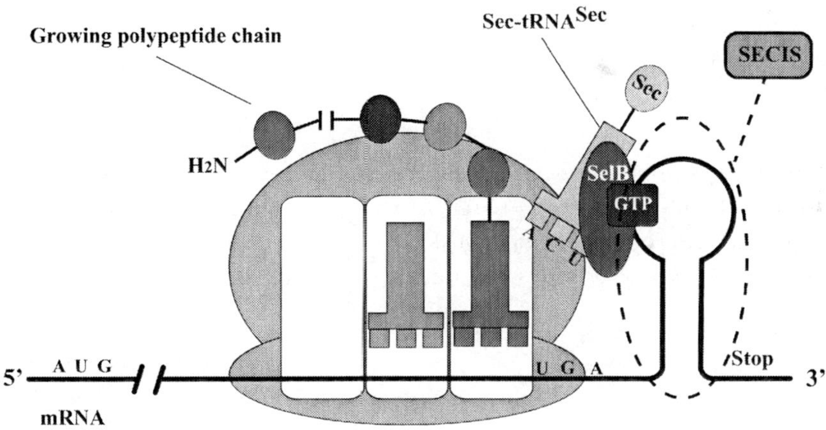

Figure VI.30: The Selenocystein insertion process. SECIS element is circled in a dashed ellipse. It is an essential partner to coordinate the insertion of Selenocystein at the proper time and place.

has been said, the SECIS element is located downstream of the UGA codon; but in eukaryotes and archaea it is located in the 3'UTR (3'-untranslated region is the region of the mRNA located downstream the "stop" signal. Then, it is a part of the mRNA that is not translated into protein), while in bacteria it is inserted in the ORF (the open reading frame is the part of the mRNA that is read by the ribosome to be translated into protein. It contains the codified information).

This is a very interesting data, since the prokaryotic SECIS element has to preserve, then, not only its secondary structure, but also the potential sequence that has to be translated[88]. In the case of eukaryotes and archaea, where the SECIS element is located in a region that will not be translated into protein, only the secondary structure must be preserved, in order to interact with the proper partners that are involved in Sec insertion. This characteristic allows eukaryotes and archaea to have more control over this element, since it can be decoupled from the codifying aspect of the nucleotides and devote them to assuring the proper three-dimensional structure needed for the role the SECIS element plays in the Selenocysteine insertion mechanism.

[88] Cf. C. ALLMANG – L. WURTH – A. KROL, «The Selenium to Selenoprotein Pathway».

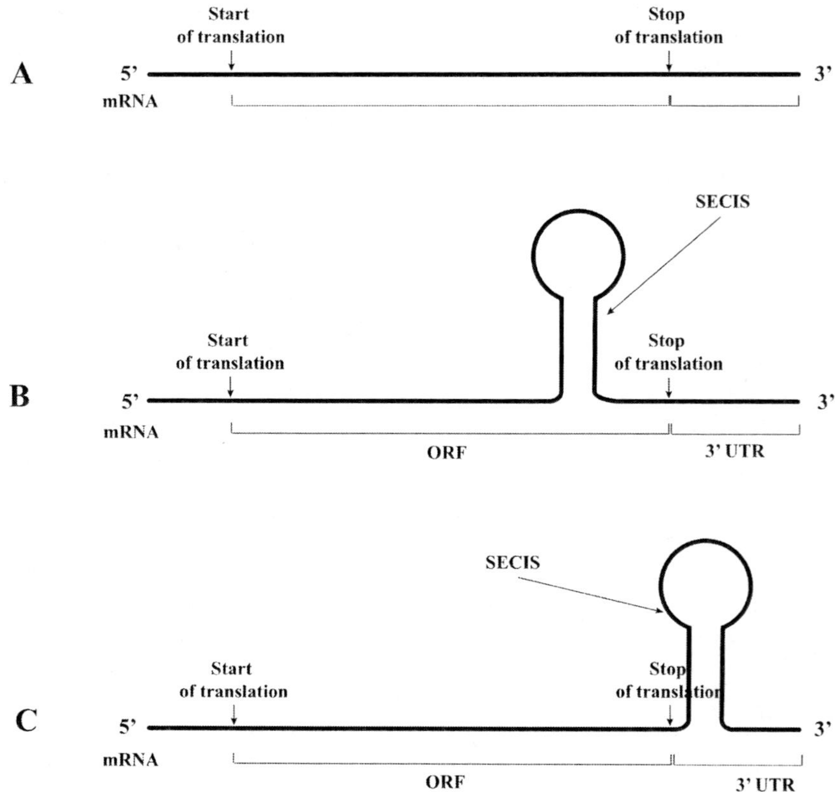

Figure VI.31: Structure of three mRNA's. A) In every mRNA can be distinguished the Open Reading Frame (ORF), that is read by the ribosome in order to be translated into protein, and the 3' Untranslated Region (3'UTR), that is not translated because the ribosome finds a stop signal before it that stops translation. B) A prokaryotic mRNA of a selenoproetin is shown. The SECIS element, indispensable for Sec insertion, is located within the ORF. Then, the sequence of the SECIS element not only has to perform the role of Sec insertion, but also its very structure has to be translated into protein. C) In the case of eukaryotes and archaea, the SECIS element is located in the 3'UTR. Then, the sequence of the element is not important, it just needs to retain the overall structure that allows it to perform its part in the Sec insertion mechanism. Furthermore, the SECIS element located in the 3'UTR can perform different Sec insertions if it were the case.

b) *Structural requirements*

If we focus on eukaryotes, we discover out two different types of SECIS elements: type 1, with an internal loop, an apical loop and two helixes; and type 2, with an additional bulge between the apical and the internal loop, that allows also the formation of a further helix (Figure VI.32). On the other hand,

archaea, has a structure that resembles the apical part of type 2 SECIS element of eukaryotes[89].

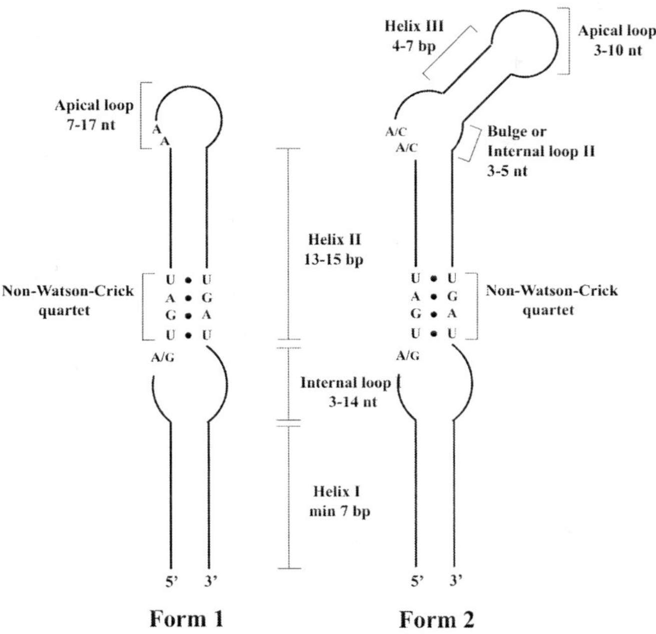

Figure VI.32: Secondary structure of type 1 and type 2 eukaryal SECIS element. A/C and A/G indicate that A is the prevalent base, but in some cases C and G can also be found, respectively. Adapted from A. Takeuchi – D. Schmitt – C. Chapple – E. Babaylova– *al.*, «A short Motif in Drosophila». Bp stands for "base-pairs" and nt for "nucleotides".

It is not well known why there are two types of SECIS elements in eukaryotes. It seems that both of them can compete for the same protein factors in order to play its role in Sec insertion. That fact points out that no specific binding factors are required for the different types of SECIS elements. Moreover, interchanging type 1 and type 2 SECIS elements do not change Sec insertion activity. It has been suggested that perhaps the extra helix of type 2 SECIS element could be a stabilizing element that nucleates the folding of the structure,

[89] Cf. C. Allmang – A. Krol, «Selenoprotein Synthesis».

favouring it thermodynamically[90]. This explanation seems plausible since both types preserve the same essential elements to retain Sec insertion activity. As can be seen in Figure VI.32, there is a consensus sequence that has been found in all eukaryotic SECIS elements. There are some bases that cannot be changed at all. Specially, two conserved adenosines in the apical loop and four non-Watson-Crick base pairs that form a quartet (5'-UGAU-3':5'-UGAU-3') located in helix II[91]. An unpaired adenosine that precedes the quartet is also essential for the proper function of SECIS element. This unusual pattern probably provides a chemical configuration for the interaction of the SECIS element with other factors[92]. Indeed, the disruption of these interactions leads to the impairment of selenoprotein synthesis[93]. It has also been demonstrated that a distance between 9 to 11 base pairs is needed for the stem situated between the non-Watson-Crick core and the conserved adenosine of the apical or internal loop[94]. Despite the

---

[90] Cf. E. Grundner-Culemann – G.3. Martin – J.W. Harney – M.J. Berry– *al.*, «Two Distinct SECIS Structures». The hypothesis of the thermodynamically favoured type 2 is supported by the fact that this type is more frequent in nature. In C.E. Chapple – R. Guigó – A. Krol, «SECISaln, a Web-based Tool» the proportion was established in 62 for type 1 and 224 for type 2.

[91] As it has been characterized in M.J. Berry – L. Banu – J.W. Harney – P.R. Larsen– *al.*, «Functional Characterization of the Eukaryotic», R. Walczak – E. Westhof – P. Carbon – A. Krol– *al.*, «A Novel RNA Structural Motif» and R. Walczak – P. Carbon – A. Krol, «An Essential Non-Watson-Crick». Further evidence for the G-A/A-G quartet instead of the A-G/G-A is found in Y. Ito – T. Mizutani, «Evidence of G-A/A-G Pair». However, a SECIS-like element has been characterized lacking the two canonical adenosines, as it could be seen in K.V. Korotkov – S.V. Novoselov – D.L. Hatfield – V.N. Gladyshev– *al.*, «Mammalian Selenoprotein».

[92] Cf. R. Walczak – E. Westhof – P. Carbon – A. Krol– *al.*, «A Novel RNA Structural Motif».

[93] Cf. R. Walczak – P. Carbon – A. Krol, «An Essential Non-Watson-Crick». In this paper the authors perform an experiment with a battery of mutants of the core sequence in order to establish the essentiality of that core. The importance of the non-canonical interactions in order to have a functional SECIS element is demonstrated. «The comparative advantage of non-Watson-Crick over Watson-Crick base pair resides is that the former present chemical groups in either the major or minor groove of the helix, available for specific interactions».

[94] Cf. E. Grundner-Culemann – G.3. Martin – J.W. Harney – M.J. Berry– *al.*, «Two Distinct SECIS Structures».

apparent indifference of an mRNA to have either one type of SECIS element or the other, it is also true that, in nature, a single selenoprotein mRNA SECIS element always adopt only one type of secondary structure[95].

Also a minimal distance between the UGA codon and the SECIS element is needed. The minimal spacing requirement is 51-111 nucleotides[96]. This distance permits the SECIS element, with the recoding machinery attached to it, to bend over in order to get closer to the UGA codon that has to be reached.

The G·A tandem found in the core of the SECIS element is not exclusive of this kind of RNA structures. It has been also found in other RNAs, like ribosomal RNAs (rRNAs) and small nuclear RNAs (snRNAs)[97]. These kinds of RNA are in close relation to protein partners and are of great importance for their proper function. The quartet adopts a particular structure known as kink-turn (or K-turn). So we can see how the adoption of non-canonical interactions between RNA base pairs, permits the exploration of RNA-protein interactions. In fact, twelve basic geometric types have been characterized in order to unify the nomenclature of the diversity of interactions between RNA base pairs[98].

It is worth noting that there is evidence to say that what is essential for the proper function of the SECIS element is not the sequence of the core but its structure. It has been demonstrated that the SECIS can cope with some mutations if a non-conserved base at position 5 in the core is able to maintain certain secondary structure interactions that preserves the final shape of the K-turn[99]. Then, «the exact sequences of the paired nucleotides are comparatively unimportant, provided that a consensus combination of length and thermodynamic stability of the base-paired structures is maintained»[100].

---

[95] Cf. D. FAGEGALTIER – A. LESCURE – R. WALCZAK – P. CARBON– *AL.*, «Structural Analysis».

[96] Cf. G.W.3. MARTIN – J.W. HARNEY – M.J. BERRY, «Selenocysteine Incorporation in Eukaryotes».

[97] Examples of K-turn in rRNA, snRNA and mRNA are shown in D.J. KLEIN – T.M. SCHMEING – P.B. MOORE – T.A. STEITZ– *AL.*, «The Kink-turn».

[98] Cf. N.B. LEONTIS – E. WESTHOF, «Geometric Nomenclature».

[99] Cf. Q. SHEN – J.L. LEONARD – P.E. NEWBURGER, «Structure and Function» and G.W.3. MARTIN – J.W. HARNEY – M.J. BERRY, «Functionality of Mutations at Conserved».

[100] H. KOLLMUS – L. FLOHÉ – J.E. MCCARTHY, «Analysis of Eukaryotic mRNA Structures», 1195.

Then, the informational sequence of the SECIS element is unimportant if compared with the three-dimensional structure it adopts. This shows us that the important question is not to maintain a certain sequential order, but an overall three-dimensional pattern that would allow the interaction with protein partners. Then, sequential information does not make the difference. Instead, the overall shape of the SECIS element can be semiotically tested to see whether it can be linked to the rest of the partners that drive Sec insertion.

The structure of the SECIS element allows me to stress something that I have pointed out previously: the "coding for" paradigm does not stand in these cases (see Chapter III, section 3.1.1). Indeed, it has been shown that the precise sequence of nucleotides that make the SECIS element is not the most important feature to perform its function in the Sec insertion event. More important than the sequence is the overall shape of the SECIS, allo-wing it to interact with the proper partners to do what needs to be done. Then, what natural selection selects is not the sequence *per se*, but a sequence that would yield a three-dimensional structure that would allow the interaction of SECIS with the rest of the partners that are needed to perform Sec insertion.

c) *The SRE*

The SRE (Selenocystein redefinition element) is a secondary structure that has been found at a conserved distance (7 nucleotides) 3' to the UGA codon[101] (Figure VI.33). SRE does not always show the same secondary structure, even if the selenoprotein mRNAs of a given eukaryotic specie are compared. The SRE secondary structure could be variable, but also the coding potential of the nucleotide sequence that shapes that secondary structure has to be preserved, as in the case of prokaryotes. This is what happens, in fact, in the bacterial SECIS element, which does not have strict secondary structure conservation among bacteria.

Those recoding elements are not as well studied as the SECIS, but they are also important elements of the recoding machinery. We can have an idea of

[101] Cf. M.T. Howard – G. Aggarwal – C.B. Anderson – S. Khatri– *al.*, «Recoding Elements». This recoding elements in the open reading frame have been identified also in the rat liver selenoprotein P mRNA, as shown in K.E. Hill – R.S. Lloyd – R.F. Burk, «Conserved Nucleotide Sequences».

the importance of this type of structures noting that a mutation in the SRE of SEPN1 (the gene that codes for the human Sel N) leads to a type of myopathy because of a decreased efficiency in Sec insertion[102]. Although the SECIS element is essential for Sec insertion, these SREs also play an important role in regulating the efficiency of this process, providing adequate surrounding conditions for the UGA codon to be recoded[103]. SRE can also have the role of slowing down the ribosome in order to give time for the SECIS element to reach the insertion locus and perform the insertion event.

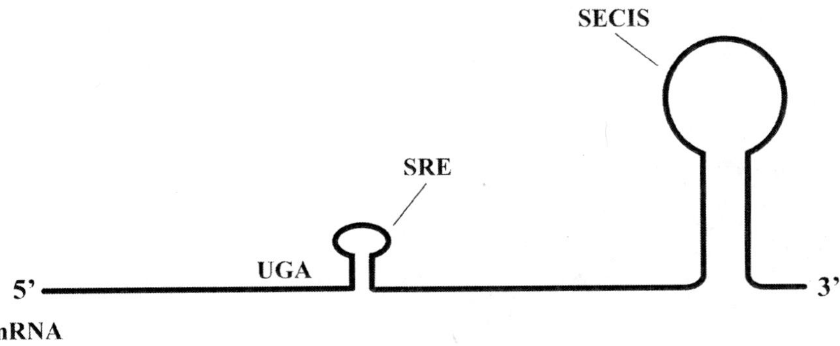

Figure VI.33: Schematic view of a SRE and a SECIS element in the same mRNA. SRE is near the UGA codon, at a certain distance, and is an important element to have a recoding event, specially for providing adequate surrounding for Sec insertion.

d) *Bacterial SECIS*

The insertion of Sec in bacterial selenoproteins depends on the stem loop secondary structure that the mRNA adopts downstream of the UGA codon[104]. Moreover, it has been shown that the stem loop has to be at a precise distance

---

[102] Cf. B. MAITI – S. ARBOGAST – V. ALLAMAND – M.W. MOYLE– *AL.*, «A Mutation in the SEPN1». Myopathy could also be caused by a point mutation in the 3'UTR of Sel N mRNA, as it is reported in V. ALLAMAND – P. RICHARD – A. LESCURE – C. LEDEUIL– *AL.*, «A Single Homozygous Point Mutation».

[103] Cf. M.T. HOWARD – M.W. MOYLE – G. AGGARWAL – B.A. CARLSON– *AL.*, «A Recoding Element». In this paper it is shown that the SRE allows the recoding machinery to use either the methylated form of the Sec-tRNA^{Sec} or the unmethylated one.

[104] Cf. F. ZINONI – J. HEIDER – A. BÖCK, «Features of the Formate».

from the UGA codon[105]. So, in bacteria, as happens also in eukaryotes and archaea, it is not only important the presence of the SECIS element downstream of the UGA codon, but it has to be located at a precise distance from it, otherwise the translation rate of the mRNA decreases markedly. However, the sequence requirements of the SECIS element are not very strict. The important thing is to preserve an overall three-dimensional structure that can interact with SelB; that is the reason why there is great sequence flexibility[106].

However, some studies point at other important elements besides SECIS. One of those is the SRE, which has been considered in the former section; but there are also at least two more features that regulate the Sec insertion activity. It has been reported that formate dehydrogenase H mRNA form *E. coli* has two additional interesting elements: a single C residue downstream of the UGA codon and six nucleotides located at 5'. So we have here not one but three regulatory elements that, depending on the availability of selenium, allow the Sec insertion or the termination of translation. Furthermore, what counts in these new regulatory elements is not its three-dimensional shape but its sequence. In fact, these two signals are both required for translation termination in conditions of selenium starvation. The C residue acts in this mechanism, but in the case of the six nucleotides placed upstream the UGA codon, it is not the sequence of them what matters, but the two amino acids they code for[107] (Figure VI.34).

---

[105] Cf. J. HEIDER – C. BARON – A. BÖCK, «Coding From a Distance» and J.T. CHEN – L. FANG – M. INOUYE, «Effect of the Relative Position».

[106] Cf. K.E. SANDMAN – D.F. TARDIFF – L.A. NEELY – C.J. NOREN– *AL.*, «Revised Escherichia Coli Selenocysteine Insertion».

[107] Cf. Z. LIU – M. RECHES – H. ENGELBERG-KULKA, «A Sequence in the Escherichia». This particular feature is also reported in J. HEIDER – C. BARON – A. BÖCK, «Coding From a Distance»: «It is demonstrated here that the correct folding of this structure [the SECIS element] as well as the existence of primary sequence elements located within the loop portion at an appropriate distance to the UGA codon are absolutely required». Also in K.K. MCCAUGHAN – C.M.D. BROWN, M. E. – M.J. BERRY – W.P. TATE– *AL.*, «Translational Termination Efficiency in Mammals», W. WEN – S.L. WEISS – R.A. SUNDE, «UGA Codon Position Affects», K.E. SANDMAN – C.J. NOREN, «The Efficiency of Escherichia Coli» and M. GUPTA – P.R. COPELAND, «Functional Analysis of the Interplay» it is showed the importance of the UGA codon context for the Sec insertion mechanism.

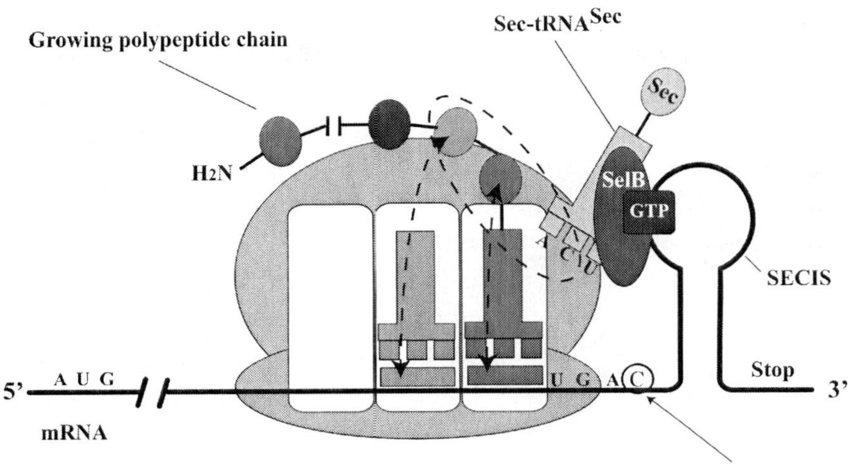

Figure VI.34: Elements of the Sec insertion mechanism in the mRNA of the enzyme formate dehydrogenase H in *E. coli*. The C residue (circled) located downstream the UGA codon and the chemical surrounding provided by the two amino acids that are codified by the six previous nucleotides (the relation between the six previous nucleotides and the two amino acids they code for is indicated with dashed arrows) are a very important element in the regulation of Sec insertion. Indeed, these elements are important to read the UGA codon as "stop", rather than as "Sec", when selenium is not available.

This is a very interesting example to see how many elements concur in order to facilitate the Sec insertion. It is difficult to see how the six nucleotides placed upstream of the UGA codon can have some incidence in the recoding event. Precisely because they are upstream the UGA codon, they cannot have any influence on it, but the two amino acids they code for can. Then, the two previous amino acids synthesized before the Sec insertion, provide a chemical environment that influence in the Sec insertion event.

Here we have an example of top-down causation. The newly synthesized amino acids do not have the capacity to insert Sec in the polypeptide chain, but they can influence that event providing the chemical conditions that would facilitate the insertion event rather than the stop one. Indeed, a single amino acid located within a polypeptide chain, cannot exert any kind of cause over the insertion of new amino acids during translation. However, if amino acids are considered not as individual items, but as providing a chemical surrounding, we see that this can influence the insertion or not of Sec in the polypeptide chain. The polypeptide chain, and more concretely the two amino acids that have been already added to it, can be seen as the higher level instance that constraints the chemical features of the lower level in order to insert Sec or finish translation.

Another important thing to be taken into account is the fact that, in prokaryotes, the location of the SECIS element in the ORF requires the dissociation of SelB and the melting ("melting" is a verb used when there is the separation of hydrogen bonds between nucleotides, as there are in the stem part of the SECIS element) of the stem-loop structure when the ribosome reads the mRNA[108]. On the contrary, eukaryotes have the SECIS element in the 3'UTR, so it has not to be melted during ribosome's read-through and SBP2 (a protein that binds the SECIS element, as we shall see) has not to be dissociated from it. These characteristics make eukaryotic Sec insertion more efficient than prokaryotic. SBP2 remains bound to the SECIS element, ready to perform another Sec insertion in another UGA codon of the mRNA (as it happens with SelP mRNA), or can wait for another translation round[109].

e) *Archaeal SECIS*

The most important difference between archaeal and eukaryotic SECIS elements is that the former does not have the non-Watson-Crick quartet (Figure VI.35 and compare with Figure VI.32). However, its Sec insertion activity has been tested and it can perform this role[110]. Astonishingly, there is one case in which the SECIS element is not in the 3'UTR but in the 5'UTR[111]. Then, we have an example of a SECIS element that is not located at the end of the mRNA, but at the beginning. Nonetheles, it participates in the Sec insertion event. It is a good example to see that the important thing to perform a certain operation is not the exact location and precise chemical components, but an overall pattern that allows the proper interactions among the elements that take part in such operation. Once again, the retaining of the binding properties is more important than the structure itself and the place where it is located.

---

[108] Cf. A. HÜTTENHOFER – J. HEIDER – A. BÖCK, «Interaction of the Escherichia Coli».

[109] Cf. S.C. LOW – E. GRUNDNER-CULEMANN – J.W. HARNEY – M.J. BERRY– *AL.*, «SECIS-SBP2 Interactions».

[110] Cf. M. ROTHER – A. RESCH – W.L. GARDNER – W.B. WHITMAN– *AL.*, «Heterologous Expression of Archaeal Selenoprotein».

[111] Cf. R. WILTING – S. SCHORLING – B.C. PERSSON – A. BÖCK– *AL.*, «Selenoprotein Synthesis in Archaea».

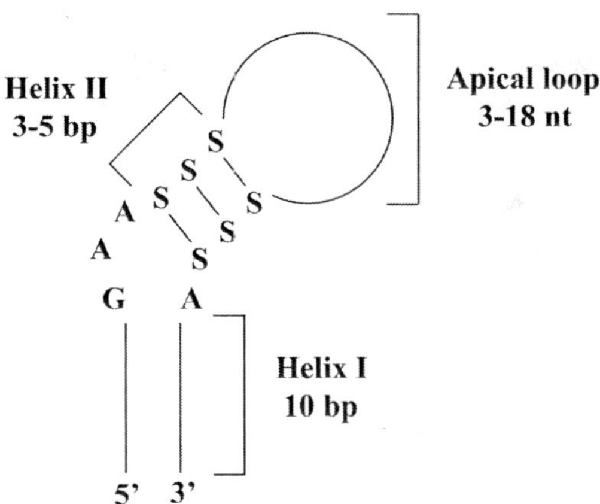

Figure VI.35: Secondary structure of archaeal SECIS element. Note that the non-Watson-Crick quartet is absent. S stands for G or C. Bp stands for "base pairs" and nt for "nucleotides". Adapted from C. ALLMANG – A. KROL, «Selenoprotein Synthesis».

3.2.3 Specific protein factors involved[112]

I have just presented the RNA elements that have a role in Sec incorporation. Let us now present the proteins that are needed in order to target Sec insertion. There are needed many proteins in order to drive the translation process. In the example I am showing, there are also the usual protein factors that are needed for protein synthesis. I just want to focus here on the specific protein factors that are needed for inserting Sec during the translation process. Those protein factors are SBP2, EFSec and ribosomal protein L30 in eukaryotes and archaea, and SelB in prokaryotes.

a) *SECIS binding protein 2 (SBP2)*

SBP2 is an 846 amino acid protein required for the translation of eukaryotic mRNA selenoproteins that binds the SECIS element[113] (as has been shown

[112] For a short characterization of the protein factors involved in Sec insertion, see J.E. SQUIRES – M.J. BERRY, «Eukaryotic Selenoprotein Synthesis».

[113] Cf. P.R. COPELAND – J.E. FLETCHER – B.A. CARLSON – D.L. HATFIELD– *AL.*, «A Novel RNA Binding Protein». We find a clear example of the importance of SBP2 for the translation of selenoprotein mRNAs in V. ALLAMAND – P. RICHARD – A. LESCURE – C. LEDEUIL– *AL.*, «A Single Homozygous Point Mutation». Here it is shown that a point mutation in the

previously in Figure VI.18). Its sequence can be divided in two main domains, as it is shown in Figure VI.36. The function of the N-terminal domain has not been yet characterized (aa 1 to 398). Figure VI.36 shows the protein in its secondary structure, not as its final folding state. Sketching a protein in such way allows to easily identify the different domains that are present and the role each of them play in the protein. The amino acids of the protein are numbered from the N-terminal end (because it begins with the amino group of the first amino acid added during translation) to the C-terminal end (which ends with the carboxyl group of the last amino acid added).

Although the N-terminal domain has not been fully characterized, a nuclear localization signal (NLS) has been found in it, suggesting a role in the shuttling of the protein through the nuclear membrane[114]. A nuclear localization signal is a short sequence that is present in the proteins that have to be sent to the nucleus. Those NLS are recognized by transporter proteins located in the nuclear membrane and then translocated into the nucleus.

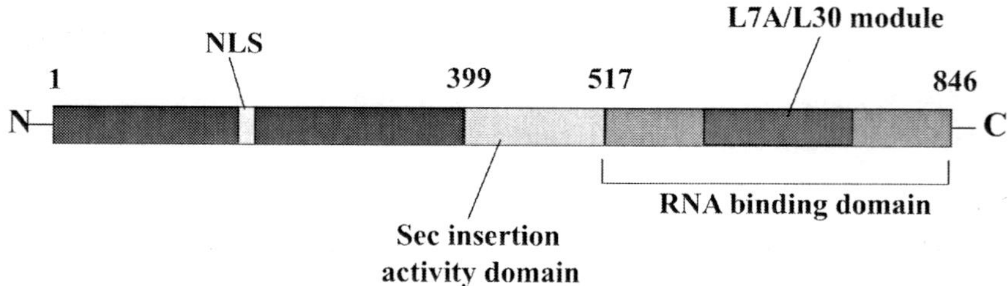

Figure VI.36: Scheme of the SECIS binding protein 2 (SBP2). Three domains can be easily identified. The N-terminal domain contains a nuclear localization signal (NLS), the central domain is responsible for the Sec insertion activity and the RNA binding domain carries the binding site for SECIS and the L7/L30 module, that is necessary for the interaction with the ribosome. Adapted from C. ALLMANG – A. KROL, «Selenoprotein Synthesis».

SECIS element of the Sel N mRNA prevents SBP2 from binding and, therefore, no translation takes place. That point mutation is the unique cause of a particular type of myophaty. Further data about the specificity of the SBP2-SECIS interaction can be found in A. CLÉRY – V. BOURGUIGNON-IGEL – C. ALLMANG – A. KROL– *AL.*, «An Improved Definition». For a genetic characterization and the map of the expression pattern of human SBP2, see A. LESCURE – C. ALLMANG – K. YAMADA – P. CARBON– *AL.*, «cDNA Cloning, Expression Pattern».

[114] Cf. L.A. DE JESUS – P.R. HOFFMANN – T. MICHAUD – E.P. FORRY– *AL.*, «Nuclear Assembly of UGA Decoding».

The C-terminal section contains two different domains: a functional domain responsible for Sec insertion activity (aa 399 to 517) and a RNA binding domain (aa 518 to 846). In this RNA binding domain a SECIS RNA binding activity and a ribosomal binding activity are located. This SECIS RNA binding activity permits SBP2 to bind the SECIS element to the non-Watson-Crick quartet and to the phosphates located all along the helix I, as it is showed in Figure VI.37. The RNA binding domain also contains the L7A/L30 module, which is also present in other proteins that interact with RNA[115]. This domain allows SBP2 to interact with the ribosome[116]. In fact, L7A/L30 is a module that can interact with the K-turn structures, which, among others, can be found in rRNAs (a family of RNAs that are part of the ribosome).

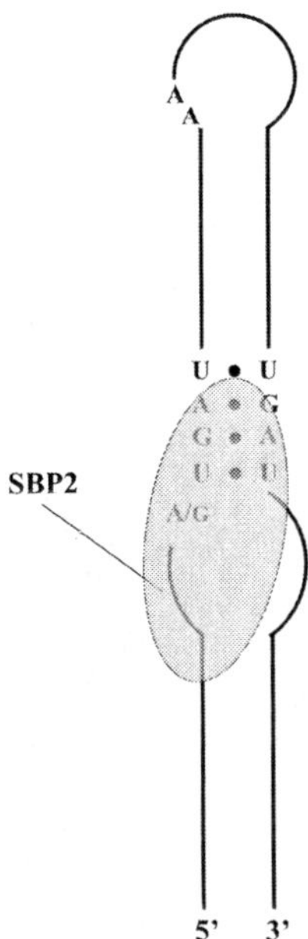

Figure VI.37: Interaction between SBP2 and a type 1 SECIS element. Adapted from A. Krol, «Evolutionarily different RNA motifs».

[115] Cf. J.E. Fletcher – P.R. Copeland – D.M. Driscoll – A. Krol– *al.*, «The Selenocysteine Incorporation Machinery» and C. Allmang – P. Carbon – A. Krol, «The SBP2 and 15.5 kD/Snu13p».

[116] Cf. K. Caban – S.A. Kinzy – P.R. Copeland, «The L7Ae RNA Binding Motif».

The Sec insertion activity (in the so called SID, for Sec Insertion Domain) is located next to the RNA binding domain. Its residues are not only responsible for the insertion of Sec in the UGA codon of mRNA, but they also play a role in the interaction with the ribosome[117].

As can be seen in Figure VI.36, the SECIS binding site and the L7A/L30 module are overlapped. This poses the question about the mechanism of action of SBP2. Does SBP2 bind the SECIS element and the ribosome at the same time? Or does it bind first one and then the other? The characterization of these binding activities leads to one Sec insertion model or other, as it shall be seen (sections 3.3.2 and 3.3.3 in this chapter)[118].

As has already been said, little is known about the function of the N-terminal end of SBP2. It has been reported that the 70% of the SBP2 is disordered, and this N-terminal domain is responsible for this unfolded state. «This finding categorizes SBP2 in the growing list of Intrinsically Disordered Proteins (IDP) and suggests that the disordered section of SBP2 may acquire a proper fold in the presence of protein partners»[119]. This is an important point where the theoretical framework proposed can be of great help. Indeed, we are faced with

---

[117] Cf. J. Donovan – K. Caban – R. Ranaweera – J.N. Gonzalez-Flores– *al.*, «A Novel Protein Domain».

[118] In S.A. Kinzy – K. Caban – P.R. Copeland, «Characterization of the SECIS», it is said that both elements cannot be bound at the same time. On the contrary, in P.R. Copeland – V.A. Stepanik – D.M. Driscoll, «Insight into Mammalian Selenocysteine Insertion» and P.R. Copeland, «Regulation of Gene Expression», it is proposed a model in which SBP2 can interact at the same time with the SECIS element and the ribosome.

[119] C. Allmang – L. Wurth – A. Krol, «The Selenium to Selenoprotein Pathway», 1420. This is a surprising issue that has pushed the biologist to reconsider what they thought that was a settled principle. «The 'structure–function relationship' has been a central dogma for decades in the protein field. As a result, an absence of folding has long been considered as an experimental artefact during the purification step and a cause of aggregation or proteolysis, rather than of biological significance. However, it is now clear that lack of folding is not synonymous of lack of biological activity. This emerged from studies like that on a cyclin-dependent kinase (Cdk) inhibitor (p21) that is unstructured in absence of its target, but becomes structured upon binding to it. Since then, a large body of other examples has been identified and the role of structural disorder in the function of RNA and protein chaperones has been reviewed», V. Oliéric – P. Wolff – A. Takeuchi – G. Bec– *al.*, «SECIS-binding Protein 2», 1003-1004.

certain domains that do not reach their final folding state until they interact with their partner. This allows for the possibility that different partners can interact with those domains, which would adopt different folding states in order to properly interact with all the possible partners. We have then an example of how the folding, and consequently the function, of a protein is something not definitely established, but that depends on the environment, on the partners that are available.

This is a very good example to see that functions are contextual. Indeed, those disordered domains that are found in some proteins, as SBP2, do not reach their final folded state until they interact with a partner. Depending on the partner they interact with, their folding, and consequently their function, would be one or other. Then, the role of those disordered domains (and, by extension, of those proteins) is contextual; it depends on what partners are in the surroundings of the protein, since the interaction with different partners would led to different ways of folding the protein they interact with.

From a semiotic point of view, those proteins are signs of referents that are not clearly established until there is an interaction of the protein with one of the partners that can be bound to it. There is a certain degree of plasticity in the binding properties of these kinds of proteins; which means that they can be sign of different referents, participating then in different operations.

b) *EFSec and SelB: special elongation factors*

As we have seen above (section 3.1), during protein synthesis the new amino acids that enter the A-site of the ribosome have to be inserted into the growing polypeptide chain, that occupies the P-site. When the tRNA has unloaded its amino acid, it passes through the E-site in order to outgo from the ribosome. Elongation factors[120] are needed in order to undergo the conformational changes required to allow each molecule to occupy its proper site at each time. They bind to the ribosome and, through GTP hydrolysis, drive the addition of the amino acids to the growing polypeptide chain. EF-Tu and EF-G are bacterial elongation factors, while EF1 and EF2 are found in eukaryotes.

In prokaryotes, EF-Tu mediates the entering of the aminoacyl-tRNA into

[120] Here I explain the main characteristics of these elongation factors, as can be seen in B. Alberts – *al.*, *Molecular Biology of the Cell*, 377-378 and B. Lewin, *Genes IX*, 167-168.

the A-site, checking that tRNA matches with the amino acid it carries. So it feeds the ribosome with the aminoacyl-tRNAs that are needed for the synthesis of the growing polypeptide chain (Figure VI.38).

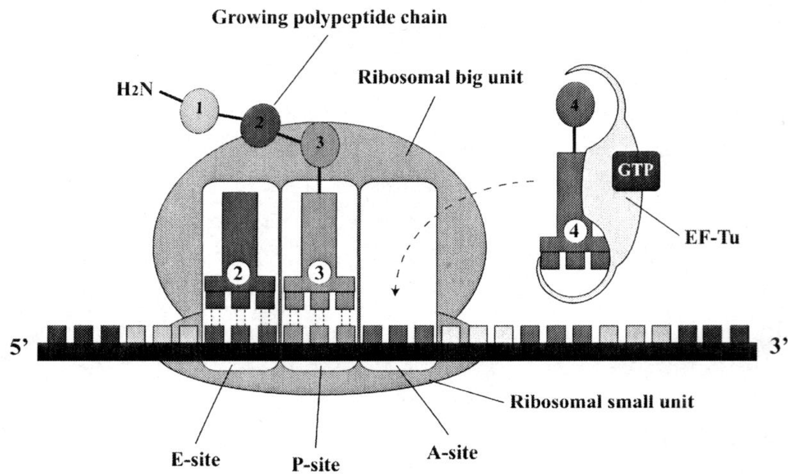

Figure VI.38: The role of the elongation factor EF-Tu in bacterial protein synthesis. EF-Tu checks that the tRNA anticodon matches with the amino acid it carries.

Another function of EF-Tu is to monitor the interaction between the tRNA anticodon and the mRNA codon at the A-site of the ribosome. It checks that the tRNA is the one that matches with the codon, making sure that the amino acid that has to be inserted is the one that corresponds with the codon that is being read by the ribosome, according to the genetic code. This proofreading is made with EF-Tu attached to the tRNA, for preventing the premature insertion of the amino acid into the polypeptide chain until it is clear that is really the tRNA that is proved the one that corresponds with the codon (Figure VI.39).

The irreversibility of this reaction is guaranteed by the GTPase activity of EF-Tu[121]. As has been said previously, GTP is one of the energetic molecules

[121] «The hydrolysis of EF-Tu-GTP is relatively slow: it takes longer than the time required for an incorrect aminoacyl-tRNA to dissociate from the A site, therefore most incorrect species are removed at this stage. The release of EF-Tu-GDP after hydrolysis also is slow, so any surviving incorrect aminoacyl-tRNAs may dissociate at this stage. The basic principle is that

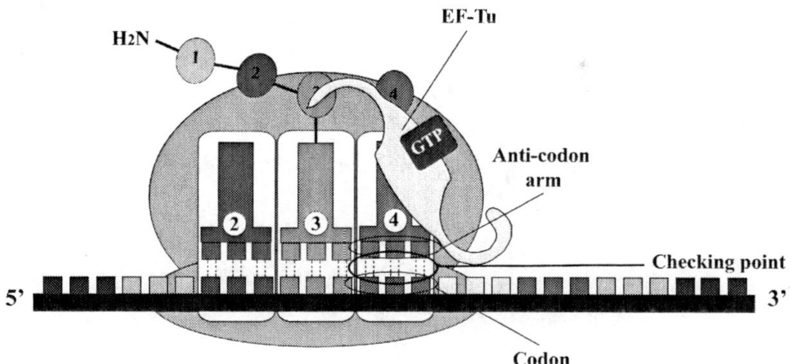

Figure VI.39: EF-Tu checks that the tRNA anticodon matches with the mRNA codon it is bound to.

used by the cell. When it has been checked that the anti-codon pairs correctly with the codon, the ribosome triggers the hydrolysis of GTP, provoking the EF-Tu release from the ribosome and allowing the polypeptide chain to react with the new amino acid in order to be inserted in the chain (Figure VI.40).

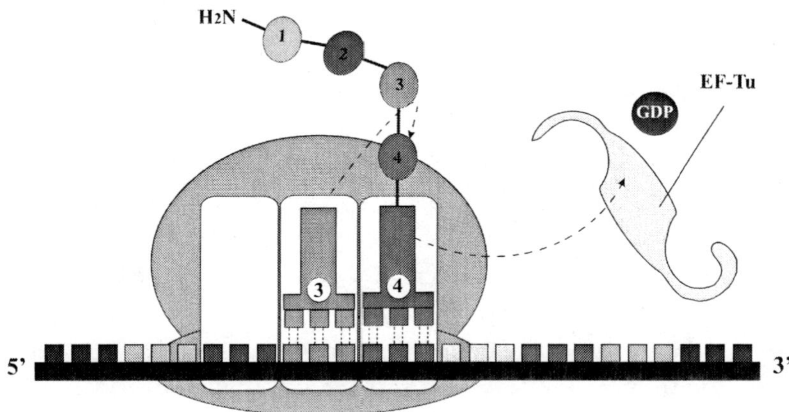

Figure VI.40: The ribosome hydrolyzes GTP into GDP, provoking the release of EF-Tu and the reaction of the growing polypeptide chain with the amino acid that is attached to the newly bound tRNA.

---

the reactions involving EF-Tu occur slowly enough to allow incorrect aminoacyl-tRNAs to dissociate before they become trapped in protein synthesis». B. LEWIN, *Genes IX*, 168.

Once the insertion of the new amino acid is completed, the ribosome has to move forward in order to read the following codon of the mRNA. The forward movement is facilitated by EF-G elongation factor (Figure VI.41).

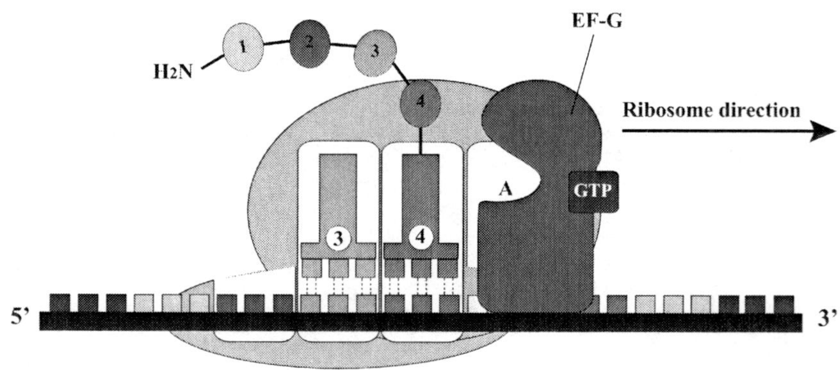

Figure VI.41: EF-G elongation factor enters the A-site and brings forward the big subunit of the ribosome to achieve the next codon that has to be read in order to translate it in the proper amino acid.

Once the big subunit of the ribosome has been placed over the next codon, the small unit follows the same direction, triggering the hydrolysis of GTP into GDP to ensure the irreversibility of the movement. Once GTP is hydrolyzed, EF-G is released (Figure VI.42).

EF-Tu and EF-G are bacterial elongation factors that are required for the adequate translation of mRNA codons. Besides those elongation factors, during Sec incorporation another elongation factor to interact with the SECIS element and to insert Sec into the growing polypeptide is also needed[122]. In prokaryotes this elongation factor has been named SelB and, as in the case of EF-Tu, is also a GTP binding protein (as it has been shown in Figures VI.12 to VI.17). Furthermore, it shows extensive homology with the N-terminal region of EF-Tu elongation factor[123]. In fact, there is high homology between the three domains

[122] Cf. C. Förster – G. Ott – K. Forchhammer – M. Sprinzl– *al.*, «Interaction of a selenocysteine-incorporating tRNA» and S. Ringquist – D. Schneider – T. Gibson – C. Baron– *al.*, «Recognition of the mRNA Selenocysteine».

[123] Cf. K. Forchhammer – W. Leinfelder – A. Böck, «Identification of a Novel Trans-

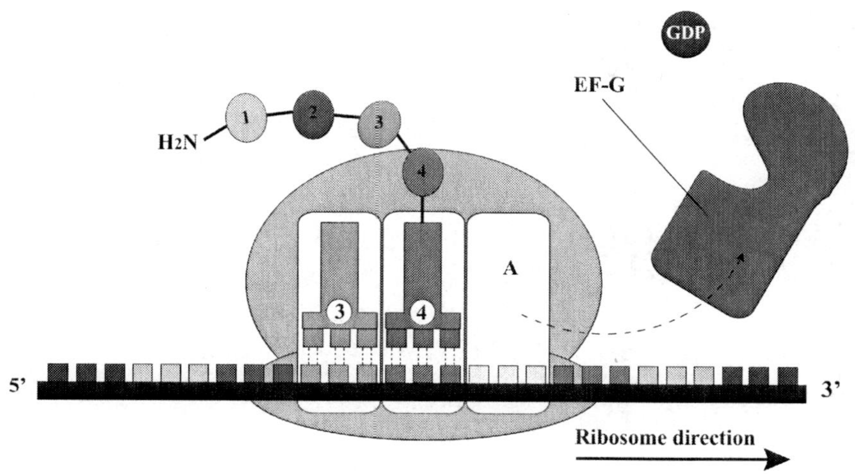

Figure VI.42: Once the whole ribosome has reached the next codon to be read, GTP is hydrolyzed to GDP and EF-G is released to allow a new tRNA entering the A-site.

of EF-Tu and three of the four domains of SelB. This region with high homology contains the binding site for the aminoacylated tRNAs (Figure VI.43).

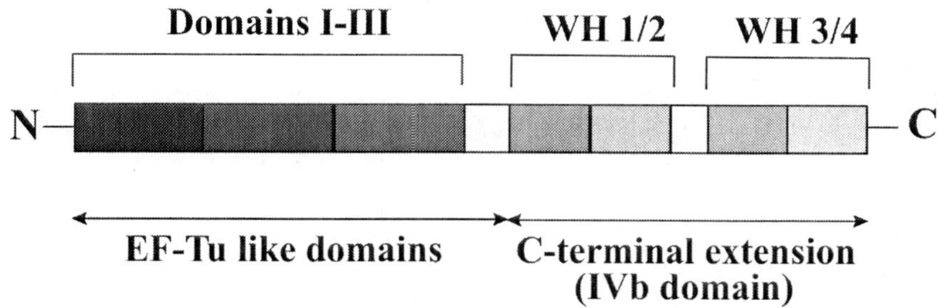

Figure VI.43: Schematic view of the domain's organization of SelB in *E. coli*. Adapted from N. SOLER – D. FOURMY – S. YOSHIZAWA, «Structural Insight».

lation Factor», K. FORCHHAMMER – K.-P. RÜCKNAGEL – A. BÖCK, «Purification and Biochemical Characterization of SELB» and R. HILGENFELD – A. BÖCK – R. WILTING, «Structural Model». It has been also reported the evolutionary relationship between this specific elongation factor and an initiator factor, as could be seen in P.J. KEELING – N.M. FAST – G.I. MCFADDEN, «Evolutionary Relationship Between Translation».

SelB binds only Sec-tRNA, but not other aminoacylated tRNAs, as does EF-Tu. A specific loop in the domain III of SelB is responsible for the interaction with Sec-tRNASec and for the measuring of its extra long acceptor arm[124]. The C-terminal domain of SelB (also called IVb), on the other hand, is not found in EF-Tu. This domain binds the SECIS element, so SelB is a bridge between the stem loop mRNA structure and the Sec-tRNASec.[125] SelB does not recognize a particular sequence of the SECIS element but its overall three-dimensional structure[126]. Four tandem winged helix motifs (WH1-4) have been identified in the C-terminal domain of SelB. The interaction between SelB and mRNA is due to these kind of motifs (Figure VI.44). This interaction acts not only as a recognition pattern between this two partners, but also has a role as a molecular switch in order to communicate with mRNA and Sec-tRNASec for the proper insertion of Sec[127].

---

[124] Cf. M. LEIBUNDGUT – C. FRICK – M. THANBICHLER – A. BÖCK– *AL.*, «Selenocysteine tRNA-specific Elongation Factor». Is interesting to note that «this loop is strictly conserved among archaea and is also present in eukarya and bacteria, but absent in EF-Tu and EF1-A, suggesting a unified tRNA^{Sec}-SelB/EFsec recognition pattern. Once again, this interaction principle represents an appealing adaptive evolution of two ligands», C. ALLMANG – A. KROL, «Selenoprotein Synthesis», 1568.

[125] In fact, SelB contains all three domains that characterize EF-Tu, but another one is added in order to interact with mRNA, as is explained in M. KROMAYER – R. WILTING – P. TORMAY – A. BÖCK– *AL.*, «Domain Structure of the Prokaryotic» and D. FOURMY – E. GUITTET – S. YOSHIZAWA, «Structure of Prokaryotic SECIS». For a three-dimensional characterization of the four domains of SelB, see M. SELMER – X.-D. SU, «Crystal Structure». For a structural study of the binding between SelB and the mRNA, see S. YOSHIZAWA – L. RASUBALA – T. OSE – D. KOHDA– *AL.*, «Structural Basis for mRNA Recognition» and T. OSE – N. SOLER – L. RASUBALA – K. KUROKI– *AL.*, «Structural Basis for Dynamic Interdomain». A kinetic study about the interaction between SelB and Sec-tRNA^{Sec} is shown in A. PALESKAVA – A.L. KONEVEGA – M.V. RODNINA, «Thermodynamic and Kinetic Framework».

[126] Cf. A. HÜTTENHOFER – J. HEIDER – A. BÖCK, «Interaction of the Escherichia Coli» and M. KROMAYER – B. NEUHIERL – A. FRIEBEL – A. BÖCK– *AL.*, «Genetic Probing of the Interaction».

[127] Cf. N. SOLER – D. FOURMY – S. YOSHIZAWA, «Structural Insight» and N. SOLER – D. FOURMY – S. YOSHIZAWA, «Molecular Switch in Tandem».

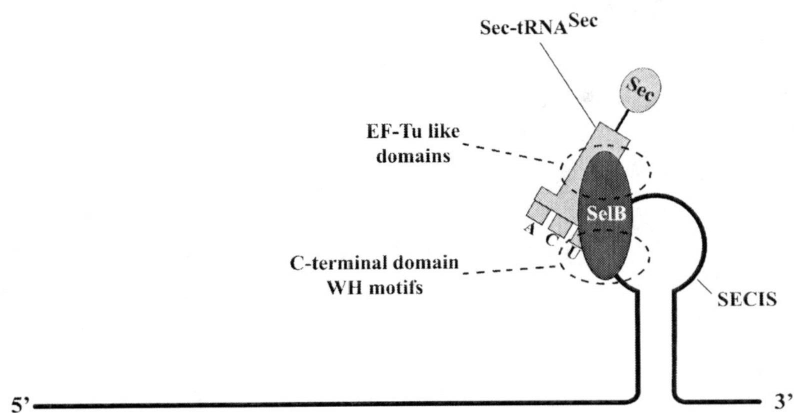

Figure VI.44: Interaction of SelB with Sec-tRNASec and the SECIS element. SelB interacts with Sec-tRNASec through its EF-Tu like domain. As EF-Tu is a general elongation factor for all tRNA's, Sec-tRNASec needs an especific elongation factor. SelB also needs to bind the SECIS element, since it is needed to insert Sec-tRNASec into its proper place in order to be added to the polypeptide chain. This interaction is done through the WH motifs.

In archaea[128] and eukarya[129], the specialized elongation factor needed for Sec insertion is EFSec[130]. Its specialized function comes from the fact that EFSec interacts with Sec-tRNASec, but not with Ser-tRNASer or Ser-tRNASec. As in the case of SelB, the N-terminal domain of EFSec shows high homology with the eukaryotic elongation factor EF1. But in contrast with bacterial SelB, EFSec does not show a SECIS binding domain; so it is unable to bind the mRNA (Figure VI.45).

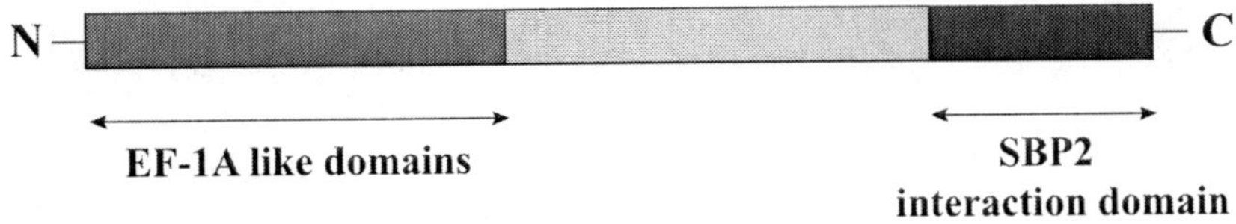

Figure VI.45: Schematic view of the N- and C- terminal domains of EFSec. N-terminal domain has high homology with elongation factor EF-1A, while the C-terminal domain allows EFSec to interact with SBP2. Adapted from C. Allmang – A. Krol, «Selenoprotein Synthesis».

---

[128] Cf. M. Rother – R. Wilting – S. Commans – A. Böck– *al.*, «Identification and Characterisation».

[129] Cf. K. Yamada, «A New Translational Elongation Factor».

[130] M. Rother – R. Wilting – S. Commans – A. Böck– *al.*, «Identification and Characterisation» and D. Fagegaltier – N. Hubert – K. Yamada – T. Mizutani– *al.*, «Characterization of mSelB».

How can EFSec then exert its role? As we have seen in eukarya and archaea, there is a SECIS binding protein (SBP2) that can bind both SECIS element and EFSec, delivering the Sec-tRNASec to the ribosome[131](Figure VI.46).

A high homology between the N-terminal domains of EF-Tu/EF1 and SelB/EFSec is expected, since all of them are translation factors. On the other hand, they have differences in their C-terminal end just to perform the peculiar elongation event that is the insertion of Sec. In the case of bacterial SelB, the C-terminal has a SECIS binding domain, while EFSec has a SBP2 interaction domain.

What is interesting of all these interactions is that what must be preserved is the organized interaction of the partners in order to allow them to perform their proper operations. In the case of eukaryotes and prokaryotes, different partners are involved, but the overall interactions between them preserve the essential features to perform a Sec insertion event.

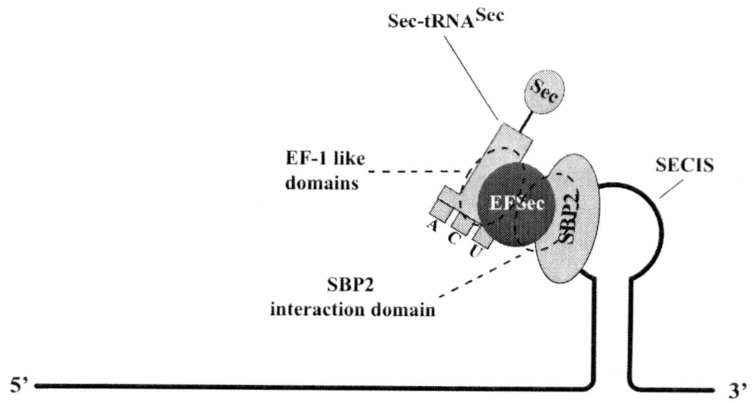

Figure VI.46: Interaction of EFSec with Sec-tRNASec and SBP2. EFSec has an EF-1 like domain, that allows the interaction with Sec-tRNASec. It does not have a domain with which directly interact with the SECIS element, but it does have a domain that binds SBP2, which in turn binds the SECIS element.

c) *Ribosomal protein L30*

This protein is only found in eukaryotes and archaea, and is an important partner of the Sec insertion mechanism[132] (see Figures VI.18 to VI.23). It has

[131] Cf. R.M. Tujebajeva – P.R. Copeland – X. Xu – B.A. Carlson– *al.*, «Decoding Apparatus».

[132] Cf. L. Chavatte – B.A. Brown II – D.M. Driscoll, «Ribosomal Protein L30».

been seen that L30 can compete with SBP2 for the SECIS element. It suggests a mechanism «in which L30 displaces transiently SBP2 to bring the SECIS RNA to the vicinity of the ribosomal A site»[133]. But the question is far from being completely understood. In fact, there is also some evidence that the L30 is far from the A site, and then, it is not known how this can match with the former data[134].

Since this protein is only found in eukaryotes and archaea, and only these two kingdoms has the SECIS element at the 3'UTR (prokaryotes has it in the ORF next to the UGA codon), it is possible that it plays an important role in directing the SECIS element to the surroundings where the Sec insertion event has to take place.

3.3 *The whole mechanism at a glance*

In the previous section I have described the different partners that interact in the Sec insertion machinery. It has been showed that many elements are required: specialized RNA structures and proteins that guide Sec insertion into the growing polypeptide chain. In this section I shall describe the whole process in eukaryotes. Once I have described the partners individually, it is time to put them together and see how they interact.

3.3.1 Supramolecular complexes

To fully understand the Sec insertion mechanism, we cannot stop at the particular characters of the elements that take part in this pathway. It is impossible to reconstruct all this biosynthetic mechanism if we only try to individuate the partners and study their singular properties. Small-Howard stresses that as follows:

> Biological processes depend on the dynamic network of interacting proteins within a cell. A full understanding of the function of an individual protein is dependent upon a broader appreciation of the proteins and/or the nucleotide partners that it interacts with in functional complexes. Ribosomal studies highlight the emergent properties

[133] C. ALLMANG – A. KROL, «Selenoprotein Synthesis», 1567.

[134] Cf. M. HALIC – T. BECKER – J. FRANK – C.M.T. SPAHN– *AL.*, «Localization and Dynamic Behavior».

of proteins and nucleic acids functioning together as supramolecular groups. Several studies have provided evidence for the existence of other supramolecular translation complexes, consisting of ribosomes, mRNA, EFs, amino-acyl tRNAs and amino-acyl tRNA synthetases[135].

Sec insertion could be considered a top-down causation event. It is something that cannot be deduced from a detailed study of the partners at play. It is an event that depends on a spatial and temporal order of steps that leads to the novelty of the insertion of an element that could not be inserted by the canonical procedure.

The fact is that the elements considered until now interact in a very precise way, forming supramolecular complexes; that is, associations of partners that are at play in order to be closer to better perform their roles. Two major supramolecular complexes have been identified in eukaryotes. The association of the SECIS element and the SBP2 forms one of them. This complex binds also the ribosomal protein L30, which mediates the incorporation process at the ribosome.

The other supramolecular complex includes Sec-tRNASec, EFSec, SECp43, SPS2 and SPS1. SPS1 is, like SPS2, a Selenophosphate synthetase, the enzyme that gives the selenium in the proper chemical form in order to synthesize Sec[136]. It seems that SPS2 interacts with SPS1 and, on the other hand, SECp43 interacts with Sec-tRNASec and EFSec[137]. The association of all these partners leads to its nuclear translocation. This association takes place through SECp43, since this protein interacts with SPS2[138]. Small-Howard summarizes the role of SECp43 as follows:

[135] A.L. SMALL-HOWARD – M.J. BERRY, «Unique Features of Selenocysteine Incorporation», 1494.

[136] However, it has been demonstrated that only SPS2 is essential for selenocysteine synthesis, as has been shown in X.-M. XU – B.A. CARLSON – R. IRONS – H. MIX– *AL.*, «Selenophosphate Synthetase 2», X.-M. XU – B.A. CARLSON – ZHANG,Y. – H. MIX– *AL.*, «New Developments in Selenium Biochemistry» and X.-M. XU – B.A. CARLSON – H. MIX – Y. ZHANG– *AL.*, «Biosynthesis of Selenocysteine».

[137] Cf. F. DING – P.J. GRABOWSKI, «Identification of a Protein Component».

[138] Cf. X.-M. XU – H. MIX – B.A. CARLSON – P.J. GRABOWSKI– *AL.*, «Evidence for Direct Roles».

Thus, SECp43 appears to coordinate the activities of selenophosphate biosynthesis, seryl- to selenocysteil-tRNA conversion, and tRNA methylation. SECp43 also coimmunoprecipitates with EFsec and enhances the association between EFsec and SBP2 in vivo, linking the factors involved in the early steps of selenocysteine biosynthesis and tRNA charging to the later steps resulting in the cotranslational incorporation of selenocysteine into selenoproteins[139]

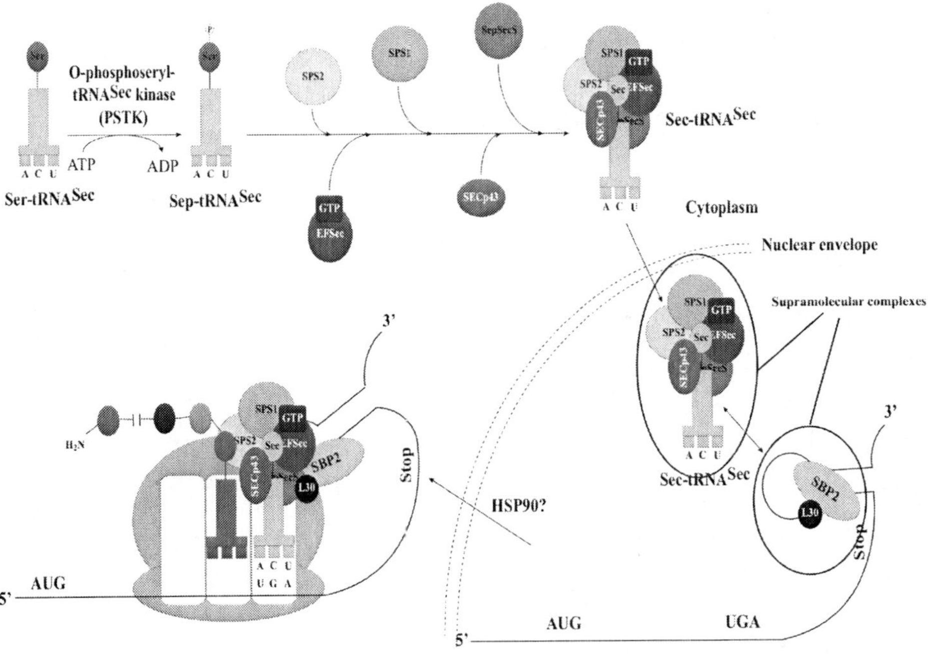

Figure VI.47: The SECIS element-SBP2 complex is formed in the nucleus. Sec-tRNASec enters the nucleus bound to many cofactors: SPS1, SPS2 and Sec synthetase (SepSecS), proteins that are required for the Sec-tRNASec synthesis. EFSec is the specialized transcription factor required for Sec insertion, and SECp43 and HSP90 have probably a shuttling activity between the nucleus and the cytoplasm. Adapted from F.P. Bellinger – A.V. Raman – M.A. Reeves – M.J. Berry– *al.*, «Regulation and Function of Selenoproteins».

As it could be seen in Figure VI.47, SECp43 binds Sec-tRNASec. EFSec binds SBP2 as well, and the entire complex enters the nucleus. In the nucleus,

[139] A. Small-Howard – N. Morozova – Z. Stoytcheva – E.P. Forry– *al.*, «Supramolecular Complexes», 2338.

EFSec may bind the SECIS element. Then the mRNA, with the complex bound to the SECIS element, exits the nucleus and goes to the cytoplasm to undergo protein synthesis.

So it can be seen that SECp43 is a protein that facilitates the assembly of the partners in order to have the first supramolecular complex, formed by Sec-tRNASec, SPS1, SPS2, SepSecS and EFSec. SECp43 also regulates the entering of this complex in the nucleus. Once the complex is in the nucleus, it can interact with the other supramolecular complex, which is formed by the SECIS element, SBP2 and (probably) the ribosomal protein L30. So SECp43 plays an important role in connecting all the partners that are necessary for the insertion of Sec, localizing their reactions in the same place.

Once the supramolecular association of the partners has been shown, two models proposed for the Sec insertion can be considered. There is no consensus regarding the mechanism of Sec insertion. Further data are needed to understand the precise steps that are performed to insert Sec during protein synthesis. If we take into account how the partners are bound and the two supramolecular complexes that are formed, two hypothetical models of interaction can be sketched.

3.3.2 First model

Based on the fact that SBP2 is found associated with ribosomes and that it cannot interact simultaneously with the ribosome and with the SECIS element, the first model proposes that SBP2 is bound to the ribosome as a previous step to Sec incorporation (Figure VI.48)[140].

In this model SBP2 is initially bound to the ribosome, and would interact with the SECIS element only if translation takes place. Then, the SECIS element would induce a change in SBP2 in order to allow Sec-tRNASec to enter the A-site[141]. It has been also suggested that SBP2 is bound to a ribosomal pool, making them competent for Sec insertion. This interaction would take place through the 28S rRNA, a RNA molecule present in ribosomes. Some ribosomes would be prepared beforehand with SBP2 in order to be able to perform a Sec insertion event. From that point of view, selenoprotein mRNAs, whose translation will be done by a ribosome that has not bound SBP2, will finish at

[140] Cf. S.A. KINZY – K. CABAN – P.R. COPELAND, «Characterization of the SECIS».

[141] Cf. K. CABAN – P.R. COPELAND, «Size Matters».

the Sec insertion codon. On the other hand, if SBP2 is previously bound to that ribosome, the Sec insertion will likely take place[142].

This model proposes a role for L30 consisting in displacing SBP2 from the SECIS element once the insertion event has finished. But further investigation has to be done since it has been found that SBP2 strongly binds SECIS, and it's difficult to explain how L30 can displace it[143]. Let us see the process according to this model (Figure VI.48 to VI.51).

- Step 1. SBP2 is previously bound to the ribosome and recruits the SECIS element.

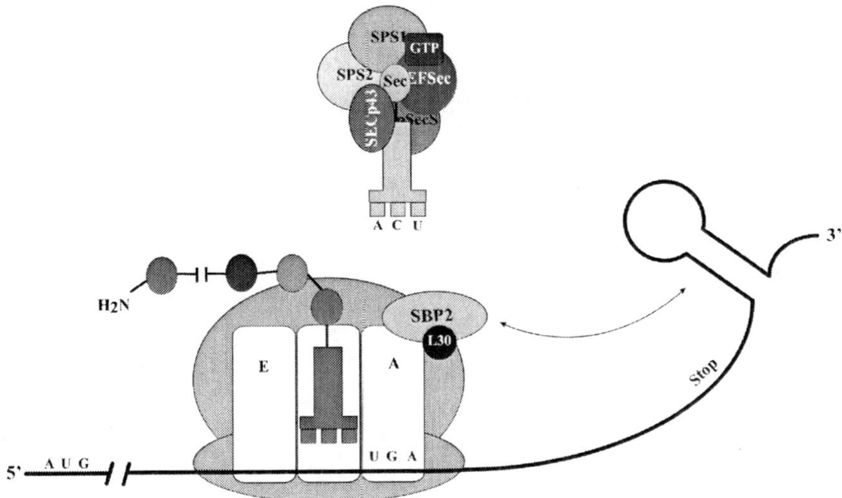

Figure VI.48: In the first model, SBP2 is bound to the ribosome, not to the SECIS element.

- Step 2. The Sec-tRNASec, with all its partners attached, interacts with the SECIS element and SBP2.

[142] Cf. P.R. Copeland – V.A. Stepanik – D.M. Driscoll, «Insight into Mammalian Selenocysteine Insertion».

[143] Competition studies about the bound strength between SECIS and SBP2 are performed in S.C. Low – E. Grundner-Culemann – J.W. Harney – M.J. Berry– *al.*, «SECIS-SBP2 Interactions».

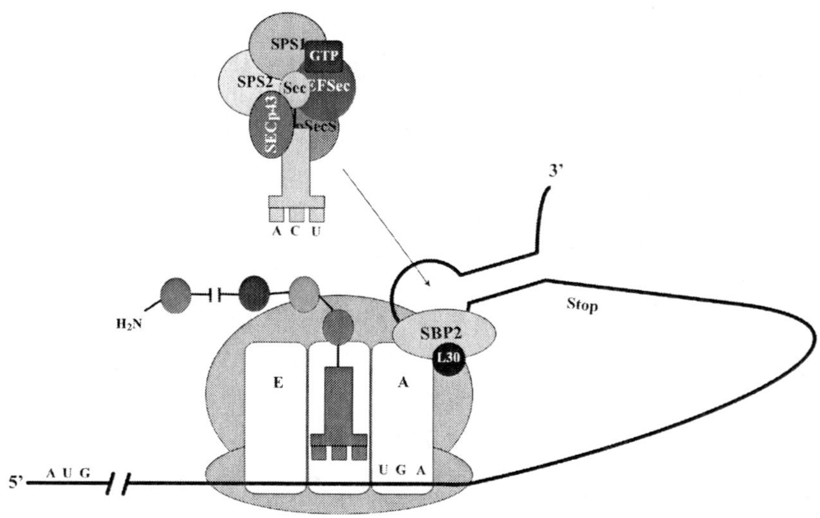

Figure VI.49: Only when the SECIS element binds SBP2 can Sec-tRNASec be recruited.

- Step 3. EFSec triggers its GTPase activity and Sec-tRNASec is inserted in the A-site.

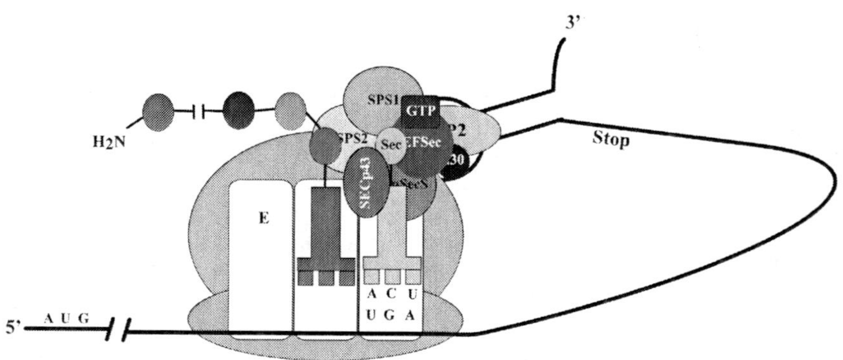

Figure VI.50: The recoding event takes place.

- Step 4. Once Sec-tRNASec has been inserted, the ribosomal protein L30 has to unbind SBP2 from the SECIS element in order to restart the system.

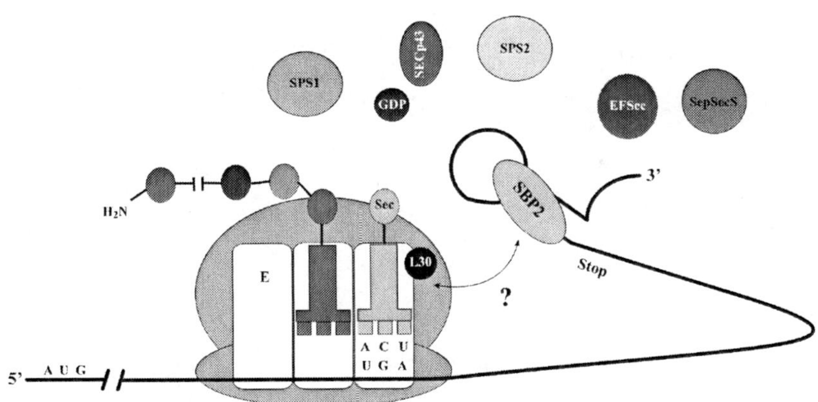

Figure VI.51: It is not totally understood how ribosomal protein L30 displaces SBP2 from the SECIS element, but something has to occur in order to reset SBP2 again in the ribosome.

3.3.3 Second model

In the second model, SBP2 is initially bound to the SECIS element (Figure VI.52). This interaction allows SBP2 to recruit the EFSec/Sec-tRNASec complex and deliver it into the A-site of the ribosome[144]. «An approaching ribosome will lead L30 to displace SBP2, the binding of L30 to the SECIS RNA inducing a more closed conformation of the SECIS K-turn. This movement triggers the release of the Sec-tRNASec and GTP hydrolysis»[145]. As in the first model, it seems that L30 is a key factor in resetting the system for another round of Sec insertion (if the mRNA has another UGA codon to be recoded into Sec).

There is evidence that the kink-turn of the SECIS element has two different conformations (open and kinked), depending on which partner it binds. The open form of the SECIS element can bind either SBP2 or L30; while the kinked

[144] For the dependent assembly of SBP2, EFSec and Sec-tRNA^{Sec}, see A.M. Zavacki – J.B. Mansell – M. Chung – B. Klimovitsky– *al.*, «Coupled tRNASec-dependent Assembly». It also occurs in bacteria, as it is reported in M. Thanbichler – A. Böck – R.S. Goody, «Kinetics of the Interaction».

[145] C. Allmang – A. Krol, «Selenoprotein Synthesis», 1568. In prokaryotes the GTPase activity of SelB depends on the SECIS binding and the presence of the ribosome, as is shown in A. Hüttenhofer – A. Böck, «Selenocysteine Inserting RNA Elements».

form binds L30 with great efficiency. In other words, the SECIS element acts as a molecular switch[146].

Let's present the second model step by step and see the difference with the former one (Figures VI.52 to VI.55).

- Step 1. In the first stage of the reaction, SBP2 is bound to the SECIS element. This assembly can recruit the Sec-tRNASec/EFSec complex.

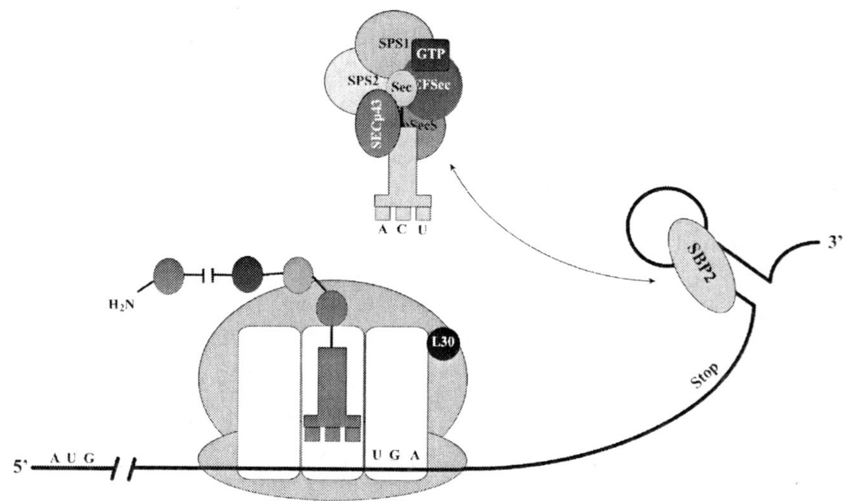

Figure VI.52: In the second model, SBP2 is initially bound to the SECIS element and recruits the Sec-tRNASec.

- Step 2. Once the SECIS element interacts with the ribosome through L30, SBP2 would be transiently displaced, in order to position the SECIS element near the A-site.

[146] Cf. D. FOURMY – E. GUITTET – S. YOSHIZAWA, «Structure of Prokaryotic SECIS» and L. CHAVATTE – B.A. BROWN II – D.M. DRISCOLL, «Ribosomal Protein L30».

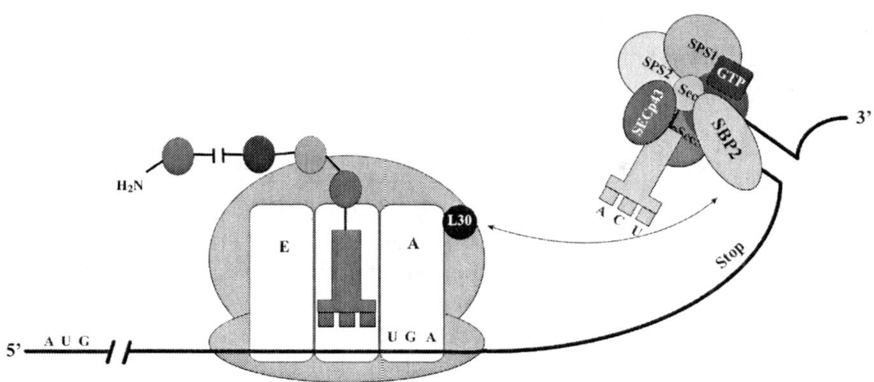

Figure VI.53: Only when Sec-tRNASec is bound with its partners to SECIS, can the complex interact with the ribosome.

- Step 3. L30 would induce the conformational change of the SECIS element (from the open to the kinked conformation). This change induces EFSec to hydrolize GTP and deliver Sec-tRNASec into the A-site. Then, we can see that the exchange between L30 and SBP2 permits the proper positioning of the SECIS element in order to insert the Sec-tRNASec at the right moment.

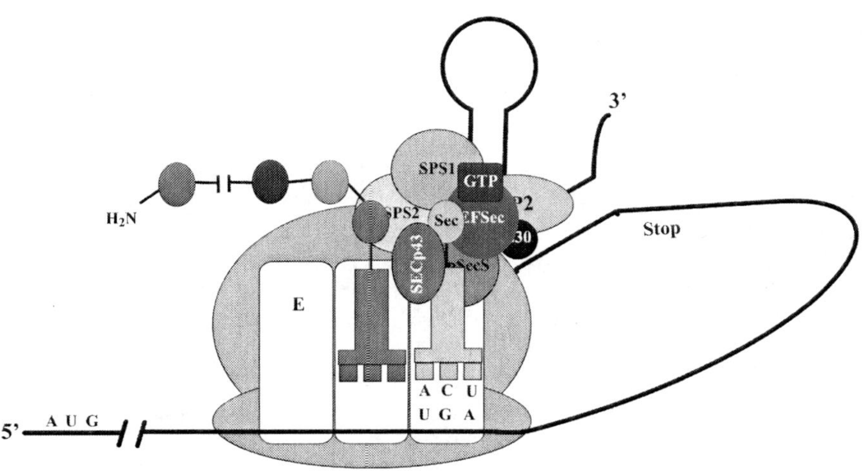

Figure VI.54: The SECIS element passes from an open conformation to a kinked, enhancing Sec insertion.

- Step 4. The SECIS element returns to its open conformation and still binds SBP2, ready to perform another recoding event.

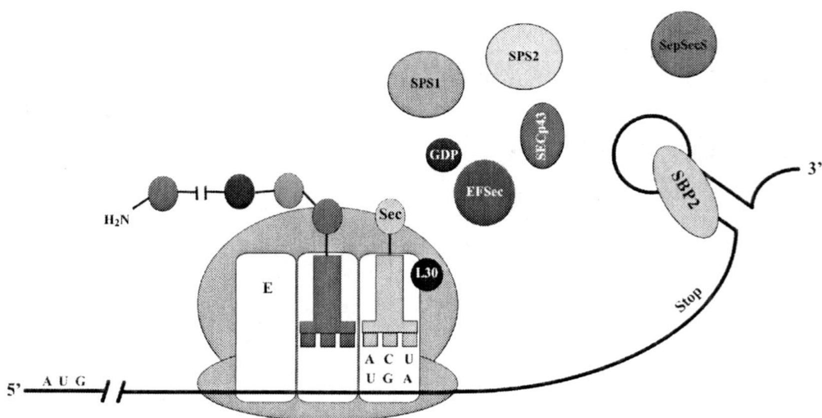

Figure VI.55: The entire complex is disassembled and SBP2 remains bound to the SECIS element, ready for another round of Sec insertion if necessary.

New data are accumulating in favour of this second model. It seems that SBP2 is not a limiting factor[147] in Sec insertion; and this matches with the fact that the proportion with the SECIS element would be 1:1. Furthermore, it is Sec-tRNASec that is the limiting factor when selenoprotein mRNA is abundant[148]. The low exchangeability of SBP2 once it binds the SECIS element responds also to an economical principle, as Hoffmann explains it:

> SBP2 may remain bound to the SECIS element and promote serial deliveries of Sec-tRNASec through multiple rounds of translation, thus promoting efficient synthesis

[147] The limiting factor in a chemical reaction is the factor whose relative quantity is the minor and, consequently, it dictates the number of reactions that can be done. In the example of Sec insertion, if we have one SBP2 for each SECIS element, all mRNAs are capable of performing a Sec insertion event. If we also assume that the ribosome is a factor that is ubiquitous in the cell, the limiting factor would be Sec-tRNA^{Sec}, since its quantity depends ultimately in Se availability.

[148] Cf. R.R. Jameson – A.M. Diamond, «A Regulatory Role for Sec tRNA[Ser]Sec».

of selenoproteins. In fact, this process is most likely required in the case of SelP, in which 10-17 selenocysteines are incorporated per polypeptide[149].

3.3.4 Competition between releasing factors and recoding machinery

Whatever the process may be, it is clear that when the ribosome arrives at the UGA codon that has to be recoded, a dynamical competition takes place between the releasing factors (proteins that release the polypeptide from the ribosome) and the recoding machinery[150]. The ribosome slows down its translation rate when it arrives in the surroundings of the UGA codon[151]. This pause could be due to the presence of a SRE and/or to the torsions that are induced in the mRNA because of the vicinity of the helix that forms the SECIS element. That pause allows a dynamical competition between the releasing factors and the Sec insertion[152]. So we can see here how the kinetics of the process and the timing of the UGA read-through are critical elements for the recoding event. Perhaps that is one of the reasons why a recoding signal should have a secondary structure, as has the SECIS element: it slows down the ribosome in order to allow the recoding machinery to play its part. Of course, other elements are needed to have such a recoding event (we have seen that many partners are needed to have a Sec insertion event), but «the timing of the secondary structure unfolding during translation may be a critical feature of competition between release factor 2 and tRNASec for decoding UGA»[153].

SelP is an example of the dynamical competition that takes place at its 10

[149] P.R. HOFFMANN – M.J. BERRY, «Selenoprotein Synthesis», 772.

[150] Cf. M.T. NASIM – S. JAENECKE – A. BELDUZ – H. KOLLMUS– *AL.*, «Eukaryotic Selenocysteine Incorporation».

[151] One way to see how translation is slowed consists in observing the polysome distribution. When the ribosomes accumulate at a certain mRNAs, it means that the translation process has been slowed. J.E. FLETCHER – P.R. COPELAND – D.M. DRISCOLL, «Polysome Distribution» and G.W. MARTIN III – M.J. BERRY, «Selenocysteine Codons Decrease Polysome Association».

[152] Cf. J.B. MANSELL – D. GUÉVREMONT – E.S. POOLE – W.P. TATE– *AL.*, «A Dynamic Competition» and E. GRUNDNER-CULEMANN – G.W. MARTIN III – R. TUJEBAJEVA – J.W. HARNEY– *AL.*, «Interplay Between Termination and Translation».

[153] J.B. MANSELL – D. GUÉVREMONT – E.S. POOLE – W.P. TATE– *AL.*, «A Dynamic Competition», 7284.

UGA codons. Since Sec insertion is an inefficient process, UGA recoding codons have to be surrounded by favourable conditions to perform that insertion instead of being terminated[154]. The base located immediately after the UGA codon is a very important parameter to determine whether there has to be Sec insertion or termination. In fact, Driscoll noticed that,

> Early studies on the role of termination in regulating Sec incorporation showed that the base following the UGA codon (the fourth base) is a critical determinant of termination efficiency, which inversely correlates with the efficiency of Sec incorporation. Termination efficiency corresponds to the hierarchy A > G > C > U, where UGAA terminates most efficiently and UGAU the least[155].

3.3.5 Knowing when to really stop

In eukaryotic selenoprotein mRNAs, the UGA codon is recoded from stop to Sec. One question that could emerge is: Then, how is the translation process terminated? If all the UGA codons of selenoproteins' mRNAs are recognized as something to be recoded, which is the true stop signal? In these cases, the open reading frame is terminated by either a UAA or a UAG stop codon. Nonetheless, there are three exceptions to this rule. In the case of mammalian selenoprotein W mRNA and *Schistosoma mansoni* glutathione peroxidase mRNA, the first UGA codon found in the ORF encodes selenocysteine, while the second UGA codon is a termination codon. In theses cases no alternative stop codon is needed, since the distance between the last UGA codon and the SECIS element is less than the minimum required, so it is physically impossible to the second UGA codon to be recoded into Sec. The third exception is the mammalian type 2 deiodinase mRNA. Its second UGA codon is seven codons from the truly stop codon (that in this case is UAA). The second UGA codon is recoded in order to incorporate Sec, but neither this residue nor the remaining seven amino acids are essential for the enzyme activity[156].

---

[154] Cf. S. Suppmann – B.C. Persson – A. Böck, «Dynamics and Efficiency».

[155] D.M. Driscoll – P.R. Copeland, «Mechanism and Regulation of Selenoprotein Synthesis», 29.

[156] Reviewed in P.R. Hoffmann – M.J. Berry, «Selenoprotein Synthesis».

4. Summing up

The presentation of a case study has been of great help to see that the informational paradigm developed in the previous chapters can be a framework to better understand the scientific results. The detailed study of the Sec insertion mechanism has revealed many points where that framework is a good interpreting instance.

Sec insertion in proteins during translation is something virtually impossible to be done according to genetic code rules. To perform such insertion, the cell acts as a whole, recruiting the suitable partners that can somehow influence the dynamics of protein synthesis to make the insertion of an amino acid that has no room in the genetic code feasible. The genetic code has established a relation between signs (the codons of the mRNA) and referents (the amino acids to which each codon stands for). There is a closed relation between codons, amino acids and the final protein that is synthesized at the end of the process.

The fact is that the cell needs to put into some proteins an amino acid that cannot have a sign in the genetic code, because all the possible signs already have their referent. The cell can perform this insertion because there is a change in the ascription of the referent to one of those signs of the genetic code. Indeed, the UGA codon that codifies for the signal "stop", has a new referent that would led it to stand not for "stop" but for "Sec".

The cell achieves this aim providing a new context for some UGA codons. All the partners involved in the Sec insertion event are used for surrounding the UGA codon with a new chemical context that would relate the sign carried by the UGA codon with the referent "Sec" instead of "stop". The elements that have been studied are proteins and RNA structures that can be considered as a higher level of organization. Those elements cannot exert a direct causality over the UGA codon to change what it stands for. UGA codon codifies for a "stop" signal because it chemically interacts with a releasing factor (a factor that stops translation). In order to avoid such termination, it is needed to provide a chemical interaction that can link the UGA codon to another referent. The whole network of elements involved is there to exert constraints over the UGA codon surroundings in order to chemically interact not with the releasing factor, but with the anti-codon of Sec-tRNASec. The interaction between the UGA codon and the Sec-tRNASec is the lower level into which the partners cannot exert their causality directly. That is why they provide the proper conditions for facilitating the chemical interaction of the lower level items that exert efficient causality (in this case the breaking and forming of chemical bonds).

It has been seen that there is no need to have a certain sequence of RNA or a certain protein sequence to have a Sec insertion event; what is important is to preserve a binding pattern between all the partners involved in order to link the different elements of the mechanism and then give a chance to Sec to be inserted.

The chemical characters of the partners involved provide the chemical conditions for efficient interaction with the protein synthesis machinery. Those chemical features exercise a bottom-up causality. On the other hand, the cell has selected those features that, in an organized manner, can be constrained in a top-down fashion to perform a reaction that they would have never perform individually or without a semiotic link between them. The cell has selected those interactions among the partners of the Sec insertion machinery that link the actions of the partners to the needs of the cell, particularly to the needs of selenoprotein's supply.

In the next chapter I shall explain the control mechanisms the cell uses to regulate all this elements according to the needs of the whole.

Chapter VII

Regulatory elements of Selenocystein insertion mechanism

1. Regulation is the key

In chapter VI the Selenocystein insertion mechanism at a molecular level was estudied. I have focused in describing the partners that take part in this process and how they interact with each other. In the present chapter I am going to study the regulation exerted over the Selenocystein insertion mechanism. I shall show that there are several points of regulation of this mechanism, and they are located at different stages: transcription, translation, activation depending on the availability of selenium, the redox status of the cell, the different affinity of the partners and the interaction with other proteins that play other roles in the cell. The cell does not have a sole control point for this mechanism, but many. The interaction of all these control systems allows the cell to have a tuned reaction depending on the interaction with the environment.

Hence, the function of a certain mechanism does not depend only on the partners that are involved in such mechanism. The other elements of the cell that regulate the availability and activity of those very partners are also very important. A molecular mechanism is inserted in the cell's whole metabolic web, and this is what allows to that mechanism to be linked to the functions and needs of the cell.

2. Hierarchical regulation of selenoproteins

«Two well-described and long-known hierarchical principles can be observed in higher mammals, i.e., not all tissues and not all selenoproteins are equally well supplied with the trace element when it becomes limiting»[1]. One of the interes-

[1] L. Schomburg – U. Schweizer, «Hierarchical Regulation of Selenoprotein Expression», 1453.

ting topics in the regulation of selenoproteins is their distinct response to the lack of dietary selenium. The low availability of selenium does not affect all tissues and all selenoproteins equally. There are some proteins whose mRNA expression decreases dramatically shortly after the selenium deficiency, while others are able to maintain the level of expression for a longer period of time[2]. The repletion of selenium also shows this hierarchical regulation. Not all selenoproteins begin to be synthesized at the same rate when selenium becomes available again. There are some selenoproteins that are synthesized first, depending on their function and the tissue where they are expressed[3].

There are many parameters that have to be taken into account to fully understand the hierarchical regulation of selenoproteins. One of these parameters is the stability of mRNA, which differs from protein to protein[4]. The stability of all mRNAs is subjected to the mechanism of nonsense-mediated decay (NMD), which consists in the degradation of those mRNAs that have a premature stop codon in its sequence (as selenoprotein mRNA has)[5]. However, other elements are thought to interact with mRNA in this type of regulation, specially the SECIS element and the UGA codon itself. In fact, the internal loop and the helix 2 of the SECIS element, which flank the kink-turn motif, are important for the efficiency of Sec insertion activity[6]. It has also been reported that the presence of an intron downstream of the UGA codon targets the mRNA for NMD when the Sec insertion process is inefficient[7]. The nature of the 3'UTR

---

[2] In K.E. Hill – P.R. Lyons – R.F. Burk, «Differential Regulation of Rat Liver» it could be seen that the mRNA levels of glutathione peroxidase decreases markedly than it does the levels of Selenoprotein P or type I iodothyronine 5'-deiodinase.

[3] Cf. X.G. Lei – J.K. Evenson – K.M. Thompson – R.A. Sunde– *al.*, «Glutathione Peroxidase and Phospholipid Hydroperoxide» and G. Bermano – F. Nicol – J.A. Dyer – R.A. Sunde– *al.*, «Selenoprotein Gene Expression».

[4] Cf. K. Wingler – M. Böcher – L. Flohé – H. Kollmus– *al.*, «mRNA Stability and Selenocysteine Insertion». In the case of SelW, selenium exerts a stabilization effect over mRNA, as is shown in Q.P. Gu – W. Ream – P.D. Whanger, «Selenoprotein W Gene Regulation».

[5] Cf. P.M. Moriarty – C.C. Reddy – L.E. Maquat, «Selenium Deficiency Reduces the Abundance».

[6] Cf. L. Latreche – O. Jean-Jean – D.M. Driscoll – L. Chavatte– *al.*, «Novel Structural Determinants».

[7] Cf. S.L. Weiss – R.A. Sunde, «Cis-acting Elements Are Required» and X. Sun –

is also a determinant character for the translational rate of selenoproteins when selenium is low[8].

Another important element is the selenium supply to the different tissues, depending on the blood concentration of Sel P and the differential uptake of these proteins by certain kind of tissues[9]. All the regulatory elements that are described in the successive sections play, in one way or other, a role in the decision of which proteins have to be expressed, where and under what conditions.

2.1 *SBP2 as a key controller*

One control step for regulating selenoprotein translation is the SBP2-SECIS interaction. Studies about the limiting elements of Sec insertion have been performed overexpressing the mRNA of selenoproteins. This overexpression yields a diminution of the Sec insertion process due to a stoichiometry problem (stoichimetry studies the proportions that are needed in order to deploy a reaction). In this situation the most important question is: what is the limiting factor? It has to be one of the partners that interact in the process. In the referred study it has been shown that the restoration of the efficiency of Sec insertion, depends on the availability of neither tRNASec nor EFSec. Only when SBP2 is overexpressed can the limitation produced by selenoprotein mRNA expression be overcome. That means that SBP2 plays an important regulatory role in selenoprotein synthesis[10]. As happens very often in biology, the limiting factor is the one that has the regulatory key of the process. A limiting factor, then, is the factor that is in minority in a reaction, making the reaction dependent on its availability. For example, if we want to synthesize water (H_2O) form oxygen (O_2) and hydrogen (H_2), we will synthesize a number of molecules of water depending on the quantity of the reactant that is lesser in quantity. If we have just one molecule of hydrogen and millions of oxygen, we would just synthesize

P.M. Moriarty – L.E. Maquat, «Nonsense-mediated Decay of Glutathione Peroxidase».

[8] Cf. G. Bermano – J.R. Arthur – J.E. Hesketh, «Role of the 3' Untranslated Region».

[9] Cf. L. Schomburg – U. Schweizer, «Hierarchical Regulation of Selenoprotein Expression».

[10] Cf. S.C. Low – E. Grundner-Culemann – J.W. Harney – M.J. Berry– *al.*, «SECIS-SBP2 Interactions».

one molecule of water, because the rest of molecules of oxygen will not have more hydrogen to react with. The lower reactant always imposes its rules.

Furthermore, competition essays have shown that SBP2 binds tightly to the SECIS element, and it does not exchange between different SECIS elements. So we can affirm that the stoichiometry of the binding between SBP2 and SECIS is 1:1. «These results predict that, under conditions in which SBP2 was limiting *in vivo*, differential affinity of the protein [SBP2] for different SECIS elements would result in the preferential synthesis of some selenoproteins over others»[11].

This sheds light about the competition between the ribosome and the SBP2-SECIS element. To prevent NMD, the SBP2-SECIS element complex has to arrive at the UGA codon before the ribosome, since it will terminate the translation if no other information arrives there in order to recode what in normal conditions will be a stop signal. So the different affinities of SBP2 for the different SECIS elements of the selenoprotein mRNAs serves as a regulatory step in the hierarchical regulation of their expression[12]. Depending on their affinities, the complex will arrive faster or slower at the UGA codon, preventing the reading of the "stop" codon or recoding it into Sec. The different affinities are seen as different signs that allow the recoding event to take place more or less efficiently.

However there is a research group that have found that there is little or no difference in the binding affinities of SBP2 for the SECIS element, so it hardly can be the only mechanism to explain the hierarchical expression of selenoproteins[13]. In any case everybody agrees that other elements have to be involved in this regulatory process.

---

[11] S.C. LOW – E. GRUNDNER-CULEMANN – J.W. HARNEY – M.J. BERRY– *AL.*, «SECIS-SBP2 Interactions», 6883.

[12] Cf. S.C. LOW – E. GRUNDNER-CULEMANN – J.W. HARNEY – M.J. BERRY– *AL.*, «SECIS-SBP2 Interactions» and J.E. SQUIRES – I. STOYTCHEV – E.P. FORRY – M.J. BERRY– *AL.*, «SBP2 Binding Affinity». In J.L. BUBENIK – D.M. DRISCOLL, «Altered RNA Binding Activity» it is shown that a mutation in SBP2 alter the pattern of expression of a subset of selenoproteins important for thyroid hormone metabolism.

[13] Cf. J.E. FLETCHER – P.R. COPELAND – D.M. DRISCOLL – A. KROL– *AL.*, «The Selenocysteine Incorporation Machinery».

2.2 *Who controls the controller?*

In the previous section it has been seen that SBP2 exerts an important role in deciding which selenoproteins have to be synthesized first in case of selenium scarcity. Here it will be showed that SBP2 activity is also regulated. Its expression varies widely between different tissues and cell types[14]. However, it has been seen that there is no correlation between mRNA and SBP2 activity; that means that mRNA levels are fairly constant in different tissues, while its activity is highly regulated and is not the same in all tissues. That means that the control point is at the translational process.

Highly conserved regions have been found in the 3'UTR of SBP2 mRNA. In the proximal region of 3'UTR a binding site for CUG-BP1 (one of the proteins that binds GU rich sequences) has been identified. In the distal region there is a long sequence where multiple proteins can be attached (Figure VII.1). One of these proteins is HuR. Both CUG-BP1 and HuR are members of the Turnover and Translation Regulatory RNA-binding protein family (TTR-RBP). These kind of proteins, bound to mRNA, can alter its stability, regulate the alternative splicing, regulate the translational process and they can also shuttle the mRNA between the nucleus and the cytoplasm[15].

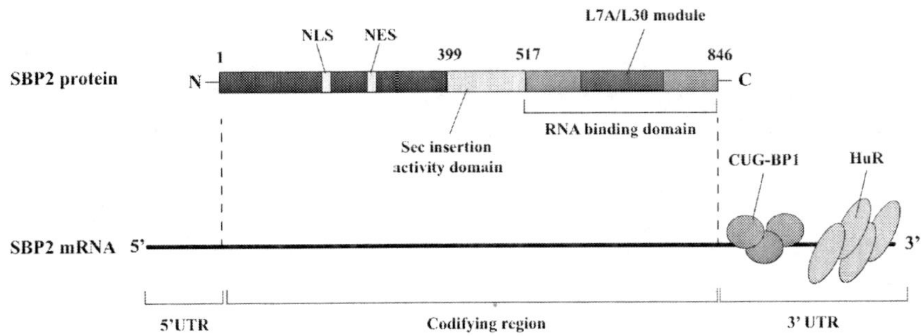

Figure VII.1: CUG-BP1 and HuR are proteins that bind the 3'UTR of SBP2 mRNA. These proteins regulate the stability and rate of translation of the mRNA.

[14] Cf. P.R. COPELAND – D.M. DRISCOLL, «Purification, Redox Sensitivity, and RNA» and P.R. COPELAND – J.E. FLETCHER – B.A. CARLSON – D.L. HATFIELD– *AL.*, «A Novel RNA Binding Protein».

[15] Cf. J. BUBENIK – A. LADD – C.A. GERBER – M. BUDIMAN– *AL.*, «Known Turnover and Translation».

Interestingly, it can be seen here that the 3'UTR has an important regulatory role. Despite being a relative extensive region of eukaryotic mRNAs, and although this could be seen as something absurd from the informational point of view, it is not at all. In fact, if no codifying information is contained in the 3'UTR, there is other type of information that is no less important. It contains the binding sites for proteins that could regulate the translation rate of the mRNA, a feature that should not be overlooked. Furthermore, there is a correlation between the 3'UTR length and the complexity of the organism[16]. The more complex an organism is, the longer the 3'UTR region of its mRNA will be. Then, a longer 3'UTR region allows more proteins to bind to it and, therefore, a tighter control can be exerted on the translation rate of that mRNA. More control proteins means more modularity and a more tuned control over the process. Again it is shown that modularity is important in order to exert tighter control over the partners to be regulated.

Despite the interesting role of SBP2 in selenoprotein synthesis and the control mechanisms through which it is controlled, there is a problem that remains unsolved. «Neither the activity nor the expression level of SBP2 is modulated by selenium»[17]. So presumably other factors have to link the regulatory control that SBP2 exerts over the selenoprotein mRNAs and the selenium availability. One of these factors is the eukaryotic initiation factor 4a3 (eIF4a3). This initiation factor binds to a subset of SECIS elements, limiting the accessibility of SBP2 and, consequently, preventing Sec insertion. eIF4a3 has a major affinity for SECIS elements with large apical loops, while it has low affinity for those with small apical loops. Interestingly, PHGPx and TR1 SECIS elements have small apical loops, so eIF4a3 does not bind them (Figure VII.2). These two selenoproteins are always privileged in the synthesis, as they are important housekeeping proteins[18].

---

[16] Cf. F. Mignone – C. Gissi – S. Liuni – G. Pesole– *al.*, «Untranslated Regions of mRNAs» and B. Mazumder – V. Seshadri – P.L. Fox, «Translational Control by the 3'-UTR».

[17] M.E. Budiman – J.L. Bubenik – A.C. Miniard – L.M. Middleton– *al.*, «Eukaryotic Initiation Factor 4a3», 480.

[18] Cf. M.E. Budiman – J.L. Bubenik – A.C. Miniard – L.M. Middleton– *al.*, «Eukaryotic Initiation Factor 4a3».

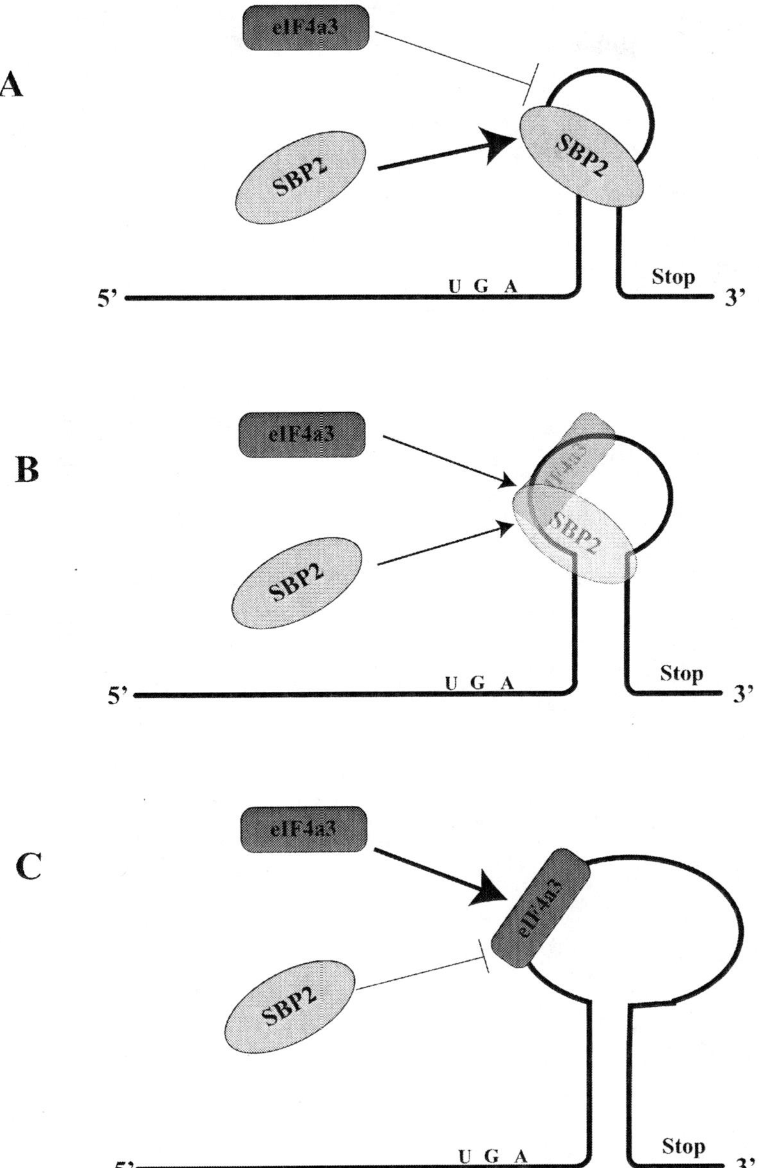

Figure VII.2: Different affinities of eIF4a3 and SBP2 to different SECIS elements, depending on the SECIS's size. A) A SECIS element of low size binds preferently SBP2. B) There is a competition of eIF4a3 and SBP2 when the SECIS element has a medium size. C) When SECIS has a big size, only eIF4a3 binds it, blocking the possibility of having a Sec insertion event.

The different affinities among those partners are used to establish an order in the translation of different selenoprotein mRNA. The whole pattern of different affinities and bindings can be seen as an icon, as a way the cell has to represent the different mRNAs at stake to itself and the priority some of them have to be translated first.

The link between the availability of selenium and the selenoproteins that have to be synthesized according to this availability takes place at tRNASec level. As it will be shown below, the two tRNASec isoacceptors are also an important regulatory point of selenoprotein synthesis.

2.3 *The role of oxidative stress*

Papp *et al.* suggest another way to regulate SBP2 activity. This regulation point depends on the oxidative stress and affects the intracellular location of SBP2[19]. Indeed, it has been shown that SBP2 contains a nuclear localization signal (NLS) and a nuclear export signal (NES) (Figure VII.1). So its relative amount in the nucleus and the cytoplasm depends on the balance between the nuclear import and export traffic, which takes place through the CRM1 pathway[20].

SBP2 is found predominantly in the cytoplasm, bound to the ribosomes. In normal conditions, the nuclear export signal is recognized by CRM1 and SBP2 is mostly exported to the cytoplasm (Figure VII.3).

When the cell undergoes oxidative stress (the term "oxidative stress" refers to an increase of molecules with high oxidative capacity, which is a threat to the proper functions of many cellular proteins and structures), some amino acids of SBP2 form disulfide bonds and glutation-mixed disulfides, yielding a change of conformation of SBP2, which does not allow it to bind SECIS. Moreover, this oxidized form, makes the nuclear import signal more recognizable. Then, SBP2 enters the nucleus and no Sec insertion takes place (Figure VII.4).

---

[19] Cf. L.V. Papp – J. Lu – F. Striebel – D. Kennedy– *al.*, «The Redox State».

[20] This pathway is one of the shuttling systems between the nucleus and the cytoplasm, and is described in M. Fornerod – M. Ohno – M. Yoshida – I. Mattaj– *al.*, «CRM1 Is an Export Receptor».

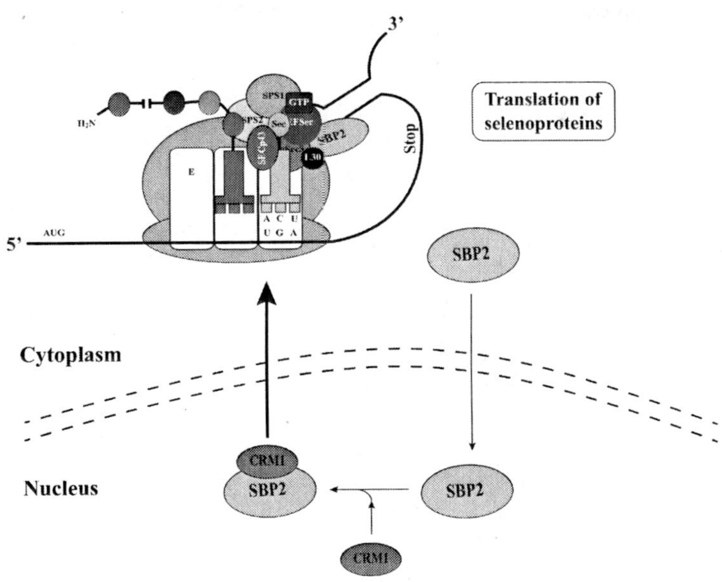

Figure VII.3: SBP2 traffic between the nucleus and the cytoplasm in normal conditions. SBP2 is mostly exported through the CRM1 pathway.

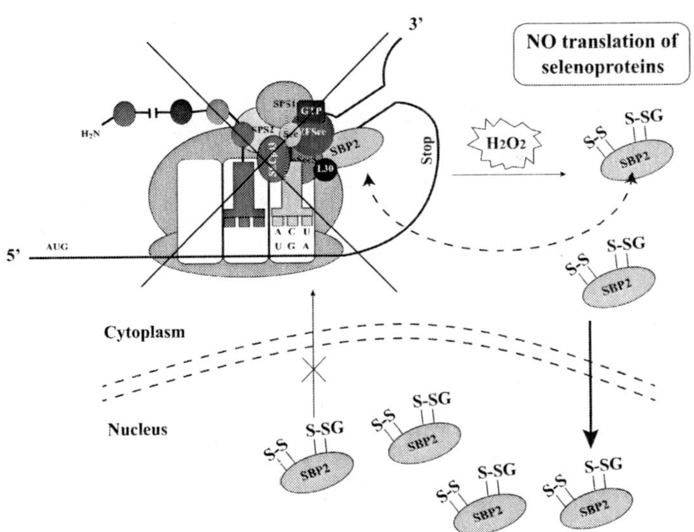

Figure VII.4: During oxidative stress, cysteine residues of SBP2 become oxidized, forming disulfide bonds (S-S) and glutation-mixed disulfides (S-SG). This disrupts the binding with the ribosome and exposes SBP2's nuclear import signal, triggering its transport to the nucleus.

Only when the cell recovers from oxidative stress can SBP2 be reduced by thioredoxin (Trx) and glutaredoxin (Grx) systems. Then, SBP2 recovers its former conformation, the nuclear export signal is available again to be recognized by CRM1 and SBP2 can exit the nucleus again (Figure VII.5).

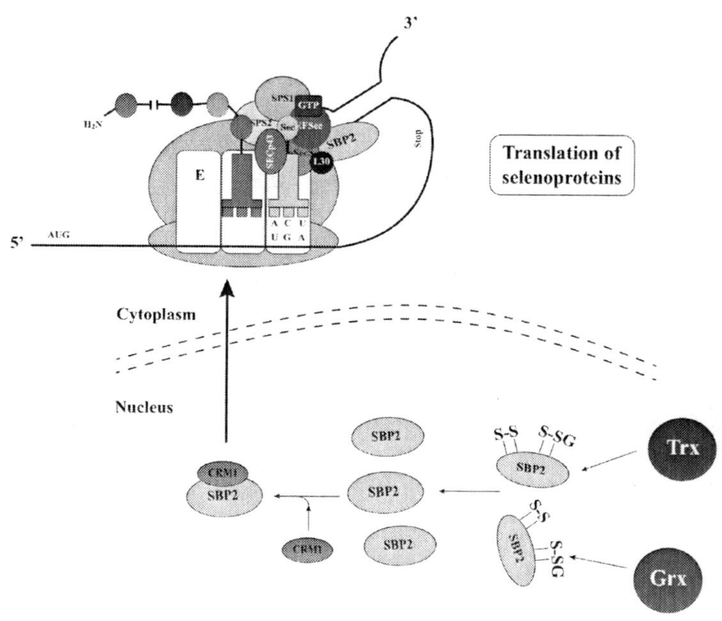

Figure VII.5: When the normal redox conditions of the cell are recovered, SBP2 is newly reduced by thioredoxin (Trx) and glutaredoxin (Grx), exposing again its nuclear export signal and shuttled back to the cytoplasm.

This regulatory role of SBP2 strongly depends on its cellular localization, so it matches better with the first model of Sec insertion that it has been previously discussed (chapter VI, section 3.3.2). In fact, these data fit in with the fact that SBP2 binds the ribosome initially rather than the SECIS element. However, it could be that an SBP2-SECIS interaction takes place when it has to be a means of transport from the nucleus to the cytoplasm, and then the SBP2 could bind the ribosomes[21]. This could be the reason why there are two possible models of

[21] Cf. L.V. PAPP – J. LU – F. STRIEBEL – D. KENNEDY– *AL.*, «The Redox State».

Sec insertion. SBP2 would definitely bind the SECIS element but, is this the first step for the Sec insertion mechanism or just a way to transport SBP2 from the nucleus to the cytoplasm and then bind the ribosomes? Further investigation is needed to clarify this point[22].

It seems paradoxical that the oxidative stress inhibits the Sec insertion mechanism, since there are many selenoproteins involved in redox reactions. It is proposed that «since Sec is an extremely redox-sensitive amino acid, it seems feasible for its translation to be restrained by oxidizing conditions»[23]; otherwise, a high oxidizing environment will inactivate its catalytic function in the core of selenoproteins.

2.4 *Transcriptional regulation.*

The eukaryotic gene of tRNASec (*Trsp*) is subjected to transcriptional regulation in eukaryotes, depending on the tissues where it is required[24]. The expression of this gene depends on the transcriptional factor STAF (Sec tRNA gene transcription Activating Factor)[25]. This protein regulates selenoprotein synthesis by both enhancing *Trsp* transcription in a tissue dependent manner and controlling tRNASec modifications[26].

Several regulatory elements have been identified for the proper transcriptional regulation of *Trsp*. It would be expected that the transcription of *Trsp*, as the other tRNA genes, would take place through RNA polymerase III[27]; but

[22] «A question that remains unanswered, however, is the role of nuclear SBP2. It was previously suggested that SBP2 may enter the nucleus to associate with selenoprotein-encoding mRNAs and facilitate their transport to the cytoplasm for translation». L.V. Papp – J. Lu – F. Striebel – D. Kennedy– *al.*, «The Redox State», 4907.

[23] Cf. L.V. Papp – J. Lu – F. Striebel – D. Kennedy– *al.*, «The Redox State», 4907.

[24] Cf. For a structural analysis of $tRNA^{Sec}$ gene, see M.R. Bösl – M.F. Seldin – S. Nishimura – M. Taketo– *al.*, «Cloning, Structural Analysis and Mapping».

[25] For a genetical characterization of the *Staf* gene, see K. Adachi – M. Katsuyama – S. Song – T. Oka– *al.*, «Genomic Organization, Chromosomal Mapping». The molecular cloning of the gene and its full sequence can be found in K. Adachi – H. Saito – T. Tanaka – T. Oka– *al.*, «Molecular Cloning and Characterization».

[26] Cf. B.A. Carlson – U. Schweizer – C. Perella – R.K. Shrimali– *al.*, «The Selenocysteine tRNA».

[27] In eukaryotic cells, each class of RNA is transcribed by one type of RNA polymerase:

there are some differences between the tRNASec gene and the other tRNAs transcription.

It was thought that the class III genes were those transcribed by RNA polymerase III because they would have a specific type of regulatory sequence to be recognized by this polymerase. The same would take place with RNA polymerase II. The first characterization of genes transcribed by RNA polymerase III showed that these types of genes have intragenic regulatory elements; that is, the sequences that regulate the gene transcription are located downstream from the start point of transcription and, obviously, these elements also carry the sequential information to be translated into amino acids. Further on it was found that some genes transcribed by RNA polymerase III have their regulatory elements entirely upstream of the starting point (as is the case of vertebrate U6 and human 7SK genes). It was intriguing to discover that these kinds of genes have a type of promoter like the class II gene have (those transcribed by RNA polymerase II). But this is not the end of the story. It was also found that there is also a type of genes transcribed by RNA polymerase III that combine both intragenic and extragenic regulatory elements. *Trsp* is an example of these genes that have combinatorial promoter elements[28] (Figure VII.6).

tRNA genes have two intragenic regulatory sequences: box A and box B[29]. But in the case of *Trsp*, transcription is regulated by an internal B box and three upstream elements: a TATA motif located at position -30, a proximal sequence element (PSE) at position -70 and a distal sequence element (DSE) at -200. These elements are also found in the U6 snRNA genes[30]. The DSE regulates the expression of *Trsp* in a tissue dependent manner[31].

RNA polymerase I transcribes rRNA, RNA polymerase II transcribes mRNA and RNA polymerase III transcribes tRNA and other small RNAs. B. Lewin, *Genes IX*. 611.

[28] Cf. O.S. Gabrielsen – A. Sentenac, «RNA Polymerase III» and I.M. Willis, «RNA Polymerase III».

[29] Cf. G. Galli – H. Hofstetter – M.L. Birnstiel, «Two Conserved Sequence Blocks».

[30] Cf. P. Carbon – A. Krol, «Transcription of the Xenopus laevis». It has also been demonstrated that the B box is not essential. In fact, the three upstream regulatory elements are enough to engage the transcription, as is reported in J.M. Park – I.S. Choi – S.G. Kang – J.Y. Lee– *al.*, «Upstream Promoter Elements».

[31] Cf. V.P. Kelly – T. Suzuki – O. Nakajima – T. Arai– *al.*, «The Distal Sequence Element».

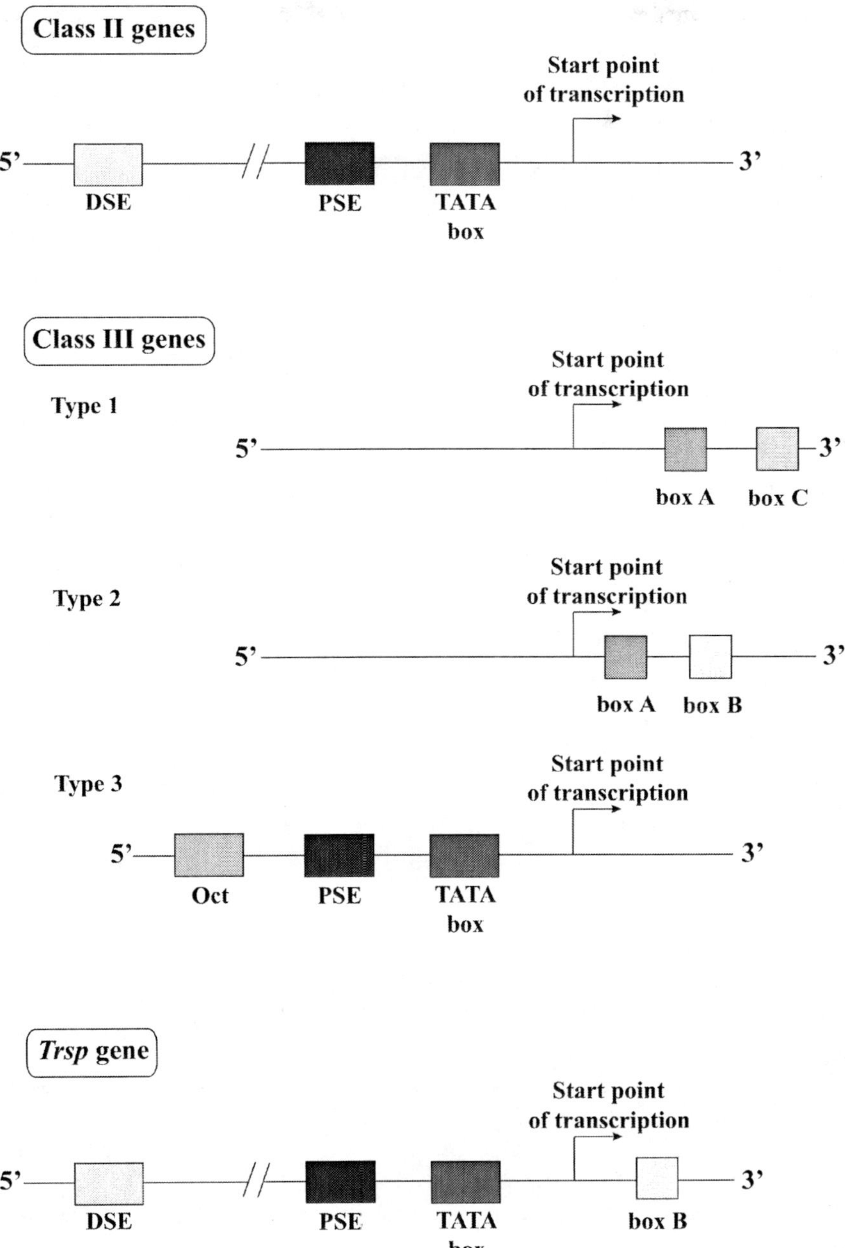

Figure VII.6: Regulatory regions of class II and III genes in eukaryotes. Note that the regulatory region of *Trsp* gene is a combination of the regulatory system used by class II genes and the type 2 of the class III genes.

It has been shown that many factors are needed to induce the transcription of *Trsp*: a TATA binding protein (TBP), a PSE binding protein (PBP)[32] and STAF (the factor that binds DSE)[33]. This is what makes the regulation of *Trsp* dramatically different in comparision with other tRNA genes, whose transcription relies on the intragenic A and B boxes.

The STAF protein contains ten zinc finger domains. It has been seen that STAF is a translation factor that acts not only in *Trsp* transcription, but also in the transcription of U6 gene. The requirements of the zinc finger interactions are different in each case, indicating that there is flexibility in the disposition of the zinc finger domains in order to recognize different promoter elements[34].

The diverse disposition of the promoter elements in the different family of genes shows us the modular elements that the genetic system uses in order to have the necessary availability of combinations to control genetic expression properly (see chapter V, section 5.2.2). The cell exchanges the order of the different elements of the gene in order to control them differently, according to the needs and the functions of those genes. It would be unaffordable to have a different mechanism of control for every single gene. The cell uses some elements and orders them in different dispositions to have a variety of control elements with few basilar items. All along the history of evolution, life has been testing different combinations of regulatory elements to respond to different situations, changing the way of controlling gene expression. When the same elements are combined in a different way, the result of the expression exerted over a gene is

---

[32] Cf. W. Meissner – I. Wanandi – P. Carbon – A. Krol– *al.*, «Transcription Factors Required». It has also seen that TBP interacts with the TATA sequence in the promoter of *Trsp*, so there is a great difference with the TATA-less promoters of the canonical tRNA genes; it is reported in E. Myslinski – C. Schuster – J. Huet – A. Sentenac– *al.*, «Point Mutations 5' to the tRNA». However, it has also been seen that TBP plays a role in TATA-less class III promoters, although the interaction is done through different surfaces from those that interact with TATA box, as it is shown in J.M. Park – J.Y. Lee – D.L. Hatfield – B.J. Lee– *al.*, «Differential Mode of TBP Utilization».

[33] Cf. C. Schuster – E. Myslinski – A. Krol – P. Carbon– *al.*, «Staf, a Novel Zinc Finger Protein».

[34] Cf. M. Schaub – A. Krol – P. Carbon, «Flexible Zinc Finger Requirement». For a detailed structural analysis of the STAF-DNA interactions, see M. Schaub – A. Krol – P. Carbon, «Structural Organization of Staf-DNA Complexes».

also different. The formal distribution of those regulatory sequences leads to different results in the expression of the genes they control.

2.5 *Translational regulation*

Let us see an example of regulation at the translational level in prokaryotes. It is very common in prokaryotes to have genes that are used in the same function to be in the same cluster. A cluster of genes is a set of genes that are transcribed in the same mRNA molecule. That means that all these genes are transcribed by the same promoter (the non-coding region of the gene that regulates its expression). *Escherichia coli,* for example, has selA and selB genes codified by the same mRNA molecule, but selC and selD, are not in the same cluster[35].

An interesting control mechanism has been found in the translation of the selA and selB genes of *E. coli.* As in other members of the prokarya kingdom, their selenopreotein mRNAs contain the SECIS element in the open reading frame, permitting the interaction with SelB in order to insert Sec-tRNASec at the UGA codon. But the intriguing question is that «a SECIS-like structure was identified in the 5' non-translated region of the selAB transcript, encoding selenocysteine synthase and SelB»[36].

This SECIS-like element forms also a quaternary complex with GTP, SelB and Sec-tRNASec. Nonetheless, the role of this complex is not that of Sec insertion but the translational regulation of selA and selB genes. In fact, when the complex binds the SECIS-like element, inhibition of the selA and selB mRNA translation takes place.

As can be seen in Figure VII.7, this is a very finely-tuned system that allows Sec-tRNASec to be synthesized only when selenium is available and when Sec-tRNASec is required for the synthesis of selenoproteins. If selenium is available and there is transcription of selenoprotein genes, Sec-tRNASec binds the SECIS element of the mRNAs, allowing for Sec insertion. In this situation, the mRNA of selA and selB is translated normally, without restriction.

When there are no selenoproteins mRNAs, the complex formed by Sec-

[35] Cf. G. SAWERS – J. HEIDER – E. ZEHELEIN – A. BÖCK– *AL.*, «Expression and Operon Structure».

[36] M. THANBICHLER – A. BÖCK, «The Function of SECIS RNA», 6925.

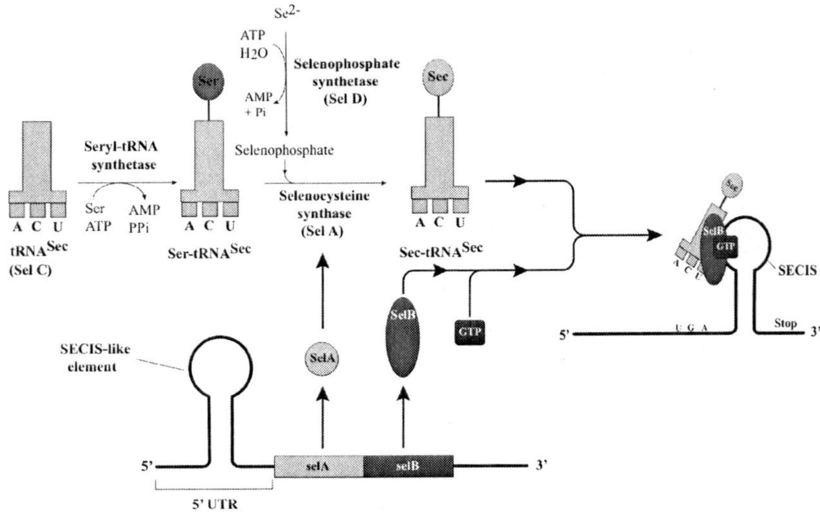

Figure VII.7: When there are selenoprotein mRNAs, Sec-tRNASec binds its SECIS elements in order to insert Sec. Then, Sec-tRNASec does not bind the SECIS-like element, allowing the normal transcription of selA and selB genes that are necessary for Sec insertion.

tRNASec, SelB and GTP, binds the SECIS-like element located at 5'UTR of the selA and selB mRNA. This binding diminishes the transcription of this mRNA sharply, repressing the expression of selA and selB (Figure VII.8).

The reason for this regulatory mechanism could be the high reactivity of Sec. A great concentration of Sec-tRNASec could enable the spontaneous hydrolysis of Sec, allowing its selenol group to react with other cell proteins. So it is important to avoid the cellular accumulation of neither Sec-tRNASec nor Sec. That is why this system is regulated in that way. When selenium is available and is needed for Sec insertion during the translation of selenoproteins, selA and selB are also translated. But when no selenoprotein synthesis is required, there cannot be accumulation of Sec-tRNASec. In that case, the selenium metabolism has to be stopped at selenophosphate and cannot go further. This negative regulation is possible since the translation of selA and selB can be repressed when no Sec-tRNASec is needed for selenoprotein synthesis. The dynamical competition between the SECIS elements and the SECIS-like element for the binding of Sec-tRNASec is the key that allows such a finely-tuned mechanism.

It is also interesting to note that the SECIS-like element located at 5'UTR

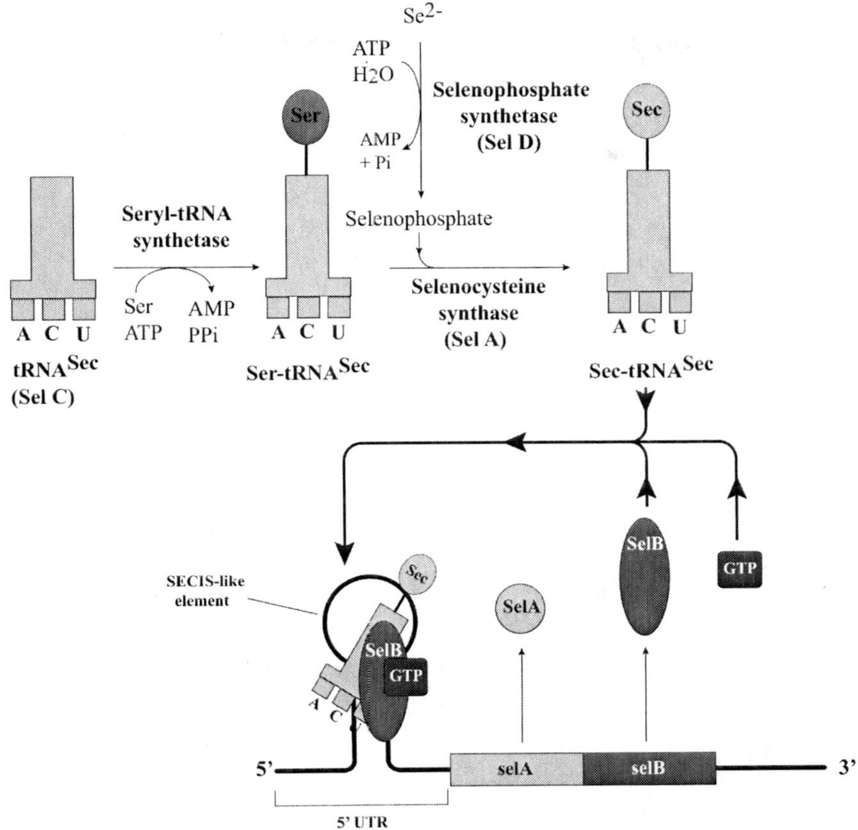

Figure VII.8: When there are no selenoprotein mRNA to bind, the Sec-tRNASec binds the SECIS-like element, preventing the mRNA translation. The levels of SelA and SelB decrease sharply.

can also perform a Sec insertion when it is inserted downstream from the UGA codon, even though with less efficiency[37]. Then, the difference is not due to the chemical nature of the partners but in the way they are linked to the whole process.

But the intriguing question here is the fact that this kind of regulation is done with the same elements that direct Sec insertion at 3'UTR. It has to be said that the sequence of the SECIS-like element is distantly related to the "true"

[37] Cf. M. Thanbichler – A. Böck, «The Function of SECIS RNA».

SECIS elements; but the point is that all its binding properties are preserved. The same quaternary complex is formed in the SECIS element located in the 3'UTR and in the SECIS-like element of the 5' end.

Despite the fact that the binding properties remain, there is a conformational rearrangement that explains the difference between the two elements. In fact, the interaction between the IVb domain of SBP2 and the SECIS-like element is different from the tight interaction that binds this domain with a "true" SECIS-element. Thanbichler explains,

> It was speculated that there exist multiple possibilities to fit RNA molecules into a binding site at the protein, although these molecules do not necessarily include all the features needed for biological activity. The SECIS-like element is a natural example for such a system: the overall binding properties are maintained, but the steric arrangement of determinants that are necessary to trigger the conformational change enabling SelB to interact with the ribosome is different, so that the ability to decode a UGA codon is drastically impaired[38].

I think that this is a very good example that shows the usefulness of the informational paradigm I have introduced in the first part of this work. The structure composed of the binding of the SECIS, the Sec-tRNASec, SelB and GTP is the same both at the 3' UTR of selenoprotein mRNAs and at the 5'UTR of the mRNA that codifies for SelA and SelB. The elements that are bound are the same in both cases, but they are used differently. Indeed, the binding to the SECIS element of selenoprotein mRNA is a sign of a referent that stands for selenium availability and then for the convenience to synthesize selenoproteins. On the contrary, when selenium is scarce and no selenoproteins can be synthesized, the same partners are bound to the SECIS-like element located in the 5'UTR of SelA and SelB mRNAs. But then, this binding pattern stands now for a different referent: now it is a sign of the absence of selenium and so it is used to block the translation of proteins (SelA and SelB) that are involved in Selenocysteine insertion.

The sign is the same in both cases, but depending on the context (presence

---

[38] M. THANBICHLER – A. BÖCK, «The Function of SECIS RNA», 6932.

or absence of selenium), it is used differently. That means that if the cell is able to use the same sign in two different contexts to have two different answers, it is capable of dealing with them semiotically, not being absolutely determined by the chemistry of the partners that are at play, but by the way the cell uses them in order to achieve its vital aims.

In figures VII.7 and VII.8, it can be considered that the elements that enter in this regulatory system are a higher level of organization. The partners that work in this regulatory system are linked in such a way that allows canalizing their chemical properties in order to achieve different results depending on the needs of the cell at each moment. Then, the whole network of items exerts a top-down causation over the individual items (the lower level instance) in order to allow them to interact in such a way that it would be for the sake of the cell at that precise situation.

Besides this particular way of translational regulation, there are also other ways to regulate the amounts of selenoprotein mRNAs. In the case of the human thioredoxin reductase, there is an AU-rich sequence in the 3'UTR that down-regulates the levels of mRNA. In general, the AU-rich regions are instability elements that facilitate the deadenylation of the poly-A, a string of A nucleotides added at the end of each mRNA. This deadenylation provokes a high turnover of the mRNA and, then, the level of expression could be tightly controlled[39]. This kind of regulation is also found in the mRNAs of growing factors and proto-oncogenes, whose expression has to be regulated strictly.

2.6 *Isoacceptors*

Another element to consider is the role that dietary selenium plays in the proportion of the two isoacceptors of tRNASec. Two tRNAs are isoacceptors if they bind the same amino acid. As it has been said, there are four post-transcriptional modifications that have to be done to tRNASec in order to be well folded and so to be functionally available. It is also known that these modifications have to occur in a certain order, the methylation at the wobble position is the last to be done (see chapter VI, section 3.2.1.d). Dietary selenium enhances this last step of the maturation pathway of tRNASec, increasing the amount of the

[39] Cf. J.R. Gasdaska – J.W. Harney – P.Y. Gasdaska – G. Powis– *al.*, «Regulation of Human Thioredoxin Reductase».

methylated form[40]. Some proteins depend on the methylated form of tRNASec, but others do not. This fact shows that the methylated isoform, which depends on the availability of selenium, exerts a regulatory role in the translation of certain selenoproteins[41].

Some selenoproteins are preferentially expressed with the methylated form of tRNASec, while others are preferentially expressed with the unmethylated one. In fact, many of the housekeeping selenoproteins use the unmethylated form[42]. This is one of the mechanisms that regulates the hierarchical expression of selenoproteins when selenium is scarce or when it becomes available again. The proteins that have to be synthesized and, therefore, have to use the little selenium that is available, use the unmethylated form. The selenoproteins that need the methylated form are not expressed in these conditions. When selenium becomes available, the methylated form begins to increase, and then proteins that depend on this isoacceptor can be synthesized.

Let us give an example of the importance of these isoacceptors as key regulators in selenoprotein synthesis. In selenium-deficient cells the expression of glutathione peroxidase (GPx1) is downregulated. But the synthesis of PHGPx and TR1 remains unaffected[43]. These latter selenoproteins use the unmethylated form of tRNASec, so their synthesis does not depend very much on selenium availability. As we have seen above, they also have a small apical loop in their SECIS element, so no inhibition of eIF4a3 takes place. We can see that the complex pattern of selenoprotein synthesis is not regulated by a single mechanism; but here we have an example of how one of the mechanisms reinforces the other. Two different regulatory elements are used to better tune the mechanism of synthesis according to Se availability.

Selenoprotein mRNAs that use the unmethylated form do not bind eI-

[40] Cf. A.M. Diamond – I.S. Choi – P.F. Crain – T. Hashizume– *al.*, «Dietary Selenium Affects Methylation».

[41] Cf. B.A. Carlson – X.-M. Xu – V.N. Gladyshev – D.L. Hatfield– *al.*, «Selective Rescue of Selenoprotein Expression».

[42] Cf. B.A. Carlson – X.-M. Xu – V.N. Gladyshev – D.L. Hatfield– *al.*, «Selective Rescue of Selenoprotein Expression» and B.A. Carlson – M.E. Moustafa – A. Sengupta – U. Schweizer– *al.*, «Selective Restoration».

[43] Cf. M.E. Budiman – J.L. Bubenik – A.C. Miniard – L.M. Middleton– *al.*, «Eukaryotic Initiation Factor 4a3».

F4a3 because they have a SECIS small apical loop, which always enhances their translation. Other selenoproteins that only have to be synthesized when the concentration of selenium is higher have big apical loops in their SECIS elements and use the methylated isoform of tRNASec, which is only available when selenium is present.

3. **Expanding the genetic code**

Since the discovering of Sec as a new amino acid, much has been written about the flexibility of the genetic code[44]. This is, in fact, a very puzzling discovery. It was known that there are a lot of amino acids that are modified *after* their insertion into the polypeptide chain; but nobody would have expected that a 21st amino acid would enter the genetic code bypassing the canonical process of protein synthesis. It is a clear example of something that has been named as recoding[45].

From the point of view of the genetic code, nothing has changed with Sec. The cell just only manages the codons it already has in order to set another amino acid that does not have its own codon. Then, the solution that the cell adopts is not to remake all the system of triplets and the amino acids linked to them from anew. What the cell makes is to take an already existing codon (UGA) and change the mRNA context in which it is inserted in order to gain a new function[46]. Also the physiological context of the cell is an important feature to be taken into account. Indeed, «such a system would provide an interesting global regulatory mechanism for gene expression. It would permit coordinated

---

[44] Cf. D. Söll, «Genetic Code: Enter a New», A. Bock – K. Forchhammer – J. Heider – C. Baron– *al.*, «Selenoprotein Synthesis». This flexibility receives also the name of a "natural expansion" of the genetic code, as it is proposed in A. Ambrogelly – S. Palioura – D. Söll, «Natural Expansion of the Genetic Code».

[45] Sec insertion is not the only example of recoding. Under this concept there are also other mechanisms of changing the meaning of the information that is carried by certain molecules. Other examples of recoding are shown in R.F. Gesteland – J.F. Atkins, «Recoding: Dynamic Reprogramming of Translation».

[46] Cf. A. Bock – K. Forchhammer – J. Heider – C. Baron– *al.*, «Selenoprotein Synthesis».

synthesis of groups of proteins during major changes in physiological conditions such as shifts from aerobiosis to anaerobiosis»[47].

This is a solution that one would not expect according to the nature of the genetic code. In fact, it is known that there are multiple codons that encode a single amino acid, but Sec insertion is a case in which a single codon, depending on the context that sorrounds it, codifies for different information: either stop or Sec[48]. It is not exactly a recodification of the codon's meaning but a change in the context the codon is in order to become a sign of a different referent.

3.1 *A context depending mechanism*

As I have already said, Sec insertion is a new trait that does not imply a rebuilding of the genetic code. We can say that the cell varies the semiotic value of a codon (UGA) changing the context of that codon dynamically. We have seen that there is a change in the mRNA context, and that fact recruits especial partners that shift the molecular context of the codon in order to be recognised as Sec rather than as a stop signal (Figure VII.9).

3.2 *Why does life have invested in Sec?*

One of the questions that arise from the discovery of Sec insertion is why life has decided to use Selenocysteine (Sec) instead of Cysteine (Cys) in certain proteins. The reason is that the presence of Sec instead of Cys in the core of selenoproteins exhibits a 10-50-fold higher activity[49]. There are cases where the

---

[47] Cf. H. Engelberg-Kulka – R. Schoulaker-Schwarz, «A Flexible Genetic Code».

[48] Cf. R. Longtin, «A Forgotten Debate». More striking is the case of *Euplotes crassus,* a ciliated protozoa, where an UGA codon can specify either Cys or Sec in the same gene, as is reported in A.A. Turanov – A.V. Lobanov – D.E. Fomenko – H.G. Morrison– *al.*, «Genetic Code Supports Targeted Insertion». The Sec or Cys insertion will depend on the availability of the functional elements.

[49] Cf. H.Y. Kim – D.E. Fomenko – Y.E. Yoon – V.N. Gladyshev– *al.*, «Catalytic Advantages Provided by Selenocysteine». Another example of a major efficiency in the activity of selenoproteins versus cysteine-containing proteins can be found in H.Y. Kim – V.N. Gladyshev, «Different Catalytic Mechanisms»; H.-Y. Kim – V.N. Gladyshev, «Methionine Sulfoxide Reduction in Mammals»; and H.-Y. Kim – V.N. Gladyshev, «Methionine Sulfoxide Reductases». In M.J. Berry – J.D. Kieffer – J.W. Harney – P.R. Larsen– *al.*, «Selenocys-

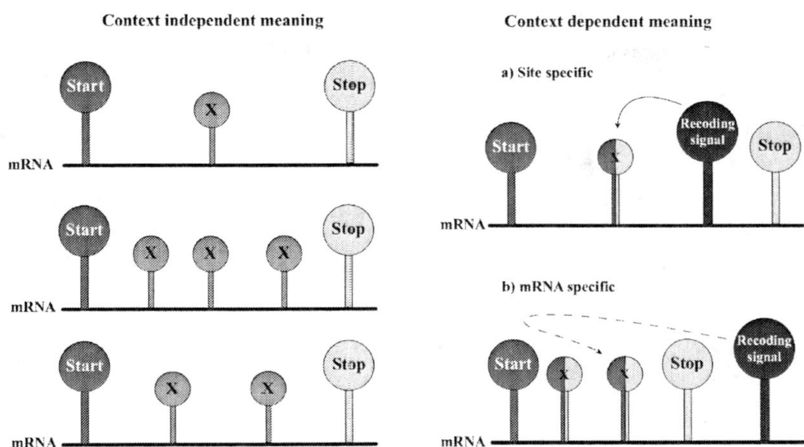

Figure VII.9: a) A new meaning can be reassigned to some stop codons independently of the context in which they are. b) But in the case of Sec insertion, the redefinition of the meaning of a stop codon depends on the context in which that codon is inserted. Adapted from J.F. ATKINS – P.V. BARANOV, «Translation: Duality in the Genetic».

increasing of activity reaches even 100-fold[50]. «The redox potential of the selenol group is about 100-150 mV more negative than that of the correspondent thiol group or cysteine»[51]. Moreover «the inherent high nucleophilicity of Sec and thereby its higher chemical reaction rate with electrophiles, as compared to Cys, seems to be a truly unique property of Sec that cannot easily be mimicked by the basicity of Cys, even within the microenvironment of a protein»[52]. That is why «nature has invested in evolving Se-dependent pathways and the specialized machinery used for Se insertion into protein»[53]. Since one of the functions

teine Confers the Biochemical Properties» it is shown that is Selenocysteine who confers to type 1 iodothyronine deiodinase the proper characteristics to convert thyroxine to T_3.

[50] Cf. G.G. KUIPER – W. KLOOTWIJK – T.J. VISSER, «Substitution of Cysteine for Selenocysteine».

[51] D. SÖLL, «Genetic Code: Enter a New», 662.

[52] E.S.J. ARNÉR, «Selenoproteins», 1296.

[53] A.V. LOBANOV – D.L. HATFIELD – V.N. GLADYSHEV, «Eukaryotic Selenoproteins and Selenoproteomes», 1425. However, some studies have established that not in all cases is Se to provide the high activity of the selenoprotein. In fact, Gromer *et al* report that thioredoxin reductases (TrxR) efficiency do not depend directly on whether the enzyme has a Cys

of selenoproteins is to neutralize dangerous oxidizing molecules, Sec is more efficient in reacting with those species than Cys.

However, it has also been reported that the higher efficiency of some selenoproteins has to pay a counterpart. In fact, in some cases, the advantage provided by having Sec instead of Cys in the catalytic core, carries also the disadvantage of a greater difficulty in returning the enzyme to the initial state in order to perform another reaction[54]. If we study the properties with which a protein that has selenium in its enzymatic core is invested with, we can understand why nature has make such an efforts to have Sec. Longtin explains some of these reasons:

> Selenium is such a good donor of hydrogen ions, and it is completely ionized at [normal blood pH levels]. Sulfur is typically the main atom involved in redox reactions [to control destructive oxygen free radicals], but selenium is so much better at it. So, it seems nature has gone to all of this trouble to incorporate it[55].

On the other hand, there are studies that support another hypothesis. Sec insertion, in some cases, would not have led to a greater efficiency of the enzymatic reaction, but to a wider variety of substrates and to the possibility to work in a broader pH range[56].

---

or a Sec in the catalytic site, but on the other amino acids that surround it. «Our findings, surprisingly, show that Sec is not necessarily required for efficient catalysis of a high Mr TrxR, and that small changes in the active site microenvironment are sufficient to yield high, selenium-independent activity. We therefore propose that Sec may not be essential for a particular enzyme reaction per se, but that this amino acid expands the metabolic capacities in terms of activity toward a wider variety of substrates and over a broader range of pH. This fact alone could explain why organisms that can rely on a continuous and adequate supply of nutritional selenium have either retained or developed selenium-dependent enzymes», S. Gromer – L. Johansson – H. Bauer – L.D. Arscott– *al.*, «Active Sites of Thioredoxin Reductases», 12623.

[54] «We propose that in MsrBs [methionine-R-sulfoxide reductases] the use of Sec is a compromise between elevated rates of Met-SO reduction and the ability to regenerate the active enzyme form by reduction with the natural electron donor». H.Y. Kim – V.N. Gladyshev, «Different Catalytic Mechanisms», 2086.

[55] R. Longtin, «A forgotten debate», 504.

[56] Cf. S. Gromer – L. Johansson – H. Bauer – L.D. Arscott– *al.*, «Active Sites of Thioredoxin Reductases».

It has also to be taken into account that the process of Sec insertion is energetically expensive: «decoding of UGA is an inefficient process and that using the third dimension of the mRNA to accommodate an additional amino acid is accompanied by considerable quantitative and kinetic costs»[57].

3.2.1 Selenoproteins and evolution

As in many other cases in biology, it is important to study the evolutionary history of selenoproteins to better understand their biological significance, why they have appeared and how the cell uses them to achieve the functions it has to deploy for its survival.

It has been shown that selenoproteins play an important function in protecting the cell against oxidative stress. The presence of selenium in the catalytic core of these kind of proteins provides them with the adequate chemical properties to neutralize oxidative species. Selenoproteins probably first appeared in low oxygen environments, as the aquatic ecosystems. Then, with the appearance of oxygen in the atmosphere, Selenocysteine would have been counterselected[58]. In fact, Selenocysteine becomes oxidized very quickly. This fits in with the fact that selenoproteins would have been highly efficient enzymes against oxidative stress. But probably because of their natural anaerobic environment, they were only able to cope with low concentrations of free radicals. Hawkes gives further data about this particular detail:

> It has been reasoned by several independent groups that the highly oxidizable Sec would have been more stable in the low oxygen environment that existed in the primordial atmosphere than it is today. This line of reasoning assumes that the rise in atmospheric oxygen some 2.4 billion years ago and known as the Great Oxidation Event created a selective pressure for replacement of Sec by cysteine. This is consistent with the occurrence of Sec incorporation via the UGA codon in all three superkingdoms of life, implying selenoproteins

[57] S. Suppmann – B.C. Persson – A. Böck, «Dynamics and Efficiency», 2284. For further information about the process' efficiency, see A. Mehta – C.M. Rebsch – S.A. Kinzy – J.E. Fletcher– *al.*, «Efficiency of Mammalian Selenocysteine Incorporation».

[58] Cf. S.C. Low – M.J. Berry, «Knowing When Not to Stop».

existed in the last universal common ancestor as much as 3.5 billion years ago, long before the Great Oxidation Event[59].

Surprisingly, it has been reported that selenoprotein synthesis is not critical for the oxidative defence in *Drosophila* since mutants lacking the special elongation factor EFSec are viable[60]. Therefore, it is difficult to justify that the selective trait that has been selected by nature was the anti-oxidative activity of selenoproteins[61]. The group that discovered this particular trait of *Drosophila* advances an hypothesis: «Thus, selenoproteins may have undergone an insect-specific adoption of novel function(s) once the components of their oxidative stress defence system became independent of selenoprotein biosynthesis»[62]. They also venture some functions that selenoproteins could have gained, as Hirosawa-Takamori explains:

> The fact that selenoprotein biosynthesis is maintained in flies suggests that following initial gene duplication events in ancestral organisms, selenoprotein-coding genes may have adopted new and possibly important, but non-vital, functions. Such functions may affect, for example, behaviour, learning and/or memory processes, which were not addressed by our present study, and may account for the continued requirement for selenoprotein synthesis once the redox homeostasis system became independent of selenocysteine-bearing enzymes during the course of insect evolution[63].

Another example has been reported in which a selenoprotein family called SelJ has not an anti-oxidative function. SelJ is a selenoprotein family found only

[59] W.C. Hawkes – Z. Alkan, «Regulation of Redox Signaling by Selenoproteins», 246.

[60] Cf. M. Hirosawa-Takamori – H.R. Chung – H. Jäckle, «Conserved Selenoprotein Synthesis».

[61] In bacteria, for example, not only the redox atmosphere is a factor for Sec selection, but also the temperature, as it is reported in Y. Zhang – H. Romero – G. Salinas – V.N. Gladyshev– *al.*, «Dynamic Evolution of Selenocysteine».

[62] M. Hirosawa-Takamori – H.R. Chung – H. Jäckle, «Conserved Selenoprotein Synthesis», 320.

[63] M. Hirosawa-Takamori – H.R. Chung – H. Jäckle, «Conserved Selenoprotein Synthesis», 321.

in actinopterygian fishes and sea urchin. Furthemore, no Cys homolog has been found in other organisms. In spite of having a redox function, this selenoproteins have a structural role in the formation of the lens of this kind of animals[64].

Comparing the different selenoproteomes one can see that the aquatic organisms have the highest content of selenoproteins. Perhaps the constant selenium supply of the aquatic medium could have been one of the factors that have pushed these life forms to take this element for their anti-redox machinery. Passing from the aquatic to the terrestrial habitat and, consequently, the scarcer availability of selenium, induced to abandonment of selenoproteins or their replace with Cys-homologs[65]. This is the case of the SelU family, whose aquatic members have Sec, while the homologous mammalian proteins insert Cys instead of Sec[66].

It is worth mentioning that even if selenoproteins are more present in aquatic organism than in terrestrial ones, vertebrates are an exception to this tendency. Indeed, if we consider plants, fungi and unicellular organisms that live in terrestrial environments, it can be seen that they have lost their selenoproteins. As it has been said, this could be because the increase of oxygen in the atmosphere made difficult to have such sensitive molecules as selenoproteins. Terrestrial organisms, in general, have found other ways to neutralize dangerous oxygen derivates. However, vertebrates do have selenoproteins. Vertebrates, and other higher organisms, are such complex that «tissue specialization, organism size and protective cover of the skin in multicellular organisms allows these organisms to control (and restrict) oxygen delivery to internal sites, thus reducing negative infuence of this gas on selenoprotein use»[67].

---

[64] Cf. S. Castellano – A.V. Lobanov – C. Chapple – S.V. Novoselov– *al.*, «Diversity and Functional Plasticity».

[65] Cf. A.V. Lobanov – D.E. Fomenko – Y. Zhang – A. Sengupta– *al.*, «Evolutio-nary Dynamics of Eukaryotic Selenoproteomes» and A.V. Lobanov – D.L. Hatfield – V.N. Gladyshev, «Eukaryotic Selenoproteins and Selenoproteomes».

[66] Cf. S. Castellano – S.V. Novoselov – G.V. Kryukov – A. Lescure– *al.*, «Reconsidering the Evolution». In A.V. Lobanov – D.L. Hatfield – V.N. Gladyshev, «Reduced Reliance» it is shown that during the evolution from fish to mammals, there was a reduction in the number of Sec insertion in SelP. There was a progressive replacement of Sec by Cys.

[67] A.V. Lobanov – D.L. Hatfield – V.N. Gladyshev, «Eukaryotic Selenoproteins and Selenoproteomes», 1427.

3.2.2 Who was first?

One question regarding the existence of Sec stills unanswered. Was Sec one of the first amino acids or, on the contrary, was it one of the latest to be added to proteins?

Let us consider Sec as one of the earliest amino acids. It should be encoded by the UGA codon that, by the way, is the "newest" of the stop codons. If so, there was an UGA codon that codifies to Sec and two stop codons: UAA and UAG. This hypothesis is supported by the fact that we can find Sec in all three kingdoms, so it would have been an early acquisition of evolution. Perhaps a high sensitivity to oxidation or to heavy metal ions led the cell to discard this amino acid and reconvert the UGA codon into stop. Then, to preserve some of the UGA codons that still coding for Sec, it was needed to mask UGA against termination, preventing the interaction with releasing factors. And, on the other hand, a special elongation factor was needed in order to recode the UGA codons that had to retain its codifying information for Sec.

Thinking of a late addition of Sec to the genetic code is a less plausible hypothesis. It would have taken place through a lateral genetic transfer. In the case of SelB it is very unlikely, as it would have needed a co-evolution of SelB and the mRNA secondary structure at which it binds. And this, since Sec is found in a high variety of enzymes (see chapter VI, section 2.3.1.), is rather difficult to maintain this hypothesis[68].

But for the consideration of this work, it does not matter when Sec insertion appeared. The important point is that the cell uses a single item to perform two different things (stop or Sec insertion) changing its surrounding conditions.

3.3 *Is Sec the only case of recoding?*

The recoding of UGA into Sec is not the only known case of recoding. Another non-canonical amino acid has been found in proteins. Its name is pyrrolysine and, as Sec, is not the product of a post-translational modification of

[68] Cf. A. Bock – K. Forchhammer – J. Heider – C. Baron– *al.*, «Selenoprotein Synthesis».

a canonical amino acid. It uses a similar system to be inserted co-translationally. In fact, Pyrrolysine is inserted also at a stop codon: the UAG[69].

It has been also guessed if there could be a 23rd amino acid. Computational research searching for new tRNAs strongly suggests that it is unlikely to find it[70]. But insertion of new amino acids is not the unique way to recode mRNA information. There are other ways to do that, like hooping, frameshifting and readthrough; strategies that I am not going to consider in this work[71].

4. **Summing up**

The Sec insertion mechanism is controlled by many different regulatory elements. All this regulatory points are used by the cell to adapt and modulate the synthesis of selenoproteins to Se availability and to the necessity of the proteins depending on the situation and the tissue where they are expressed.

The partners involved in such control are sensitive to redox conditions, and Se availability, and this makes possible that the different regulatory interactions that take place, can be linked together to achieve an important goal: to express the proper proteins at the proper time and place. It is the linkage of those regulatory elements with the needs of the cell, more than their detailed chemical mechanism, that makes them useful to become members of the controlling instance of selenoprotein synthesis.

As it can be seen from the evolutionary examples mentioned above, selenoproteins are not committed to be used only in redox metabolism. If an organism has other proteins that are responsible to maintain the redox status of the cell, selenoproteins would be used to do other works, as we have seen in *Drosophila*. The cell is the highest instance, the self, which adapts the items it has according

[69] Cf. M. Ibba – D. Söll, «Genetic Code: Introducing Pyrrolysine», J.F. Atkins – F. Gesteland, «The 22nd Amino Acid», and Y. Zhang – P.V. Baranov – J.F. Atkins – V.N. Gladyshev– *al.*, «Pyrrolysine and Selenocysteine».

[70] Cf. A.V. Lobanov – G.V. Kryukov – D.L. Hatfield – V.N. Gladyshev– *al.*, «Is There a Twenty Third».

[71] Cf. O. Namy – J.P. Rousset – S. Napthine – I. Brierley– *al.*, «Reprogrammed Genetic Decoding» and R.F. Gesteland – J.F. Atkins, «Recoding: Dynamic Reprogramming of Translation». All these recoding strategies have been also reported in Archaea, as is reviewed in B. Cobucci-Ponzano – M. Rossi – M. Moracci, «Recoding in Archaea».

to its needs. If there are some proteins that can perform a function better than others, the latter ones can be used to do another thing.

This lead us to consider the concept of function not as something linked to the chemistry of the molecules, but to the way the cell deals semiotically with them in order to achieve the goals necessary for its survival. The role of a protein can change from an evolutionary point of view, but the functions that an organism has to perform in order to keep itself alive, are always the same.

In these two chapters of the third part, I have mainly focused on the scientific data to explain the case study. In some places I have made some reference to the philosophical framework to see that the concepts presented in the first and second part have a correspondence with the empirical data. In the last chapter I am going to collect the main points of what has been already said and try to see the pertinence of the proposed framework to better understand the case study.

Chapter VIII

Application of the informational paradigm to the case study

1. **Informational paradigm and our case**

In this last chapter I want to recapitulate the main questions of what has been said and to point out how the conceptual framework sketched in the first and second part of this work can help to better understand the case study described in the third. I shall show that when the informational paradigm is applied to biological data, when it is used as a framework to interpret the molecular mechanisms of life, new considerations come out that could help for having a more complete picture of the case under study.

I am not proposing a sort of mixture between the philosophical and the scientific method. Neither do I want to propose a new way of making science through a philosophical method. I know that both the object and the method of philosophy and science are different. Science has to keep on doing science as it has been doing. Once science has obtained the results of a certain process or object, then philosophy can take these data and interpret them according to a wider paradigm than the scientific one. My guess is that the informational paradigm can provide this wider framework to better understand the concept of function in biology.

2. **The irreducibility of biological processes**

The most important issue that has been discussed in the first part of this work (chapters I and II) is whether biology has to be considered a science on its own or just a special case of physics and chemistry. There have been many attempts to reduce biology to those sciences. Reductionism acknowledges that biology is a very complex science, of course; there are many biological processes and structures that are very sophisticated and imply the interaction of many different and complex molecules. However, in the end, the reductionistic point of view thinks that this is just a matter of size, a question of complexity. If we can divide biological objects into their more fundamental parts, then we would be able to discover that everything is explainable with the laws and features of its most basilar items.

However, it has been shown that it is not possible to achieve a reduction of biology to physics and chemistry. It has been shown (chapter I, section 3.3.5) that it is not possible even to reduce thermodynamics to statistical mechanics. The reason is that, in one hand, thermodynamic processes are irreversible, and so a strict correspondence between the final and the initial state of a system cannot be established. Thermodynamics introduces a feature in the system that cannot be reduced to classical physics and chemistry because those latter sciences study reversible processes. On the other hand, it is impossible to fully establish the initial conditions of a biological system, because it is an open system that exchanges energy and matter with the environment and is far from equilibrium. The initial conditions of such systems can be only framed statistically, but it is not possible to determine exactly which those conditions are.

From a biological point of view, the cell is always changing its initial conditions, since all the chemical reactions that take place inside the cell, provide a new framework for the forthcoming ones. The cell, then, can be seen as a dynamical system where the initial conditions are always changing due to the irreversibility of the processes that take place in it.

However, the irreducibility of biology to physics and chemistry is not only due to the thermodynamic consideration of biological processes. Indeed, there is another reason that makes life an irreducible issue; it is the way it deals informationally with the molecules that characterize it. The way the cell deals with its own molecules shows that the use of them is not reduced to their physical and chemical properties. Those molecules can be used not only as instruments to perform biological reactions, but also as signs that vehicle information about referents that the cell does not control directly. In other words: biology is irreducible to physics and chemistry because it deals with molecules not only as a source of efficient causality but also as formal constraints that can influence efficient causality.

From that point of view, the survival of the cell does not depend on the preserving of certain chemical properties, but on an informational relation between its molecules and structures that allow the cell to somehow represent the environment, the functions that have to be performed and the way to perform them.

The first thing that must be noticed in the case study is that it is a clear example of where the whole is more than the sum of its parts. Indeed, if we analyze carefully the genetic code, we shall see that there are only 20 amino acids. All the cellular coding machinery is suited to manage those 20 amino acids to be used during translation of mRNA into protein. tRNA, triplets, and the enzymes that charge every amino acid to its proper tRNA, make a closed system that has achieved the important role of translating every three nucleotides of DNA in an amino acid. As it has been shown (Chapter VI, section 3.1), there is no room for a 21st amino acid in the genetic code. The genetic code cannot accommodate a new amino acid because all the combinations of mRNA triplets and amino acids are occupied.

However, it is well known that we can find in the proteins more amino acids than the canonical 20. When we analyze the sequence of many proteins, other amino acids are found. This fact could have an easy explanation, since there are some amino acids that are modified *after* having being inserted into the polypeptide chain. But this is not the case of our example. Selenocysteine (Sec) is not an amino acid that is found in some proteins because of the post-translational modification of a canonical amino acid. Sec is co-translationally inserted into the protein. That means that the cell has to somehow find the way to ascribe to Sec a tRNA and replace one of the triplets in use.

The new way the cell finds to insert Sec into some proteins is due basically to two features. On the one hand, the cell has achieved a certain degree of complexity to combine its modular structures and try new solutions for new challenges. Emergence is the process through which life has achieved such complexity. On the other hand, these complex structures and processes have to influence and constrain the processes that take place at the chemical level, providing new boundary conditions to canalize certain properties of the most basilar levels of organization. This causal influence of the higher levels of organization over the lower ones is known as top-down causation. Emergence, then, is the result of the history of the whole universe, with its growing complexity and organization, while top-down causation is, instead, referred to how life deals with this emerged structures in order to deploy the functions that are needed to survive.

2.1 *Emergence: how novelty appears in biology?*

Emergence can be defined as the way in which novelty arises out of a system and is not reducible to the sum of the parts of that system. The novelty emerged is not an extra element put into an already existing system, but it comes from a new organization of it. The concept of emergence is normally related to how novelty has emerged during the history of the universe. Emergence points at the novelty of natural processes from a philogenetic point of view; that is, how this novelty has arisen in time from the "Big Bang" until now.

The first author to use the notion "emergence" was George Henry Lewes (see Chapter II, section 3.1). Stuart Mill previously used the distinction between "homophatic" and "heterophatic" effects when he talked about the composition of causes. Homophatic effects are due to the vectorial sum of the forces at play, and then they are reducible to more elementary causes; and heterophatic effects are not the result of a vectorial sum of forces. Lewes calls Mill's heterophatic effects emergence. Lewes pointed out that there is no possibility of decomposing what has emerged into its constituent's parts. Emergence makes reference to the way those parts are organized.

In the Selenocysteine case study, it can be seen that the result of Sec insertion is not due to a straightforward application of the rules of the genetic code. It is not the vectorial sum of the elements that compose the genetic code that allows Sec to be inserted into the growing polypeptide chain. Indeed, Sec is inserted because the rules of the genetic code are used far beyond their "homophatic" effects.

The notion of emergence was enriched with the contribution of Lloyd Morgan on the concept of levels of reality. According to Lloyd Morgan, the notion of novelty in nature has to be explained through the levels of organization in which nature is structured and that have been progressively arisen all along the history of the universe. New levels of reality emerge from the lower ones because a new relation between the lower level items takes place. Less complex items establish new relations between them, allowing a more complex higher level to emerge.

From that point of view, the emergence of such a complex mechanism as that of Sec insertion would not have been possible without the previous emergent events that allowed the apparition of the levels of reality that are involved in that mechanism. It is impossible to think in a mechanism such as Sec insertion without complex elements such as the ribosome and the SECIS element; elements that, in turn, have appeared due to the organization established between their basilar elements (proteins and mRNA); which in turn are the result of a particular relation of their fundamental items (amino acids and nucleotides). Finally, amino acids and nucleotides would have not appeared if inorganic matter, at a certain time, had not been combined in such a way and under certain conditions to allow it to cross the line between non-life and life (Figure VIII.1).

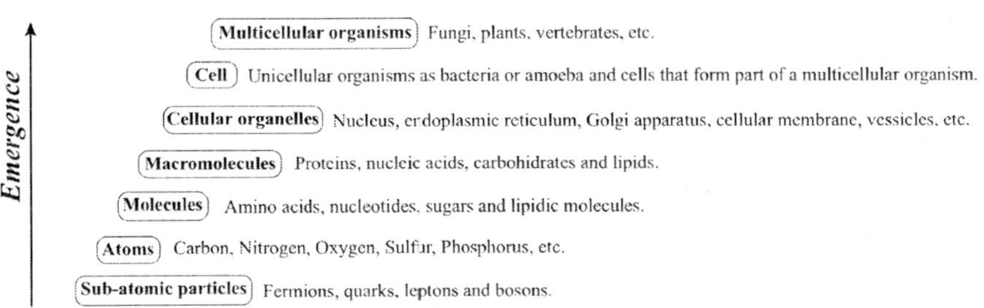

Figure VIII.1: All along the history of the universe, new levels of organization have emerged thorough different combinations of the immediate lower level items. Emergence, then, is a philogenetic aspect of reality.

C. D. Broad goes further and says that the newly emerged properties are non-deducible from the items that compound them. Examining the lower level items would never give us, *a priori*, the key to understanding how those very items are *exactly* organized in the next level of reality. The reason is that the novelty of emergence is not due to the introduction of a different material cause from that of the lower level, but just a particular organization of the already existing elements. The elements present in the Sec insertion mechanism are the same that were present in the origins of life: amino acids and nucleotides; but they have been organized all along the history of the universe in such complex structures that have allowed the emergence of novel features when those structures were co-ordinately organized. If the individual elements that are involved in the Sec insertion mechanism were studied individually, it would be difficult to deduce how they would interact to perform such an event. Once the mechanism is noticed, *a posteriori*, how the elements have interacted it can be traced back, but the *way* they have interacted cannot be deduced previously.

Broad also introduces a very interesting distinction when he talks about intra-ordinal laws and trans-ordinal laws. The former are those that are at work among the items of the same level of organization, while the latter are those laws that rule the interaction between different levels. I think that Broad's intra-ordinal laws can be understood as efficient causality, as the chemical interactions between molecules of the same level of organization. On the other hand, trans-ordinal laws are the formal constraints that higher levels of organization would exert over the lower ones. Broad's trans-ordinal laws can be, from a certain point of view, what later was called top-down causation. The novelty emerged exerts a different causality, apart from the efficient one. This formal causality is due to the constraining influence by the higher level of the possibilities the lower level items can adopt.

2.1.1 Emergent monism

My guess is that the more adequate philosophical approach to understand biology is the emergent monism. I do not understand monism as a sort of materialism, but as a philosophical tenet that acknowledges that there are in nature some common characters, some recurrent commonalities that point to a unique principle. Monism holds that variety in nature is not due to the introduction of different principles or spooky entities. Novelty in nature is not due to dualisms or other sorts of doctrines that postulate the need for strange entities or laws to explain that novelty. Nature has to be understood here not as what is available to our scientific method, but as a conception of reality that has an ultimate unity that goes beyond the epistemological approaches of individual sciences. That is the role philosophy can offer to natural sciences.

The notion of emergence is what allows us to talk about variety and novelty in a monistic world. Novelty is due to the complex organization of nature, yielding new structures and complex behaviours. According to the notion of emergence as Lloyd Morgan understands it, the whole history of the universe is an emergent process, where matter is constantly constructing complex structures. The novelty does not come from the adding of a different matter, but from a new order conferred to the already existing one.

Indeed, if universe is studied carefully, it is easy to individuate different levels of organization that emerge through the combination and complexification of simpler items that pertain at the immediately lower level. From a reductionistic point of view, the most organized structure of the universe is made of the same stuff of the most basilar elements. Then, the emerging complex reality does not depend on the adding of an already non-used element, but on a different way of using the ones are at hand; that is, depends on its formal structure.

2.1.2 Emergence and evolution

In the case study it can be seen that the organization of the items involved in the Sec insertion event constraints the possibilities of how the elements at work can interact. It has been shown that the partners involved provide a suitable chemical environment in order to facilitate the insertion of Sec instead of ending the translation. This trans-ordinal law cannot be deduced from the efficient causality each item can contribute due to its chemical properties, but is something that emerges because of a proper interaction between those very items.

The case study considered in the latter two chapters can be seen as a moment of this emergent evolution that characterizes the development of the universe. When life finds itself with the challenge of facing some dangerous environmental situations, it tries to find a solution with the items it has at hand. Evolution tests different combinations of molecules to find a solution to the problem at stake. This is how emergent evolution works. Indeed, during evolution we can see that life is more and more complex. Then, from a philogenetic point of view, emergence has yielded, through the history of evolution, novelties that have emerged from new combinations of elements to look for new solutions for the challenges of life.

As has been shown (Chapter VII, section 3.2.1), it seems that the shift from the aquatic medium to the terrestrial and the consequences it has had for selenoproteins is an example of those difficult situations life has had to face. One of the most important functions of the cell is to keep itself free from dangerous chemicals that can damage the structure of its proteins and other molecules. Some of these chemicals are oxygen derivates. When an organism is in an aquatic medium, the concentration of those dangerous chemicals is not very high. Then, it makes sense to have a family of proteins very sensitive to those molecules, since then they can be neutralized as soon as they appear. It seems that this is the reason why selenoproteins are more abundant in aquatic organisms. Then, the transition to a terrestrial habitat and the oxidizing atmosphere made selenoproteins a system of defence that became saturated very quickly; and then the dangerous oxidizing molecules could cause great damages to the cellular structures. Then, other systems have been used by terrestrial organisms to afford this challenge, and selenoproteins have been diverted to other functions, different from the ones they performed in aquatic environments. This is an example of teleology. The cell does not bother about which system to use to afford a certain problem. The important thing is to constantly perform those functions that are crucial for the cell's survival (one of which is to maintain the redox equilibrium, neutralizing the dangerous oxidizing molecules). The environment has constrained the cell to find new solutions to keep on living. The cell then, has had to readapt its internal structure, to link the environmental cues that convey information on the presence of dangerous molecules to another biochemical network that better achieves the aim of controlling the redox status in terrestrial environments.

2.2 *Top-down causation: how does the cell manage the novelty emerged?*

Top-down causation can be seen as the ontogenetic face of emergence. As emergence is a process that has been at work since the very beginning of the universe, it is a philogenetic process which has to be seen in all its temporal perspective through the different generations of organisms to study its development. On the contrary, top-down causation has to be seen as the result of the emerged reality at work in a given organism. Emergence has lead to more and more organized structures, and top-down causation is the way those complex structures can exert some influence over the distribution of the lower level items that are at the basis of those structures. Then, it can be seen that teleology is not only a characteristic of evolutionary processes, such as emergence, that strive to find the most suitable solution to a given challenge; but it is also a feature of biological processes in a given organism.

The notion of emergence helps us to understand how novelty has appeared in our universe. All along the history of the universe and the history of life, there are new structures and organizational patterns that have yielded more and more sophisticated ways to face the challenges of life. Emergence, then, is a philogenetic concept that shows that the universe is not a simple aggregate of its basilar elements, but an ordered structure where each of its levels is the result of the proper organization of the lower level. The level emerged is not just the sum of the features of the lower one, but the proper formal organization that allows novelty to arise.

When causality is considered, it is frequently thought that efficient causes are the ones that do all the work. In biology, that means that in the end, what really matters for life to go ahead is to break and form chemical bonds among the most basilar elements that form biological systems: the molecules. The reductionist programme thinks that everything in biology can be explained if we are able to identify which chemical bonds have been formed or broken.

I do not reject the importance of efficient causation in biology. However, I maintain that it is not the only cause at work in biological systems. There are formal constraints that do not act directly in the forming and breaking of chemical bonds, but that *concur* to influence how and when efficient causality has to be performed.

2.2.1 Causality in biological systems

The reductionist paradigm only admits efficient causation as the unique source of change in biological systems. Only if the forming and breaking of chemical bonds can be traced can we talk about true causality. The real work performed in a cell, from a reductionistic point of view, is only done at the chemical level. This point of view leads to admit that the features of any given organism can be explained by just considering the chemical interaction of the biological macromolecules that are involved in those features. The "coding for" paradigm (Chapter III, section 3.3.1) defends this tenet, since it maintains that the overall features of a given character can be fully explained by the genes that are involved in those features. It does not pay attention to the way genetic products (the proteins) interact with each other and with the environment, neither does it take into account the fact that the only thing that is coded by the gene is its correspondent mRNA, because the three-dimensional protein that results from the mRNA translation is no longer coded information.

The informational paradigm, for its part, does not want to leave out efficient causation. The interactions at the chemical level are taken into account, but they are only part of the story. Efficient causality is not the sole source of causation. As has been shown (Chapter II, section 5), causality in biological systems is richer than the efficient cause.

The cell, with its own structure of cellular membrane, nucleus and the other organelles, exerts the first formal constraint over all the elements and processes that take place within it. The spatial organization of the cell imposes certain constraints on everything that happens in the space delimitated by the cellular membrane. If the cell is considered to be the highest level of organization (at least it is in the case study), it can be considered that the cell exerts a top-down causation constraint over the lower level items that form that very cell.

The organization of all those structures leads to the ontogenetic appearance of formal constraints that control the way the partners have to be organized in order to achieve the cell's aim. What nature has acquired through the emergence of new structures all throughout the history of evolution is ontogenetically used by a singular organism that has inherited it from the emergent process (Figure VIII.2).

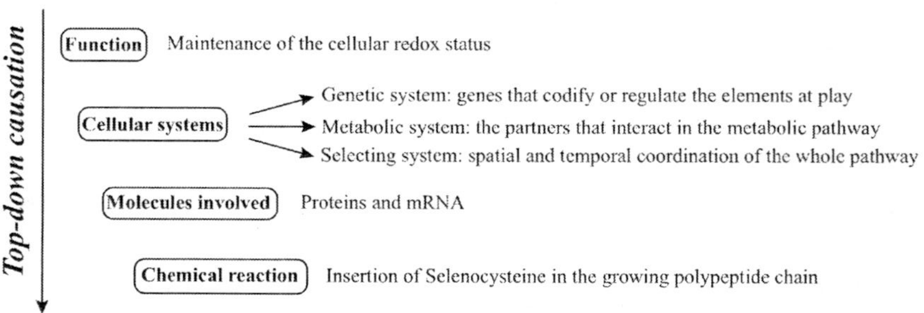

Figure VIII.2: Schematic view of top-down causation from the functional level to the chemical one.

Efficient causality is important in biology, but it has to be placed with other kind of causes that Auletta has called dynamic causes[1]. When there is breaking and forming of chemical bonds, there is an exchange of energy and momentum between molecules of the same level of organization. In the case study, the efficient causality can be seen every time there is a chemical interaction among the different partners. Besides efficient causality, there is also another kind of dynamic cause, which is *circular causality*. It can be seen in the feedback loops that characterize some steps of Sec insertion's biochemical pathway (see below), establishing a relation between the items that allows the whole pathway to be treated as a unit by the cell. Feedback loops give semiotic closure to the system (Chapter IV, section 4.3).

Final causes can also be considered in that example. Teleonomy is a kind of final cause that is characterized by the evolution of the system when it is conveyed to a local attractor. The attractor canalizes the system to achieve the final state, which can be achieved by different means. In the case study, the Sec insertion mechanism can be considered as a teleonomic structure, a local

[1] Cf. G. Auletta, «How Many Causes Are There?».

attractor configured to achieve a final state that consists in introducing a Sec amino acid into the growing polypeptide chain when certain conditions concur. The way to do that could have been very different. Other partners could have performed an equivalent biochemical operation that would have lead to the insertion of Sec. Perhaps it could have been a different mRNA structure, different from the SECIS element, that orchestrates all the partners involved in such an operation. The important thing for the cell is to find a way to insert an amino acid in a way that is different from those of the canonical amino acids of the genetic code. This could have been achieved by different means, by different equivalence classes, but in the end, the attractor has to be the final state of the inserting of Sec.

The teleonomical system can regulate the evolution of its partners to achieve the final state to which it is attracted. This can be seen when the Sec insertion mechanism is compared in eukaryotes and prokaryotes. The two main differences between the Sec insertion mechanism in eukaryotes and prokaryotes is that in the former the SECIS element is located in the 3' UTR and there is an extra protein factor in the process (EFSec and SBP2). Prokaryotes have only one protein factor (SelB), and the SECIS element is located in the very sequence of the protein, in the open reading frame (ORF), which implies certain restrictions for its structure since the SECIS element is part of the mRNA that has to be translated into protein. It can be seen that the system in eukaryotes is more complex than the one of prokaryotes (as usually happens in biology). The system has been improved in eukaryotes, since more partners allow for a more tuned control over the process. According to the fact that teleonomical systems can regulate the evolution of its partners, the changes and new partners introduced into eukaryotes can only be allowed if they concur to the achievement of the final state of the system, if they concur to the attractor; that is, for a Sec insertion event. Any new element that could be recruited for the Sec insertion mechanism is tested by the mechanism to see if it helps the mechanism to better reach its final state or not. The attractor, then, regulates the partners that can concur to achieve its final state.

2.2.2 Boundary conditions and formal constraints

One of the things that has often been repeated in this work is that life does not use any special matter that is different from that found in the non-living world. The difference is in the way life deals with matter and the structures made with it. The cell as a whole can influence upon its structures and molecules. This influence, of course, cannot be exerted by violating the physical closure of the world, and higher levels of organization (as the cell as a whole) cannot have a direct causal role (efficient causality) over the lower ones (as can be the molecular level). Then, the cell has to somehow provide the adequate conditions to influence the dynamics of its elements in order to steer them to perform the reactions needed for cell survival. Dynamic causality and formal constraints act together to achieve what they cannot get separately.

The case study of Selenocysteine can be seen as an orchestrated strategy the cell uses in order to change the chemical environment of certain UGA codons to have a Sec insertion event. All the factors involved in the Sec insertion are devoted specifically to putting Sec in the surroundings of the UGA codon to have a greater chance of action than that of the releasing factors, which would stop translation at that point. The cell cannot act directly upon that UGA codon, preventing it from being a stop event; but it can use the genetic, the metabolic and the selecting systems to somehow constraint certain molecules to perform the needed insertion event.

We have seen that there are many elements that concur to the changing of the environmental conditions that would lead to a change in the semiotic value of the UGA codon. Basically, those elements are additional proteins and mRNA structures that help that recoding event to occur. There are also other elements that have a key influence on increasing the probabilities of having a Sec insertion. These elements are neither special proteins nor mRNA elements. As we have shown (Chapter VI, section 3.3.4), in the very mRNA there are elements that have been selected in order to make the Sec insertion more feasible. One of these elements is the relation between the amino acid placed immediately after the UGA codon and the probability of a Sec insertion. There is a scale of propensities that is as follows: A > G > C > U. The immediately downstream nucleotide, then, can be selected in order to modulate the propensity of an UGA codon to be a stop codon or a Sec insertion codon. Upstream of the UGA codon SREs have also been characterized. These are elements that slow down the ribosome in order to allow the insertion machinery to be prepared and have enough time to place all its elements in the vicinity of the UGA codon.

I have to point out that there is an astonishing third element that has to be taken into account, since its influence plays also an important role in deciding how to manage the UGA codon. I am talking about the six nucleotides that are immediately upstream of the UGA codon. It is interesting that those nucleotides do not have any special role in the recoding event. Indeed, the ribosome would have passed over them when it reaches the UGA codon, so they can do little to influence the ribosome. But the two last amino acids inserted in the polypeptide chain, which are the result of the translation of these six nucleotides, have an important role in Sec insertion. Indeed, since they are the two last incorporations into the polypeptide chain, they are attached to the last tRNA and are located in the P-site of the ribosome (see Figure VI.34). They can exert some influence on the surroundings of the A-site of the ribosome, facilitating or obstructing the Sec insertion machinery.

All these elements considered together constitute what Michael Polanyi called the boundary conditions. They do not have an efficient causal role on the insertion event (only the tRNASec and the peptydil-transferase activity of the ribosome have that causal capacity) but they are the conditions of possibility for Sec to be driven in the proper place to be inserted into the polypeptide chain. All these features act together to canalize Sec at the proper place and time.

3. Information control

3.1 *Spatial and temporal order in biological processes*

The cell has many molecules in its inner space, and is also sensitive to the molecules that are in the environment. It would be impossible to deal equally with all those stimuli that constantly impact the cell or are inside it. The cell has to distinguish chemical and physical impacts from useful information. Not all elements have the same relevance for the cell. All of them carry some information (at least the information that has been needed to univocally specify them from the Shannon's paradigm point of view), but not all carry useful information for the cell.

The cell, then, has to deal differently with the elements at play to use their features in order to link the reactions they perform to the interests of the cell as a whole. The cell has to get access to the proper informational aspects of its items in order to allow them to be inserted in a certain pathways. The problem of information control is a question of information access (Chapter II, section 5.2).

Feedback control is one of the ways the cell has to control the dynamics of its components (Chapter V, section 5.1). These feedback loops allow the cell to have a control over the whole process, since different parts of the process are linked in order to have some information exchange among them. Some properties of certain items are selected to carry out this interaction. The binding properties between two items can be used to establish such a regulatory role. However, the binding is not a sufficient condition at all, since the reaction provoked by that binding has to be somehow linked to the goal the process seeks. Then, the important thing is not to bind molecules but to do it in the frame of a teleonomical process.

In the case study there are very interesting examples of feedback loops. One of them is located at the translational level (Chapter VII, section 2.5) of the genes involved in Sec insertion in *E. coli*. As it could be seen in Figures VII.7 and VII.8, *selA* and *selB* genes are located in the same cluster (in the same mRNA strain). That means that those genes are transcribed and translated together. It would be worth remembering that *selA* gene codifies for SelA protein, which is the one responsible for the transformation of Ser-tRNASec into Sec-tRNASec. On the other hand, *selB* gene codifies for SelB protein, which interacts with the SECIS element.

The important question in this example of regulation is that the mRNA of *selA* and *selB* has a SECIS-like structure in its 5'UTR. This SECIS-like structure binds exactly the same elements that the SECIS structure located at the 3'UTR of selenoproteins binds. It binds the same elements to perform a Sec insertion, but it cannot perform any insertion at all because there is no UGA codon in the vicinity. However, when all the elements are bound to that SECIS-like element (Sec-tRNASec, SelB and GTP), it avoids *selA* and *selB* mRNA translation, blocking the access of the ribosome to the mRNA.

SECIS-like element retains all the binding properties of the SECIS element, but its binding leads to a different result. If SECIS element leads to have a Sec insertion event, the SECIS-like element, when linked to the same partners, leads to an inhibition of the *selA* and *selB* mRNA translation. This feature is used by the cell to establish a feedback loop between the synthesis of SelA and SelB and their requirement depending on selenoprotein's mRNA quantities.

When selenoproteins mRNAs are present, the Sec-tRNASec-SelB-GTP complex is bound to the SECIS elements of those mRNAs in order to perform Sec insertion. When there are no more selenoprotein mRNAs, Sec-tRNASec would accumulate, being a danger for the cell, since the hydrolysis of that molecule could yield a very reactive selenol group. Then, when there are no more mRNA's to bind, the complex binds the SECIS-like element located at the 5'UTR of *selA* and *selB* mRNA, blocking its translation and preventing further synthesis of Sec-tRNASec.

The same overall binding pattern is used as a sign of two different referents. In the SECIS element is used to perform a Sec insertion event, while in the 5'UTR it is used to prevent translation when there are no more mRNAs and, consequently, no more Sec-tRNASec is needed. Then, this is an example of information control, since the same sign is used differently, depending of what the cell needs and what the environmental selenium status is.

Another example of a feedback loop is that shown in eukaryotes by the availability of SBP2 depending on the redox status of the cell. As it has been shown (Chapter VII, section 2.3), SBP2 is sensitive to the redox status of the cell. This feature makes SBP2 a protein whose cellular location depends on the redox balance. The presence of oxidizing agents confines SBP2 to the nucleus, blocking selenoprotein synthesis. When the cell recovers its redox balance, SBP2 is exported from the nucleus to the cytoplasm and can exert its role in selenoprotein synthesis. Then, an environmental cue has incidence on the spatial location of the elements that are related to the cellular response to those very cues. The organization of the cell is canalized to respond to the environmental redox status, in this case exerting an information control over the spatial localization of SBP2.

3.2 *Chemical features that are signs that stand for another thing*

In the case study many examples can be found where the important thing of a given partner to do its work is not the exact sequence or its detailed chemical structure but the preservation of certain overall patterns that allow that partner to interact with other molecules in the Sec insertion event. The cell does not need molecules of a precise structure, but molecules that can somehow interact with the rest of the partners involved in the pathway in order to perform the operation at stake correctly; in this case, the insertion of Sec into the growing polypeptide chain. Different interactions can be considered to show the importance of a global pattern instead of the chemical details of these interactions. Some of the interactions found in eukaryotes are showed in the following table:

INTERACTION	INTERACTING ELEMENTS	SEMIOTIC RELEVANCE
1) 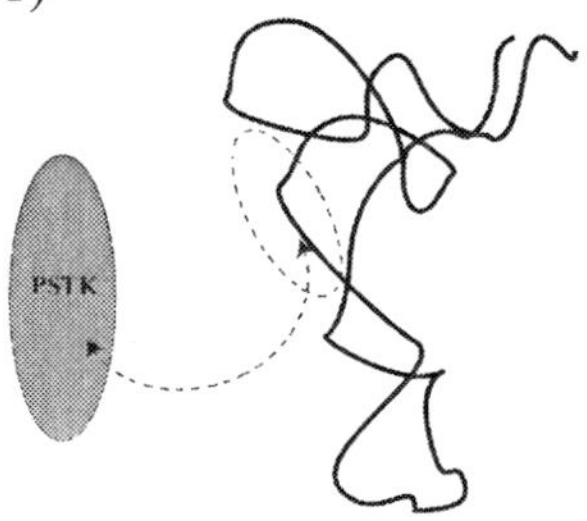	PSTK enzyme tRNA^{Sec}	The D-arm of tRNA^{Sec} is different from the D-arm of canonical tRNAs, being a discriminatory trait to be differentially used by the mechanism.
2) 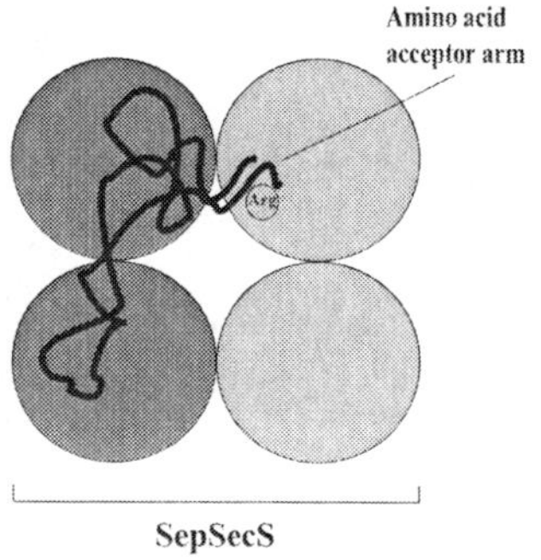	SepSecS enzyme tRNA^{Sec}	The longer amino acid acceptor arm of tRNA^{Sec} is what allows that molecule to be recognizable by SepSecS enzyme, which is the responsible of the last step that yields Sec-tRNA^{Sec}.
3) 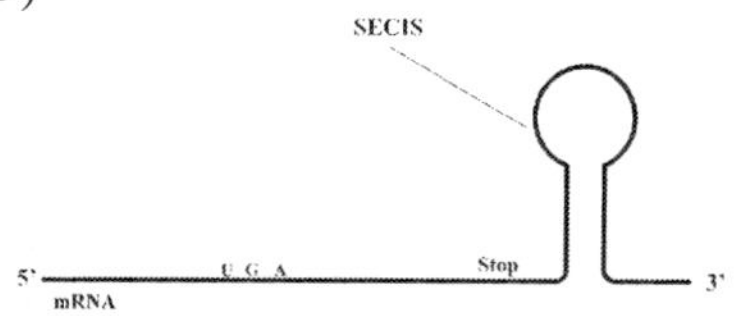	SECIS spatial localization	The localization of SECIS downstream of the UGA codon is essential to allow its bending in order to put closer the elements at play.
4) 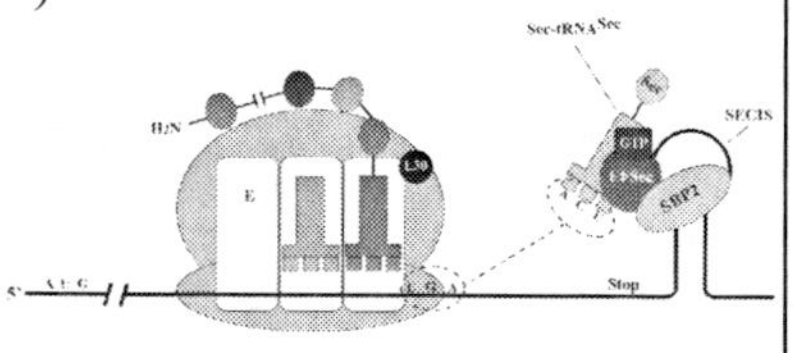	Ribosome EFSec GTP SBP2 Sec-tRNA^{Sec} SECIS	The binding among all these partners is what provides the proper chemical context to have a Sec insertion event.

Table 1: Interactions that take place during the Sec insertion event and that can be treated semiotically by the cell.

As has been already stressed, there has to be some discriminating features of tRNASec in order to be treated differently to enable it to carry a Sec amino acid to enter the Sec insertion machinery (Chapter VI, section 3.2.1.c). In eukaryotes and prokaryotes there have to be different traits that allow the differentiation between tRNASec and tRNASer in order to be processed in the proper way. The point is that the different features are not equally used by the cell. Indeed, since in eukaryotes the different D arm of tRNASec is used as the recognition trait for PSTK (Table 1.1), in the case of archaea the discriminating feature that PSTK recognizes is the acceptor arm of tRNASec. This example shows us that the most important thing for the achievement of that modification is not the fact of having a certain feature or not, but just to have the possibility to use *any* feature that makes the difference in order to treat tRNASec and tRNASer differently. PSTK has to find some anchor point in order to perform its reaction, no matter if the binding is done in a way or another (Figure VIII.3).

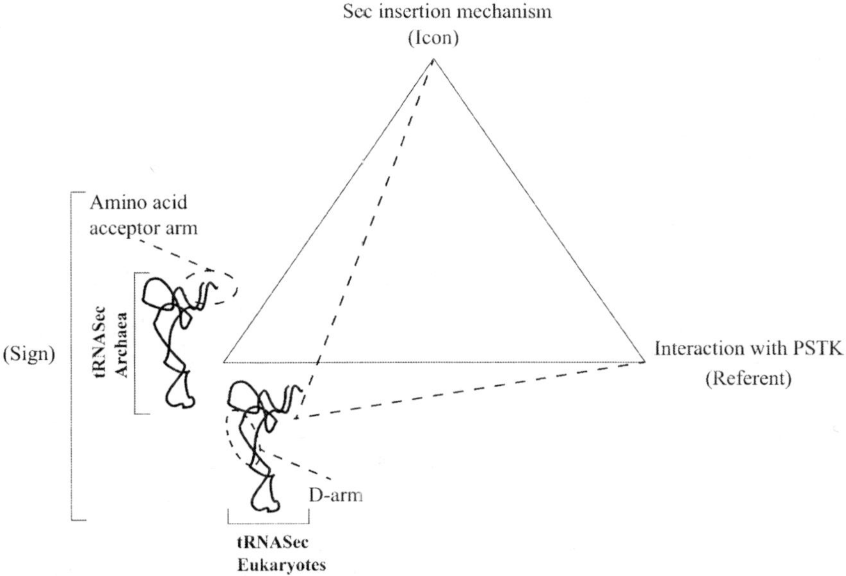

Figure VIII.3: Chemical properties are treated as signs that stand for a referent. The recognition of tRNASec by PSTK is done through a different chemical structure in archaea (amino acid acceptor arm) and in eukaryotes (D-arm). Then, the linkage of the recognition of tRNASec by PSTK in order to be inserted in Sec metabolism has to be done according to a semiotic relation, not to a chemical necessity.

This is a clear example to show that the cell is not determined by the chemistry of its molecules and, hence, that the reductionistic approach to biology is misleading. The cell does not use molecules *only* according to its chemical features. The composition of the most basilar elements of the cell does not commit it to play the same role all the time. In the case of the recognition of tRNASec by PSTK, an eukaryotic organism and an archaeal one do it differently. Eukaryotes take the D-arm as sign of the recognition of PSTK (the referent), which would allow the molecule to enter the metabolic network of the Sec insertion mechanism (the icon). On the contrary, archaea does not use the D-arm as the discriminating feature but the amino acid acceptor arm. Then, it is not the chemical structure of the items used that matters for the performance of a certain function, but its semiotic linkage; that is, to be related to a referent that has to be inserted into the whole network of the metabolic process and, by extension, in the cell as a whole to concur to its functions.

Another different feature of tRNASec is that it has an amin oacid stem arm longer than the one that the canonical ones have. The cell takes advantage of this difference and allows the interaction with SepSecS, a key enzyme in the synthesis of Sec-tRNASec (Table 1.2). In Chapter VI, (section 3.2.1.c) other differences that allow the cell to deal with these partners differently are explained.

The spatial localization of the SECIS element is another paramount feature for the Sec insertion process. This element has to be localized at a certain distance from the UGA codon, permitting its bending in order to constrain and put together all the items that are needed to have a Sec insertion event (Table 1.3). It has been shown that the chemical structure of the SECIS element has not to be very strict. But now we are in front of another important property of this item, the relative location with respect to the UGA codon. This parameter has nothing to do with a specific sequence, but just to a relative position among those elements.

Then, the important point here is to have a proper relation among those items. It is worth noting that the SECIS element has a very conserved core in the helix II. However, mutations on this core does not necessary commit SECIS element to inactivity, since other compensatory mutations can maintain the overall shape of the structure. More important than the precise elements of the sequence are the distances that have to be preserved between the UGA codon and the SECIS element (51 to 111 nucleotides), and between the non-Watson-Crick core and the apical or internal loop (9 to 11 nucleotides) (see Figure VI.32). Then, in this example it can be seen that it is not enough to have a SECIS element to perform a Sec insertion event, but it has to be properly located in relation with the UGA codon. This relative position of the elements is something that does not depend on the chemical nature of the partners, but on the possibility of the SECIS element to bend over the UGA codon to put all the elements needed physically close. It is not the sequence that matters, but the capacity of putting the elements together in the space and that these elements, with their appropriate interaction, would introduce a novelty that never would be achieved if the elements were not related in that proper way.

The relative position between the mRNA and the SECIS element exerts a formal constraint over the elements that would participate in the Sec insertion event. They do not determine that the insertion of Sec will take place every time an UGA codon is found. The connection between these two structures opens a space of possibilities, a place where the interaction among different partners could take place (Figure VIII.4). Then, it can be seen that constraints do not have an efficient causal power, but they only canalize the range of possible interactions in order to make more feasible the encounter between those items that would perform the chemical reactions through which the efficient cause is at work.

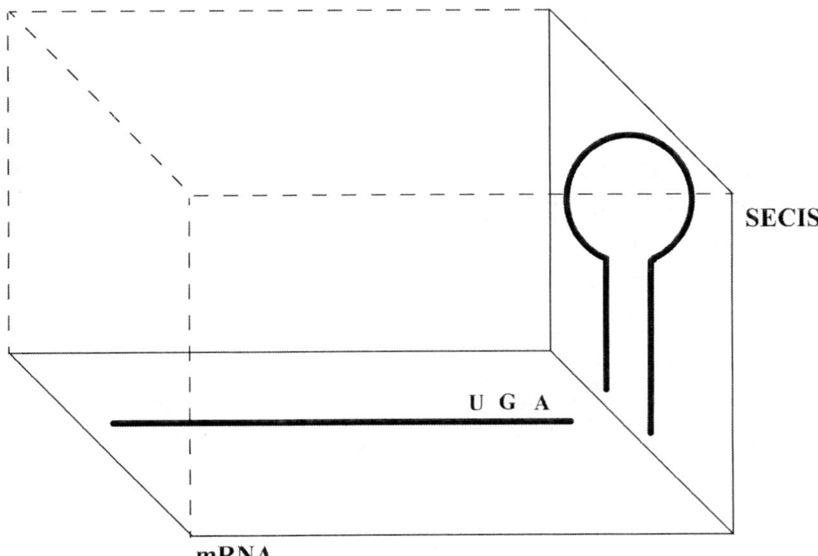

Figure VIII.4: The mRNA and the SECIS element are the same RNA string. However, they are related in such a way that the relative orientation of these two pieces of string allows for the opening of a three-dimensional space (in dashed lines) that permits the local interaction of all the partners that concur to the reading of the UGA codon not as an "stop" signal but as a "Sec" one.

As has been previously said (Chapter II, section 5.1), constraints not only reduce the number of possibilities, but they also make room for new possibilities. In the case of the relation between the mRNA and the SECIS element, it is very clear. Indeed, the relation between two bidimensional items (both, the mRNA and the SECIS are nucleotide string) opens the possibility to the interaction of new elements in a three-dimensional space.

The binding among the different partners showed in Table 1.4, links the items in a coordinated way in order to properly interact with the UGA codon to insert Sec instead of stopping translation. The structure depicted by the whole network of interactions among all these partners is an icon, a structure that represents and performs what the cell is looking for; to have the insertion of a new element that would be of great interest for a better functioning of the cell.

3.3 *Semiotics in biology*

In chapter IV semiotics was introduced as a concept that makes a difference in the informational framework. Semiotics can be understood as the way the cell links the chemical properties of biological partners and reactions to its functional needs. Molecules have to interact and their features have to be selected to play in the metabolism network according to some patterns that allow them to coordinately cooperate to the aims of the cell as a whole. Not all possible interactions are allowed, but just the narrow window that permits the partners to properly interact for the sake of the cell. That is why we can say that those interactions take place semiotically, according to certain aims that have to be achieved, not depending exclusively on the chemical nature of the items involved in achieving those aims.

One of the things that the cell has to do in dealing properly with external stimuli is to link them with an inner network of cellular structures that allow it to adequately respond to those stimuli. The cellular structures have to be somehow tuned to react in the proper way to the meaningful stimuli that come from the outside, but also to those that come from the inside as a result of the metabolic pathways performed there.

In Chapter VII, section 2, I have presented the hierarchical regulation of selenoprotein expression. It has been shown that, when selenium is available, not all selenoproteins are synthesized at the same time. There is a hierarchical organization that dictates which proteins have to be synthesized first. The organization of the Sec insertion mechanism has to be structured in such a way that allows the differential expression of selenoproteins according to the hierarchy of importance that those proteins have for the cell. The cell, then, has to be organized in such a way that it can properly respond to the selenium availability. From a certain point of view, the cell has to construct an organization that somehow represents the external availability of selenium. It is achieved with a variety of binding affinities between the different elements that take part in the Sec insertion event. This can be done due to complexity, allowing the cell to establish different levels of organization, exerting top-down causation from the higher one.

SBP2 has a different binding affinity for different SECIS elements (due to the size of their apical loop, as has been shown in Chapter VII, section 2.2). This difference on the binding, makes some SECIS elements more easily bound by SBP2 than others, making the mRNAs that carry those SECIS elements a greater chance to have a Sec insertion event. I think that, according to Peirce, this diverse binding affinity is an icon, which stands for the orderly patterns in which the proteins have to be synthesized when selenium is scarce. The complex relation of the different binding affinities of the proteins involved and the diverse SECIS elements of selenoprotein mRNAs, the different use of the isoacceptors, and any other element that can directly or indirectly concur to enhance or restrict Selenocysteine insertion, establishes a network of relations among all those elements that can be considered as an icon, a representation of the orderly interaction of all those elements.

Another element that concurs to this hierarchically organized selenoprotein synthesis is the competition between SBP2 and the eukaryotic initiation factor eIF4a3 for binding SECIS (see Figure VII.2). If eIF4a3 binds the SECIS element, then Sec insertion would be prevented. On the contrary, if SBP2 binds SECIS, Sec insertion would take place. It has been shown that eIF4a3 binds mostly SECIS elements with a large apical loop, and its binding to those with a small apical loop is less feasible. Then, those selenoproteins whose mRNA's SECIS elements have small apical loops, would be synthesized always, since eIF4a3 cannot bind them very well. Empirical data confirms this point, since housekeeping selenoprotein mRNA have always SECIS elements with small apical loops.

There is another element that concurs in this hierarchical expression of selenoproteins: the two isoacceptors of tRNASec. As has been explained (Chapter VII, section 2.6), there are two isoacceptors of tRNASec, which means that they are the same molecule, they perform the same reaction, but there are different modifications in their nucleotides. In fact, the two isoacceptors differ only in the methylation of position 34. This methylation takes place when selenium is available. If it were scarce, the unmethylated form would be predominant. Then, those selenoproteins that need the unmethylated form to perform Sec insertion, would be always synthesized, regardless the amount of selenium present. In other words: their synthesis would always be privileged. The small amount of selenium in the cell would be always used for the synthesis of those selenoproteins. On the contrary, those proteins that use the methylated form of tRNASec would be synthesized only when selenium reaches a certain concentration in the cell.

The important question is that both mechanisms are related to each other. Those selenoproteins that have small apical loops in the SECIS element also use the unmethylated form of tRNASec, allowing the protein to be synthesized even if only small quantities of selenium are present. Both mechanisms are used with the same goal and reinforce each other to make sure that the proteins that have to be always synthesized would effectively be.

Summing up, it can be said that all this patterns of different affinities and the role of the two isoacceptors with their differential dependence of selenium availability, are used by the cell to represent the selenium status inside it and to synthesize selenoproteins according to the importance they have for the cell. This entire pattern of interactions could be considered as a sign, as an icon that links the availability of selenium with the importance of selenoproteins for the cell. The cell uses this pattern to interpret, according to selenium availability, which selenoproteins have to be synthesized and which can wait until selenium levels rise a little more. Is what Jablonka called the organization of the receiver in order to properly respond to a given stimulus (Chapter III, section 3.2.1).

From that point of view, it can be said that semiotics in living systems is precisely what we see in living systems. Organisms are alive because they have narrowed down a huge amount of possible interactions among their molecules to select the interacting properties that would benefit the cell. Living organization is the most patent example of the semiotic selection life has done in order to have a net of partners that could maintain the exigencies of life. Other chemical interactions could be possible, but those that have been selected are semiotically linked to the whole network of molecules in the cell.

That is the reason why the cell as a whole would automatically select any new element that enters it, for example a new protein that appears by mutation. The accommodation of the new item in the already existing network of molecules would mean that the cell could deal with that item in order to somehow concur to any of its functions; or at least without disturbing any of them.

Semiotics, then, is a relation, a process oriented to an aim. It is something that cannot be studied as the structure of a protein or the sequence of a gene, but is something that is paramount for the cell to keep itself alive. The triadic relations established by Peirce between sign, referent and icon govern this semiotic process. Every item of the cell (a molecule or a process) is linked to the referent it is referred to. And this reference, this respect, is established by the icon, which manages the semiotic process to its own sake. From that point of view, the cell can be seen as an icon, a structure that includes all the relations that can be established among its elements, whose maintenance is the most important aim those elements are devoted to.

4. **Functions**

4.1 *The function as an act of the cell*

Function is not a category that can be ascribed to an item or a group of items. Function is a dynamical category, an act that emerges as the result of a non-reductive approach to a system. If we think in the main functions an organism has to perform in order to keep itself alive, it will be seen that is not possible to ascribe each of these functions to a given item or group of items in the cell. The function is not localized in a precise part, nor is the result of the sum of the activities of the pieces of the system. The function is a non-localized act that emerges from the interweaving of bottom-up and top-down causations.

Function is something at work, something that happens, not just a possibility among others in a given system; otherwise the system will die. It is often forgotten that an organism must perform its vital functions constantly. It is not possible to stop its activity and test new ways of performing such functions. The test must be done without stopping its functioning. That is the role of degeneracy in nature. We talk about degeneracy in a given metabolic pathway or any other structure when «several subnetworks can give rise to the same outcome and therefore in certain conditions display the same functionality, but in other conditions the same network can also display different functionalities, which shows that function is related to structure but in a loose way»[2]. The cell tests alternative paths for some operations to see if there is some other ways to better deploy functions. The cell tries to find functional equivalence classes of operations.

The concept of function in biology has to do with the viewpoint of the whole as prior to its parts. This is an approach that is very pertinent in biology, since biological systems can be understood as wholes that have ontological precedence to their parts. This can be said because of the renewal of matter in organic systems. As McLaughlin says,

> a system that remains the same only insofar as it reproduces itself by renewing and replacing its own parts is also temporally prior to (many of) these parts can thus, without backward causality, be held to be causally responsible for the existence and/or properties of these parts. Thus, the key concept in explicating the sense of functional explanation will be the concept of the self-reproduction of a system[3].

Another question that directly influences the category of function is that of causality. A reduced conception of causality that sees only efficient causes, does not allow embracing the category of function from a wider point of view.

Then, is it always misleading to talk about the function of an enzyme, a signalling pathway or a receptor? Is difficult to talk about their functions *per se*. I would rather say that they have a function when considered in a hierarchical organization in order to link its activity to that of the higher levels of the organism. RNA polymerase, for example, is an essential enzyme to transcribe DNA into mRNA. This chemical reaction is performed, but the chemical reaction could only be seen as a function when it is understood as a step towards protein synthesis, which is a higher level of consideration where it is easier to see that it has an important role for cell function.

The concept of function I am dealing with is not restricted to the performance of an item according to its structure and chemical composition. There could be a protein that performs a reaction very efficiently that is disturbing for the organism's function. On the other hand, there could be a mechanism that is not 100% efficient but that contributes to the general function of the cell.

---

[2] G. AULETTA, *Cognitive Biology*, 262.

[3] P. MCLAUGHLIN, *What Functions Explain*, 13.

Take, for example, the role of DNA polymerase. It is the enzyme that duplicates DNA during cell division. This enzyme is not fully efficient, and it makes some errors during DNA synthesis. Far for being a problem, this inefficacy is one of the sources for genetic variation, and a precious material upon which natural selection can work.

The question is not only restricted to the chemical reaction or activity a certain item has; it is also important to see how it is integrated in the hierarchy of levels of the organism. When considering a protein *per se*, we can only say that it makes an operation, or a single reaction, or it has a binding activity, etc. Only when we see how this activity is linked and amplified in order to affect the entire cell through its levels of organization, we can say that it contributes to a certain function and, as shorthand can we say that it has that very function.

4.2 *Design versus tinkering*

Our case study helps us to better understand the concept of function and to see that the functional approaches made by the etiological and the causal-role analysis (Chapter V, section 3.1 and 3.2) are not enough to understand biological functions. The causal-role analysis focuses on the role a given item play to cause or concur to a certain function. But it has been seen that the function cannot be totally dependent on the chemical properties of the item; it has also to do with the chemical environment provided by higher levels of organization when they exert top-down causation over that very item. The effect a trait has on an organism does not only depend on the efficient causality it can exert; it does not depend on its capacity of creating or break chemical bonds with other molecules. It also depends on how higher levels regulate and modulate its activity.

On the other hand, the function cannot only depend on the recent evolutionary history of the trait, as the "Modern History" approach claims. A trait can be recruited for a different function than that it has been involved in until now. Indeed, it has been shown (Chapter VII, section 3.2.1) that selenoproteins probably appeared in aquatic environments, in a moment where the atmospheric conditions where rather anaerobic. In those conditions it is very useful to have proteins that are very sensitive to any reactive oxygen specie. Since those species would be rather rare in such a reductive environment, it is necessary to have a very sensitive protein to react against those molecules, even if they appear in low concentration. That is why Sec would have been selected to be in the enzymatic core of such anti-redox proteins.

However, after the Great Oxidation Event, such highly sensitive proteins would be rapidly saturated with oxidation species; and that may be the reason why selenoproteins were then counterselected. It could also be the fact that the transition of an aqueous environment to a terrestrial one made selenium availability an important problem, and this could be also a further reason that lead to selenoproteins to no longer concur to redox equilibrium. In those organisms, as has happened with insects, the redox status of the cell has been put in charge of some other proteins that protect the cell against reactive species. There is a family of selenoproteins, the SelU family, that in aquatic organism has Sec, while the terrestrial mammals have changed Sec by Cys. Probably the latter have committed to having less efficient proteins, but there it is more feasible to find Cys in the terrestrial medium than Sec. The proteins of this family that has Cys in the case of terrestrial organisms can exert its anti-oxidant role, although it probably will not be done in such an efficient way. But the important question, as has been stressed, is not the efficiency of a single step but the convenience of the overall process to be linked to the functions of the cell.

In the case of actinopterygian fishes and the sea urchin, there is a family of selenoproteins, named SelJ, that have a structural role. Then, we have a selenoprotein that is used in a function different from that of redox stability. The presence of Sec does not commit a protein to a certain function, but it has to be evaluated by the cell as a whole to see how it can better fit into the whole metabolic network. This alternative use of selenoproteins in aquatic environemnts can be done because previously those organisms solved the important function of keeping the redox status of the cell under control by other means. Then, the selenoproteins that were responsible of that function in aquatic environments, now can be used for other functions when the redox status is maintained by other mechanisms in terrestrial environments.

These examples show us that the cell does not design its proteins in order to perform certain functions. On the contrary, there are certain functions that have to be performed, whatever the way of performing them would be. The cell takes advantage of the items it has at hand; it can also assay different combinations of the modules of the different systems (metabolic, genetic and selecting). Combinations of those different elements allow the cell to look for new or better solutions for its constant functional challenges. That is the reason why the cell seems more a tinkerer than a designer.

Susan Oyama and her notion of ontogeny of information make sense here. The cell makes its own information, selecting the interactions that can be linked to an overall goal. The important features of an item are not decided beforehand, but they have to somehow be tested in the whole network of cellular interactions.

4.3 *The creativity of nature*

Then, having studied the Selenocysteine insertion mechanism, it can be said that a new feature has emerged. The insertion of Sec into the growing polypeptide chain was something neither predictable nor deducible from the detailed study of the genetic code. Moreover, the way through which Sec is introduced in the growing polypeptide chain cannot be reduced to the mode used by the 20 canonical amino acids. Sec is introduced through another system. Of course, it would eventually be inserted through a tRNA, but the previous steps are more complex and require more partners than usual.

The emergent novelty in this case is that the cell has found a way to give a new context to an already existing item of the genetic code (the UGA codon) in order to stand for a different referent than the one it usually has. It seemed that there was no room for novelty there, but the cell takes a new habit, as Peirce would say, and has managed the elements it had at hand to achieve the novelty that was needed to keep on performing a certain function.

The new result that the cell has achieved is not due to a rearrangement of the already existing elements of the genetic code. There have not been new assignments to leave room to the incoming member. The new feature of the genetic code is not a consequence of the genetic code, but of other elements that the cell recruits in order to achieve what the genetic code cannot perform on its own. Then, we can say that there is not only the appearance of a new emergent element that is inserted in the genetic code, but that there has to be a higher level instance that has to find the way to perform that insertion in a controlled way.

The phenomenon of Sec insertion as an emergent issue gives some evidence that allow biologists to have some prejudices against ontological and epistemological reductionism. Epistemological reductionism is overruled because knowing the laws that constitute the genetic system is not enough to derive the rules governing Sec insertion. We can formulate it the other way round: the Sec insertion mechanism cannot be reduced to the laws of the genetic code. The cell uses a mechanism of insertion that, without violating the genetic code, takes advantage of it in order to yield a result that goes far beyond its intrinsic possibilities.

The case study constrains us to admit the insufficiency of what ontological reductionism claims. The success in the Sec insertion event does not depend mainly on the chemical properties of the elements involved, but on the timing that slows down the ribosome, bends over the SECIS element and puts the elements that provoke the insertion of Sec in the proper place instead of stopping translation. Of course, there are chemical reactions and different binding affinities, but all these characteristics have to be properly canalized in order to achieve the aim.

This consideration leads us to see also the incapacity of causal reductionism to account for the overall causes that take part in the insertion event. Causal reductionism claims that there are only efficient causes that can do *real* work (Chapter I, section 3.1). In the case of biology, this causality would be due to the breaking and forming of chemical bonds. According to causal reductionism, everything in a cell (and in organisms in general) can be fully explained appealing to chemical interactions and reactions.

However, we have seen that something more than the breaking and forming of chemical bonds is needed to have a Sec insertion event. All the elements involved are recruited in order to create a chemical environment around the UGA codon to diminish the probability of having a stop event and increase the probability of having a Sec insertion one. Once the UGA codon is in the proper chemical environment, the Sec insertion event would take place. The interesting point is to see how the cell provides that suitable environment and canalizes the already existing elements in order to get a novel characteristic that cannot be reduced just to the sum of the chemical properties of the elements that are involved. The whole environment provided, the new context offered, is the higher level of organization, conformed by the relation of the partners that are at play. However, they cannot exert an efficient causality over the chemical bonds of the single elements of the system, since the breaking and forming of chemical bonds pertains to a lower level of organization, the chemical one. The organization of the higher level is done in a way that exerts some constraints that facilitate the proximity of the lower level items to react among them. The causality of the higher level consists in a top-down causation that provides the adequate context in order to constrain those chemical reactions of the lower level that are important for the survival of the cell as a whole. Top-down causation is not efficient causation, but concurs to *certain* efficient causation.

Figure VIII.5 shows the elements that have been already studied and that play some role in Sec insertion. Those elements are represented as macromolecules, as big structures, without taking into account the chemical details of their constitution. The ribosome is formed of proteins and ribosomal RNA; the mRNA, with its SECIS element, is a string of nucleotides; SBP2 and EFSec are proteins that are composed of a string of amino acids; Sec-tRNASec is a molecule formed of a tRNA and the Sec amino acid attached to it; and GTP is an energetic molecule that is transformed to GDP to use its energy for the Sec insertion event.

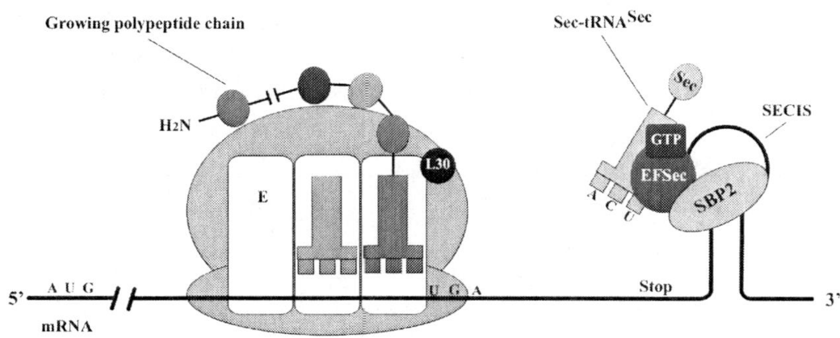

Figure VIII.5: Molecules that are needed to have a Sec insertion event. All those partners are macromolecules that, in any way or other, concur to insert Sec where normally there is a "stop" signal for translation.

The spatial organization of those macromolecules and the sequence of the interaction among them is the higher level of organization that constrains all these elements to concur to a certain effect. The effect that is looked for is that of enhancing the base-pairing of the UGA codon with the anti-codon of Sec-tRNASec instead of interacting with the releasing factor that would stop translation. Then, the whole macromolecular machinery that is at play has the aim of providing the UGA codon a contextual environment that facilitates the interaction with Sec-tRNASec rather than with the releasing factor. The base-pairing established in such a way is something that pertains to a lower level of organization. It is not a binding among macromolecules, but a very specific interaction between three nucleotides of the mRNA and three nucleotides of Sec-tRNASec. This sort of interaction pertains to the purely chemical level, and the interactions among the partners at play are steered to have that effect.

Of course, it can be objected that the interaction among macromolecules (the binding of SBP2 with SECIS, the interaction of the ribosome with the mRNA, etc.) are also chemical interactions, but they took place at a higher scale simply because the partners are bigger. This is true. However, the question is not the scale of the chemical interactions but the effect to which they concur. Indeed, the important point of the interaction between the UGA codon and the Sec-tRNASec is not the interaction *per se*, but the fact that this interaction, because it takes place in the context of the ribosome, triggers a peptydil-transferase reaction that allows the Sec amino acid carried by Sec-tRNASec to be inserted into the growing polypeptide chain. That is the reaction to which all of this mechanism concurs to: insert an amino acid that is not present in the genetic code. This is the lower chemical level over which the whole mechanism exerts top-down causation. The ultimate level the cell has to concur to exert a causal effect is the making or disrupting of chemical bonds. It is at that level that the efficient causality of the cell is at play (Figure VIII.6).

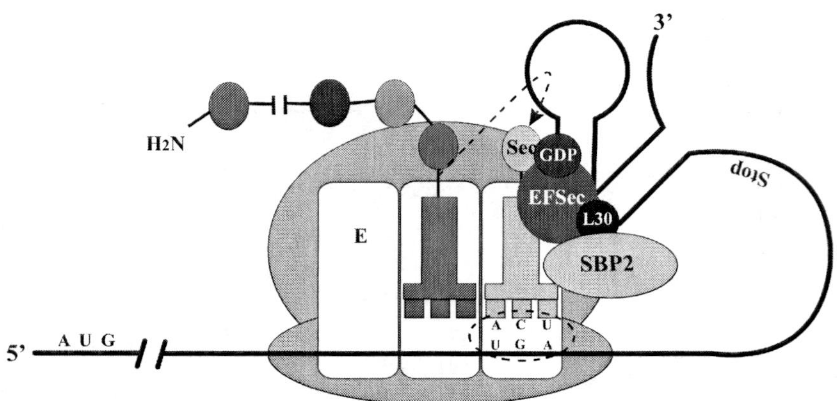

Figure VIII.6: The overall interaction between the molecules concurs to establish a precise chemical interaction between the UGA codon and the Sec-tRNASec, which in turn triggers the forming of a chemical bond between Sec and the growing polypeptide chain. Key chemical interactions are indicated with dashed lines.

The modularity of the different elements involved permits the cell to test solutions for the new situations to which selenoproteins would be a good response. It has been shown (Chapter VII, section 2.4, especially Figure VII.6) that genetic modularity is important for the proper regulation of *trsp*, the gene that codes for tRNASec in eukaryotes. There is a new organization of transcriptional components that makes *trsp* gene to be a sort of mix of class II and class III genes. New combinations of transcriptional regulatory elements have been found in order to have a proper transcriptional control over that gene. The regulatory elements of *trsp* gene are the result of a novel organization of already existing elements that have a regulatory role in other type of genes. The cell tests a new organization of items to obtain a new way of controlling genetic expression. Novelty is not achieved through the introduction of another regulatory element, but thanks to a formal re-organization of what the cell has at hand. The new way of controlling that gene is due to a new relation, not to a new material cause.

The modularity of the selecting system is also important for the proper performance of Sec insertion. When the role of oxidative stress in the cellular localization of SBP2 was shown (Chapter VII, section 2.3), it was clear that the translational regulation uses the different localization of SBP2 in the nucleus or the cytosol to exert a control point. The cell, then, uses its own structures to constrain the elements involved in our case study to be recruited and active in the proper place, the nucleus or the cytoplasm.

Perhaps the clearest case of modularity is that founded in the metabolic system, since many examples could be found in the case study. All the protein partners and mRNA elements that interact between them can be considered as modules that the cell has at hand; and they can be used because the organized interaction among them can result the insertion of Sec in its proper place.

Nature does not act as a designer, trying to find the piece that best fits in a given process. The cell cannot stop living to test which possible variant of a molecule would be better for the process it is involved. Life consists of a constant performing of vital functions that are crucial for the cell survival. There is no option of stopping the processes to see what the best solution is. The cell just lives, and when there is some challenge from the environment, it uses what it has at hand, the modules (proteins, RNAs, etc.) that are already involved in some functions or recruits partners that have had no function up until then or other partners that in the past had other functions. The cell then, works more as a tinkerer than as a designer.

The ultimate criterion to recruit a partner for a new process is the possibility to fit the already existing metabolic, genetic and selecting systems. Then, if the new partner can be inserted and controlled by them, it would be semiotically used by the cell to achieve a certain function. Many different molecules can be used by the cell, but they have to interact with the already functioning ones and have to be linked to a process that is for the sake of the cell.

Then, from that point of view, biological evolution can be understood as «the result not only of the occurrence of genetic mutations (happenings), but also of the capability of organisms to integrate the occurred mutation in a complex network of control mechanisms that is able to canalize its possible effects or contributions to the overall metabolic system»[4].

5. **Summing up**

The informational paradigm shows that the notion of function is something that has to be affirmed from the cell as a whole. It makes no sense to talk about the function of a protein or any other molecule separated from the rest of the metabolic, selecting and genetic systems. The cell uses each molecule according to how it fits into the rest of the network.

Then, more than a precise structure, what really matters is that the molecule could establish appropriate relations with the rest of elements in order to be used by the cell for its own sake. The cell, seen as an icon, is the final criterion of the overall linkage among its items, which support each other in certain respects, performing the adequate reactions to fulfill the teleological goal of the cell, which is its functioning. From that point of view, the cell could be considered prior to its parts, since its parts are integrated in the cell, depending on the contribution they can make to the whole.

All the complexity of the organization of the cell is devoted to canalize the different structures and elements to provide contexts that will enhance some reactions that would be important for the functioning of the cell. The cell, at least in unicellular organisms, is the highest organizational level that constraints its lower levels of organization in a top-down fashion to concur to certain effects that take place at the chemical level. The whole cell structure is a formal constraint that canalizes its elements to have certain efficient causes.

For doing that, the cell does not exert a control over the detailed chemical structure of its molecules and structures. The cell just tries to find combinations of items that link its functions to environmental variables. That is the system that will make sure that inside the cell there is a representation of the environment and an appropriate way to respond to any variation that would be crucial for the cell functioning. No mind is needed, nor symbolic thinking; just the cell and its network of chemicals, whose organization establishes the semiotic relevance that each point of this network has for the cell.

---

[4] G. Auletta, *Integrated Cognitive Strategies*, 133.

Conclusion

All throughout the previous pages I have tried to explain the concept of function as the touchstone to understand life as an irreducible process. It would have been very difficult to defend that tenet if I would have limited the exposition just to a conceptual comparison of the different notions of function. Of course, a philosophical reflexion on the concept of function in biology and other terms related to it is needed. However, it would have been misleading to confine this work solely to a conceptual investigation into how functions are understood.

Since the very beginning this work has been conceived as a contribution to the dialogue between philosophy and science. From that point of view, it is very important to reflect on philosophically, taking into account the data provided by empirical sciences. The third part of this work is especially devoted to that very reason to describe a biological example in detail, to explain it with philosophical categories that would help to better understand what happens in the biological world.

1. Emergence in biological systems

Emergence shows that there is a history of the universe in general and of life in particular, that has lead to the appearance of new structures and behaviours. The concept of emergence is linked to that of levels of organization. When life is observed in detail, different levels of organization from atoms, to molecules, to macromolecules, to cells, organisms and ecosystems can be identified. This layered structure of life has been emerging all throughout the history of the universe, through a philogenetic process. Different motifs and patterns of organization can be identified, and life uses those recurring structures and behaviours to increase its complexity and hence better respond to the challenges of the environment.

It can be affirmed that life is a phenomenon of increasing complexity, that has been evolving into structures and patterns of organization leading to systems that have a tight interconnection among their items.

Novelty in nature is the result of an emergent process of new structures and behaviours that are the result of new combinations of already existing items. Then, novelty in life is not due to the use of different elements from those that can be found in the inorganic world. Furthermore, life uses even fewer elements. The emergence of something new is made through a new organization, a new relation among the elements; hence, the novelty of life is due more to a formal aspect than to a material one.

Then, it is not possible to reduce a biological system to its physical and chemical characteristics, since such reduction does not take into account the relations that can be established among the items of that system and the novelty it could emerge due to those relations. Studying every single item of the Selenocysteine insertion mechanism would never led us to understand how those items interact between each other in order to perform such insertion. There is something in the mechanism that is not reducible to its items. To fully understand it, the relations established in space and time and the semiotic linkage of them to the function of the cell as a whole have to be taken under consideration.

2. Top-down causation

The novelty emerged does not only have to do with the structures that appear due to a new organization of the systems' elements, but also with the causal powers with which those higher structures can constrain the dynamics of the lower level items. However, a higher level cannot exert efficient causality over a lower one. Then, the causal powers of the former over the latter have to be carried out through a different kind of causality. This causality is known as top-down causation and consists in the formal constraints higher levels of organization exert over the lower ones in order to concur to certain efficient effects. Higher levels cannot have a direct and efficient incidence over the lower ones, but can constraint the lower level items. They can canalize them in order to facilitate certain causal efficiency among them. It has been shown in the case study, since the molecules involved in Selenocysteine insertion cannot by themselves perform such insertion. However, they can provide the appropriate surrounding conditions in order to facilitate an insertion event rather than a stop one. Those structures constrain the lower level items to facilitate certain reactions, to higher the probability of interaction among them.

Causality in biology, hence, cannot be reduced to the individual chemical reactions of the cell's molecules. Every molecule, of course, has chemical features that allow it to bind other molecules and react with certain chemicals; but the function cannot be considered as an arithmetic sum of all those individual reactions.

The cell is an organized system that exerts some causal influence due to that very organization. The cell uses all the Selenocysteine insertion machinery showed in the third part of this work to find a new way to perform a certain function. For that proposal, some molecules have been selected that can establish a binding pattern among them in order to link the novelty introduced by a perturbation to the function that has to be deployed. In the case study, it means that the cell finds the way to insert an amino acid that has special features to better deploy the antioxidant activity of the proteins that carry such amino acid.

3. **Information**

The concept of information applied to biology can be very useful to better understand how the cell works and how it interacts with the environmental stimuli. When information enters the discourse, it is normally thought that it has to do with carrying something from one place to another. Then, what it is straightforwardly accepted is that the cell receives information from the environment and from its genes and it reacts accordingly to what has been received. This characteristic, even if it is true, does not tell the whole story.

I maintain that the notion of information in biology is wider than the one we normally have when we think about information as the communication between a sender and a receiver. I propose to extend Shannon's theory of communication to Peirce's triadic semiotics, where there is no longer the dyad sender-receiver but the triad sign-icon-referent. Then, items and processes are not just information being transferred from one place to another but signs that stand for a referent in certain respect, to which they are linked through an icon.

This way of seeing biology is richer than the reductionist paradigm. Semiotics shows that the physical and chemical features of biological items are not the only important characteristics to be taken into account. Indeed, besides the chemical properties of a biological item, the use the cell makes of it is important and, consequently, the respect under which the cell treats that item as a sign. The biological role of a molecule is not then limited to its physical and chemical activity, but to the features that enter into play due to the relation that every molecule establishes with other ones.

The semiotic aspect of biological molecules allows the cell to control information. The cell cannot check every single chemical feature of all its molecules, but can exert an information control over the semiotic features of certain molecules; controlling the referent those molecules (or pieces of molecules) stand for. This control is exerted by the cell, which takes into account the necessity to establish a linkage between environmental and internal cues, and the functions that have to be performed.

The icon, the net of relations among signs and referents, is a semiotic network. This network establishes a connexion between the information and the function. It is a semiotic network that somehow represents what happens outside the cell. For example, it has been seen that the relative amount of the methylated and unmethylated forms of tRNASec depend on the selenium availability. When other elements are considered, such as the size of the SECIS element, the different affinities of SBP2 to different SECIS elements and other features of the rest of the items involved in the Sec insertion event, it can be seen that the network of interactions among all these elements is very complex. All of them can be seen as an icon, a set of items and relations that establish a semiotic correspondence between what happens outside the cell and the proper response the cell has to make to face the situation and keep it alive.

4. **Function**

Following what has been said, the concept of function has to be understood not as the sum of the individual activities of the items involved in that function, but as something that happens. The function is an act of the cell, something that has to be constantly performed in order to maintain the cell alive. The most important requirement of a function is not that of having a certain kind of molecule to perform that function, but to have such a relation among the molecules in order to link the internal and environmental cues to the performance of the function.

In the example studied, it can be seen that the cell uses a very interesting mechanism to insert one amino acid that is not in the genetic code. Saying more precisely, the cell assigns a new referent to a sign that previously had another one. All the molecular machinery used to insert Selenocysteine is devoted to change the referent to which the UGA codon is normally linked, which is a stop event.

Then, the UGA codon stands for a certain referent depending on the context in which it is placed. The UGA codon is not a univocal sign, but it points to a different referent when it is inserted in a new relation of items (an icon). It is not strictly necessary for the Sec insertion to have the set of molecules that are found in the example. Life could have found another way to perform such insertion. The crucial point is that life needed, in a certain moment, to use the amino acid Selenocysteine to be part of certain proteins. The mechanism used to perform that insertion has not been selected because of its chemical features, but due to the capacity it has to link the availability of selenium semiotically to the synthesis of selenoproteins and to the functions those selenoproteins concur to.

The cell can be seen as an icon that integrates in itself all the possible sign-object relations, selecting only those that help to maintain the integrity of the cell. The survival of the cell is the teleological end towards which all the biological pathways tend to.

Functions are not absolutely determined by the chemical features of the most basilar items of the cell (carbohydrates, lipids, nucleotides and amino acids); they are rather modulated by the higher levels of organization of the cell that exert formal constraints over those basilar items, using them semiotically in order to link them in an iconic network where everything is connected to the aims of the cell. This connection is not due to the chemical properties of the items involved, but to the semiotic use the cell does of those very items.

I'm convinced that the concept of information and function is a way to recover the Aristotelian notion of formal and final cause respectively. Life is not just an extension of the material world, but a phenomenon that is characterized by dealing semiotically with its biological items to perform the functions that life needs to keep it alive.

Initials and abreviations

3'UTR	3' Untranslated region
A	Adenine
ATP	Adenosine triphospate
bp	Base pair
C	Carbon (atomic element) or Cytosine (nucleotidic base)
Ca	Calcium
Cl^-	Chloride
Co	Cobalt
Cr	Chromium
C-tal	Carboxyl terminus of a protein
Cu	Cooper
Cys	Cysteine
DIO	Thyroid hormone deiodinase
DNA	Desoxyribonucleic acid
DSE	Distal sequence element
EF1	Elongation factor 1
EF2	Elongation factor 2
EF-G	Elongation factor G
EFSec	Elongation factor for Sec
EF-Tu	Elongation factor Tu
eIF4a3	Eukaryotic initiation factor 4a3
ER	Endoplasmic reticulum
Fe	Iron
FR	Far red (spectrum of light)
G	Guanine
GPX	Glutathione peroxidase
GTP	Guanosine triphosphate
GDP	Guanosine diphosphate
H	Hydrogen
Hb	Haemoglobin

His	Histidine
i.e.	For example
I	Iodine
i^6A37	6-isopentenyl-A37
IDP	Intrinsically disordered proteins
mcm^5U34	5-methylcarboxymethylmethyluridine
mcm^5Um34	5-methylcarboxymethylmethyluridine-2'-O-methyl ribose
Mn	Manganese
Mo	Molybdenum
mRNA	Messenger RNA
NaCl	Sodium chloride
NADPH	Nicotinamide adenine dinucleotide phosphate
Ni	Nickel
NMD	Nonsense mediated decay
N	Nitrogen
n	Note
Na^+	Sodium
N-tal	Amino terminus of a protein
NES	Nuclear export signal
NLS	Nuclear localization signal
NSEP1	Nuclease sensitive element binding protein 1
nt	Nucleotide
O	Oxygen
ORF	Open reading frame
PBP	PSE binding protein
Pi	Organic phosphate
ψ	Pseudouridine
PSE	Proximal sequence element
PSTK	O-phosphoseryl-$tRNA^{Sec}$ kinase
Pu	Purinic base (A or G)
Py	Pyrimidinic base (C or U)
R	Red (spectrum of light)
RNA	Ribonucleic acid
rRNA	Ribosomal RNA
ROS	Reactive oxygen species
S	Sulphur
SBP2	SECIS binding protein 2.
Se	Selenium
Sec	Selenocysteine
SECIS	Selenocysteine inserting sequence

Sec-tRNA^{Sec}	tRNA^{Sec} loaded with Sec
SepSecS	O-phosphseryl-tRNA:selenocysteinyl-tRNA synthase
Sep-tRNA^{Sec}	O-phosphoseryl-tRNA^{Sec}
Ser	Serine
SID	Sec insertion domain
snRNA	Small nuclear RNA
SPS1	Selenophosphate synthetase 1
SPS2	Selenophosphate syntetase 2
SRE	Selenocysteine redefinition element
STAF	Sec tRNA gene transcription activating factor
T	Thymine
TBP	TATA binding protein
tRNA	Transfer RNA
tRNA^{Sec}	Transfer RNA of Sec
TRXR	Thioredoxin reductase
TTR-RBP family	Turnover and translation regulatory RNA-binding protein
U	Uracil
Zn	Zinc

Biological Glossary

3' Untranslated region (3'UTR)

Non-coding region of the mRNA molecule that extends from the stop codon to the end of the molecule. The triplets of this region do not codify for amino acids.

Alpha helix

Folding motif found in the structure of many proteins. Amino acids are folded forming a right-handed helix.

Aminoacyl-tRNA-synthetase

Enzyme that attaches a given amino acid to a proper tRNA molecule in order to be inserted in the polypeptide chain during protein synthesis, according to the rules of the genetic code.

Amino acid

The basic unit of proteins. It is an organic molecule that contains an amino, a carboxyl group and another chemical group that differs among the 20 amino acids found in nature.

Anti-codon

Three nucleotides located in one of the arms of the tRNA molecule that are complementary to a codon of an mRNA molecule.

Archaea

Organisms that lack nucleus (such as Prokaryotes) but their genetic material is similar to that of Eukaryotes. Their metabolic pathways are related to those of bacteria.

Adenosine three phosphate (ATP)

Nucleotide formed by an adenine, a ribose and three phosphate groups. It is a molecule used by the cell to store and use energy.

Beta-sheet

Folding protein motif consisting in different sections of the polypeptide chain running alongside each other, like forming a sheet of paper.

Catalytic core

Part of an enzyme that is directly involved in a chemical reaction.

Cellular membrane

The bilipidic membrane that surrounds a cell.

Chaperon proteins

Family of proteins that help other proteins to reach their final folded state.

Chloroplast

Cellular organelle found in plants and some algae where photosynthesis is performed.

Chromatine

Complex of DNA and the proteins that help to its package, found in the nucleus of the eukaryotic cell.

Codon

Every three nucleotides found in DNA or mRNA that codify for an amino acid in a growing polypeptide chain.

Cofactor

Inorganic ion required for the proper activity of an enzyme.

Cytoplasm

The content of a cell that is confined within the cellular membrane.

Desoxyribonucleic acid (DNA)

Polypeptide formed by a linear sequence of covalently bounded nucleotides, where the cell stores its codified information.

Elongation factors

Family of proteins that facilitate both the transcription form DNA to mRNA and the translation from mRNA to protein.

Endoplasmic reticulum (ER)

Inner cellular structure found in the cytoplasm attached to the nucleus where lipids and proteins are synthesized.

Enzyme

Protein that performs a specific chemical reaction.

Epigeny

Developmental process consisting in phenotypic changes that are not the result of changes in the DNA sequence.

Epistasis

Interaction effect between two or more genes whose result differs from the sum of their effects taken separately.

Eukaryotes

Uni- or multicellular organism that has a defined nucleus where its genetic material is separated from the rest of the cell.

Exon

Segment of an eukaryotic gene that codifies for a tRNA, rRNA or mRNA molecule. The exons in a gene are found next to non-coding regions named introns.

Gene

DNA region that produces an mRNA that would be translated into protein, or another functional molecule as rRNA or tRNA.

Genetic code

Set of correlations among every triplet of the RNA and the 20 amino acids they code for.

Genetics

The study of the genes, their function and their heredity.

Genotype

The set of genes of a given individual.

Glycosilation

The addition of one or more sugar molecules to a protein or a lipid.

Golgi apparatus

Cellular organelle formed by flattened compartments where proteins and lipids are modified and sorted out.

Gradualism

Evolutionary tenet that holds that large differences in phenotypes are due to the accumulation of many slightly different intermediate states.

Housekeeping proteins

Proteins required in every kind of cell, no matter what its specialized activity is.

Intron

Noncoding region of an eukaryotic gene that is transcribed into mRNA, but then is removed through the splicing process and, hence, cannot yield a functional RNA molecule or a protein.

Isoacceptors

Two or more tRNA molecules that accept the same amino acid.

Limiting factor

The proportionally less abundant item in a reaction where many partners are involved. As it is the less abundant, it conditions the total amount of reactions that can be performed, regardless the abundance of the other partners.

Mendelian genetics

Rules of genetic inheritance discovered by Gregor Mendel. Those rules established a relation between the phenotype and the genotype of certain characters when DNA had not already been discovered.

Messenger RNA (mRNA)

RNA molecule that is the result of the transcription of a gene and specifies the sequence of a protein when it is translated by the ribosome.

Mitochondrion

Cellular organelle of eukaryotic cells where almost the entire production of ATP takes place.

Molecular genetics

Genetical approach to heredity based on the detailed molecular characterization of the partners involved in the expression and transmission of genes.

Nonsense mediated decay (NMD)

Control mechanism that looks for mRNAs that contain premature stop codons in their sequence and degrades them before being translated into proteins that would be aberrant.

Nucleotide

Purine or pyrimidine base linked to a ribose or desoxyribose sugar, which in turn is linked to one or more phosphate groups. The polymerization of nucleotides yields DNA and RNA molecules.

Ontogeny

The developmental process and stages of an organism from the fertilized zygote until its death.

Open reading frame (ORF)

A continuous mRNA molecule free from stop codons that can potentially be translated into protein.

Organelle

Cellular compartment that has a specific function inside the cell.

Phenotype

Set of observable characters that are the result of the expression of the genotype in an individual.

Phylogeny

The history of an organism or group of organisms from an evolutionary perspective, showing the linkage and dependence on common ancestors.

Polypeptide chain

Linear amino acid polymer that is the result of the translation of an mRNA molecule. The polypeptide chain needs to be properly folded to become a functional protein.

Prokaryotes

Single-cell organism whose genetic material is not enclosed inside a nucleus.

Protein

Macromolecule made of a linear polypeptide chain of amino acids linked through peptide bonds and folded in a three-dimensional structure to perform a certain reaction. Proteins are the major constituents of cells.

Punctualism

Evolutionary tenet that holds that major differences between organisms are due to rapid changes that take place in short periods of time, followed by long periods of little change.

Releasing factors

Family of proteins that stops protein synthesis in the ribosome and facilitates the release of the already synthesized polypeptide chain.

Ribosomal RNA (rRNA)

Any of the RNA molecules that form part of the ribosome and participates in the synthesis of the polypeptide chain.

Ribosome

Macromolecular complex formed by proteins and rRNA that translates the linear information of the mRNA into the linear sequence of the polypeptide chain.

RNA

Polymer formed by ribonucleotides linearly linked.

Transcription

Process through which a DNA sequence is copied into a complementary mRNA sequence.

Transfer RNA (tRNA)

RNA molecules that recognize a specific triplet of an mRNA and also carry the corresponding amino acid to which that triplet code for. They are key molecules in the process of translation of mRNA into protein.

Translation

Process through which a polypeptide chain is synthesized following the instructions of an mRNA molecule. This process takes place in the ribosome, where every three nucleotides of mRNA are recognized by a tRNA and the amino acid that binds it is transferred to the growing polypeptide chain.

Triplet

Set of three nucleotides. When the triplet codifies for an amino acid, it is also named codon.

Bibliography

Adachi, K. — Katsuyama, M. — Song, S. — Oka, T.— *al.*, «Genomic Organization, Chromosomal Mapping and Promoter Analysis of the Mouse Selenocysteine tRNA Gene Transcription-activating Factor (mStaf) Gene», *Biochemical Journal* 346 (2000) 45-51.

Adachi, K. — Saito, H. — Tanaka, T. — Oka, T.— *al.*, «Molecular Cloning and Characterization of the Murine Staf cDNA Encoding a Transcription Activating Factor for the Selenocysteine tRNA Gene in Mouse Mammary Gland», *J. Biol. Chem.* 273(15) (1998) 8598-8606.

Agazzi, E., «Introduction», in E. Agazzi, ed., *The Problem of Reductionism in Science,* Dordrecht 1991, VII-XVIII.

———, «Reductionism as Negation of the Scientific Spirit», en E. Agazzi, ed., *The Problem of Reductionism in Science,* 1991, 1-29.

Alberts, B. – Johnson, A. – Lewis, J. – Raff, M.– *al.*, *Molecular Biology of the Cell,* New York 20085.

Allamand, V. — Richard, P. — Lescure, A. — Ledeuil, C.— *al.*, «A Single Homozygous Point Mutation in a 3' Untranslated Region Motif of Selenoprotein N mRNA Causes SEPN1-related Myopathy», *EMBO reports* 7(4) (2006) 450-454.

Allen, C. — Bekoff, M., «Biological Function, Adaptation, and Natural Design», *Philosophy of Science* 62(4) (1995) 609-622.

Allmang, C. — Carbon, P. — Krol, A., «The SBP2 and 15.5 kD/Snu13p Proteins Share the Same RNA Binding Domain: Identification of SBP2 Amino Acids Important to SECIS RNA Binding», *RNA* 8(10) (2002) 1308-1318.

Allmang, C. — Krol, A., «Selenoprotein Synthesis: UGA Does Not end the Story», *Biochimie* 88(11) (2006) 1561-1571.

Allmang, C. — Wurth, L. — Krol, A., «The Selenium to Selenoprotein Pathway in Eukaryotes: More Molecular Partners than Anticipated», *Biochimica et Biophysica Acta. General Subjects* 1790(11) (2009) 1415-1423.

Amberg, R. — Mizutani, T. — Wu, X.-Q. — Gross, H.J.— *al.*, «Selenocysteine Synthesis in Mammalia: an Identity Switch from tRNASer to tRNASec», *Journal of Molecular Biology* 263(1) (1996) 8-19.

Amberg, R. — Urban, C. — Reuner, B. — Scharff, P.— al., «Editing Does Not Exist for Mammalian Selenocysteine tRNAs», *Nucleic Acids Research* 21(24) (1993) 5583-5588.

Ambrogelly, A. — Palioura, S. — Söll, D., «Natural Expansion of the Genetic Code», *Nature Chemical Biology* 3(1) (2007) 29-35.

Araiso, Y. — Palioura, S. — Ishitani, R. — Sherrer, R.L.— al., «Structural Insights into RNA-dependent Eukaryal and Archaeal Selenocysteine Formation», *Nucleic Acids Research* 36(4) (2008) 1187-1199.

Araiso, Y. — Sherrer, R.L. — Ishitani, R. — Ho, J.M.— al., «Structure of a tR-NA-dependent Kinase Essential for Selenocysteine Decoding», *Proceedings of the National Academy of Sciences* 106(38) (2009) 16215-16220.

Ariew, A., «Platonic and Aristotelian Roots of Teleological Arguments», in A. Ariew – R. Cummins – M. Perlman, ed., *Functions. New Essays in the Philosophy of Psychology and Biology*, Oxford 2002, 7-32.

Aristoteles, *Metafísica*, Madrid 1994.

Arnér, E.S.J., «Selenoproteins—What Unique Properties Can Arise with Selenocysteine in Place of Cysteine?», *Experimental Cell Research* 316 (2010) 1296-1303.

Arthur, J.R. — McKenzie, R.C. — Beckett, G.J., «Selenium in the Immune System», *Journal of Nutrition* (2003) .

Atkins, J.F. — Baranov, P.V., «Translation: Duality in the Genetic Code», *Nature* 448(7157) (2007) 1004-1005.

Atkins, J.F. — Gesteland, F., «The 22nd Amino Acid», *Science* 296 (2002) 1409-1410.

Auletta, G. — Ellis, G.F.R. — Jaeger, L., «Top-down Causation by Information Control: from a Philosophical Problem to a Scientific Research Programme», *Journal of the Royal Society Interface* (2008) 1-15.

Auletta, G., «A Paradigm Shift in Biology?», *Information* 1 (2010) 28-59.

———, *Cognitive Biology: Dealign with Information from Bacteria to Minds*, Oxford 2011.

———, «How Many Causes Are There?», *21mo Secolo. Scienza e tecnologia* 6 (2008) 41-48.

———, *Integrated Cognitive Strategies in a Changing World*, Roma 2011.

Ayala, F.J., «Introduction», *Studies in the Philosophy of Biology*, London 1974, vi-xvi.

———, «Teleological Explanations in Evolutionary Biology», *Philosophy of Science* 37(1) (1970) 1-15.

Barbieri, M., «Has Semiotics Come of Age? and Postscript», in M. Barbieri, ed., *Introduction to Biosemiotics. The New Biological Synthesis*, Dordrecht 2008, 101-113.

Baron, C. — Böck, A., «The Length of the Aminoacyl-acceptor Stem of the Selenocysteine-specific tRNA(Sec) of Escherichia coli is the Determinant for

Binding to Elongation Factors SELB or Tu», *J. Biol. Chem.* 266(30) (1991) 20375-20379.

BARON, C. — HEIDER, J. — BÖCK, A., «Mutagenesis of selC, the Gene for the Selenocysteine-inserting tRNA-species in E. coli: Effects on in Vivo Function», *Nucleic Acids Research* 18(23) (1990) 6761-6766.

BARON, C. — WESTHOF, E. — BÖCK, A. — GIEGÉ, R.— *AL.*, «Solution Structure of Selenocysteine-inserting tRNA(Sec) from Escherichia coli. Comparison with Canonical tRNA(Ser)», *Journal of Molecular Biology* 231(2) (1993) 274-292.

BATTERMAN, R.W., *The Devil in the Details*, Oxford 2002.

BECKERMANN, A., «Introduction», in A. BECKERMANN – H. FLOHR – J. KIM, ed., *Emergence or Reduction? Essays on the Prospects of Nonreductive Physicalism*, Berlin 1992, 1-21.

BECKNER, M.O., «Vitalism», *The Enciclopedia of Philosophy*, 7 and 8, 253-256.

BEHNE, D. — KYRIAKOPOULOS, A., «Mammalian Selenium-containing Proteins», *Annual Review of Nutrition* 21 (2001) 453-473.

BELLINGER, F.P. — RAMAN, A.V. — REEVES, M.A. — BERRY, M.J.— *AL.*, «Regulation and Function of Selenoproteins in Human Disease», *Biochemical Journal* 422(1) (2009) 11-22.

BERENDA, C.W., «On Emergence and Prediction», *Journal of Philosophy* 50(9) (1953) 269-274.

BERG, J.M. – TYMOCZKO, J.L. – STRYER, L., *Biochemistry*, New York 20076.

BERGMANN, G., «Holism, Historicism, and Emergence», *Philosophy of Science* 11(4) (1944) 209-221.

BERMANO, G. — ARTHUR, J.R. — HESKETH, J.E., «Role of the 3' Untranslated Region in the Regulation of Cytosolic Glutathione Peroxidase and Phospholipid-hydroperoxide Glutathione Peroxidase Gene Expression by Selenium Supply», *Biochemical Journal* 320 (Pt 3) (1996) 891-895.

BERMANO, G. — NICOL, F. — DYER, J.A. — SUNDE, R.A.— *AL.*, «Selenoprotein Gene Expression During Selenium-repletion of Selenium-deficient rats», *Biological Trace Element Research* 51 (1996) 211-223.

BERRY, M.J. — BANU, L. — CHEN, Y.Y. — MANDEL, S.J.— *AL.*, «Recognition of UGA as a Selenocysteine Codon in Type I Deiodinase Requires Sequences in the 3' Untranslated Region», *Nature* 353(6341) (1991) 273-276.

BERRY, M.J. — BANU, L. — HARNEY, J.W. — LARSEN, P.R.— *AL.*, «Functional Characterization of the Eukaryotic SECIS Elements which Direct Selenocysteine Insertion at UGA Codons», *EMBO Journal* 12(8) (1993) 3315-3322.

BERRY, M.J. — KIEFFER, J.D. — HARNEY, J.W. — LARSEN, P.R.— *AL.*, «Selenocysteine Confers the Biochemical Properties Characteristic of the Type I Iodothyronine Deiodinase», *J. Biol. Chem.* 266(22) (1991) 14155-14158.

BEURTON, P. – FALK, R. – RHEINBERGER, H.-J., ed., *The Concept of the Gene in*

Development and Evolution. Historical and Epistemological Perspective, Cambridge 2000.

Bickhard, M.H. – Campbell, D.T., «Emergence», en P.B. Andersen – C. Emmeche – N.O. Finnemann – P.V. Christiansen– al., ed., *Downward Causation*, 2000, 322-348.

Bock, A. — Forchhammer, K. — Heider, J. — Baron, C.— al., «Selenoprotein Synthesis: an Expansion of the Genetic Code», *Trends in Biochemical Sciences* 16 (1991) 463-467.

Boorse, C., «A rebuttal on functions», in A. Ariew – R. Cummins – M. Perlman, ed., *Functions. New Essays in the Philosophy of Psychology and Biology*, Oxford 2002, 63-112.

Bösl, M.R. — Seldin, M.F. — Nishimura, S. — Taketo, M.— al., «Cloning, Structural Analysis and Mapping of the Mouse Selenocysteine tRNA([Ser]Sec) Gene (Trsp)», *Mol Gen Genet.* 248(3) (1995) 247-252.

Bösl, M.R. — Takaku, K. — Oshima, M. — Nishimura, S.— al., «Early Embryonic Lethality Caused by Targeted Disruption of the Mouse Selenocysteine tRNA Gene (Trsp)», *Proceedings of the National Academy of Sciences* 94(11) (1997) 5531-5534.

Brauckmann, S., «On Genes, Cells, and Memory», *Semiotica* 127(1-4) (1999) 151-168.

Brigelius-Flohé, R. — Kipp, A., «Glutathione Peroxidases in Different Stages of Carcinogenesis», *Biochimica et Biophysica Acta* 1790 (2009) 1555-1568.

Brown, K.M. — Arthur, J.R., «Selenium, Selenoproteins and Human Health: a Review», *Public Health Nutrition* 4(2B) (2001) 593-599.

Bruni, L.E., «Cellular Semiotics and Signal Transduction», in M. Barbieri, ed., *Introduction to Biosemiotics. The New Biological Synthesis*, Dordrecht 2008, 365-407.

Bubenik, J. — Ladd, A. — Gerber, C.A. — Budiman, M.— al., «Known Turnover and Translation Regulatory RNA-binding Proteins Interact with the 3' UTR of SECIS-binding Protein 2», *RNA Biol.* 6(1) (2009) 73-83.

Bubenik, J.L. — Driscoll, D.M., «Altered RNA Binding Activity Underlies Abnormal Thyroid Hormone Metabolism Linked to a Mutation in Selenocysteineinsertion Sequence-binding Protein 2», *J. Biol. Chem.* 282(48) (2007) 34653-34662.

Budiman, M.E. — Bubenik, J.L. — Miniard, A.C. — Middleton, L.M.— al., «Eukaryotic Initiation Factor 4a3 is a Selenium-regulated RNA-binding Protein that Selectively Inhibits Selenocysteine Incorporation», *Molecular Cell* 35(4) (2009) 479-489.

Buller, D.J., «Function and Design Revisited», in A. Ariew – R. Cummins – M. Perlman, ed., *Functions. New Essays in the Philosophy of Psychology and Biology*, Oxford 2002, 222-243.

Bunge, M., «The Power and Limits of Reduction», in E. Agazzi, ed., *The Problem of Reductionism in Science*, 1991, 31-49.

Burk, R.F. — Hill, K.E., «Selenoprotein P: an Extracellular Protein with Unique Physical Characteristics and a Role in Selenium Homeostasis», *Annual Review of Nutrition* 25 (2005) 215-235.

———, «Selenoprotein P-expression, Functions, and Roles in Mammals», *Biochimica et Biophysica Acta. General Subjects* 1790(11) (2009) 1441-1447.

Burkard, U. — Söll, D., «The Unusually Long Amino Acid Acceptor Stem of Escherichia coli Selenocysteine tRNA Results from Abnormal Cleavage by RNase P», *Nucleic Acids Research* 16(24) (1988) 11617-11624.

Caban, K. — Copeland, P.R., «Size Matters: a View of Selenocysteine Incorporation from the Ribosome», *Cellular and Molecular Life Sciences* 63(1) (2006) 73-81.

Caban, K. — Kinzy, S.A. — Copeland, P.R., «The L7Ae RNA Binding Motif is a Multifunctional Domain Required for the Ribosome-dependent Sec Incorporation Activity of Sec Insertion Sequence Binding Protein», *Molecular and Cellular Biology* 27(18) (2007) 6350-6360.

Carbon, P. — Krol, A., «Transcription of the Xenopus laevis Selenocysteine tRNA(Ser)Sec Gene: a System that Combines an Internal B Box and Upstream Elements also Found in U6 snRNA Genes», *EMBO Journal* 10(3) (1991) 599-606.

Carlson, B.A. — Moustafa, M.E. — Sengupta, A. — Schweizer, U.— al., «Selective Restoration of the Selenoprotein Population in a Mouse Hepatocyte Selenoproteinless Background with Different Mutant Selenocysteine tRNAs Lacking Um34*», *J. Biol. Chem.* 282(45) (2007) 32591-32602.

Carlson, B.A. — Novoselov, S.V. — Kumaraswamy, E. — Lee, B.J.— al., «Specific Excision of the Selenocysteine tRNA[Ser]Sec (Trsp) Gene in Mouse Liver Demonstrates an Essential Role of Selenoproteins in Liver Function», *J. Biol. Chem.* 279(9) (2004) 8011-8017.

Carlson, B.A. — Schweizer, U. — Perella, C. — Shrimali, R.K.— al., «The Selenocysteine tRNA STAF-binding Region is Essential for Adequate Selenocysteine tRNA Status, Selenoprotein Expression and Early Age Survival of Mice», *Biochemical Journal* 418 (2009) 61-71.

Carlson, B.A. — Xu, X.-M. — Gladyshev, V.N. — Hatfield, D.L.— al., «Selective Rescue of Selenoprotein Expression in Mice Lacking a Highly Specialized Methyl Group in Selenocysteine tRNA», *J. Biol. Chem.* 280(7) (2005) 5542-5548.

Carlson, B.A. — Xu, X.-M. — Kryukov, G.V. — Rao, M.— al., «Identification and Characterization of Phosphoseryl-tRNA[Ser]Sec Kinase», *Proceedings of the National Academy of Sciences* 101(35) (2004) 12848-12853.

Carlson, B.A. — Yoo, M.H. — Sano, Y. — Sengupta, A.— al., «Selenoproteins Regulate Macrophage Invasiveness and Extracellular Matrix-related Gene Expression», *BMC Immunology* 10:57 (2009) .

Castellano, S. — Lobanov, A.V. — Chapple, C. — Novoselov, S.V.— al., «Diversity and Functional Plasticity of Eukaryotic Selenoproteins: Identification and Characterization of the SelJ Family», *Proceedings of the National Academy of Sciences* 102(45) (2005) 16188-16193.

Castellano, S. — Novoselov, S.V. — Kryukov, G.V. — Lescure, A.— al., «Reconsidering the Evolution of Eukaryotic Selenoproteins: a Novel Nonmammalian Family with Scattered Phylogenetic Distribution», *EMBO reports* 5(1) (2004) 71-77.

Chalmers, D.J., «Strong and weak emergence», in P. Clayton – P. Davies, ed., *The Re-emergence of Emergence. The Emergentist Hypothesis from Science to Religion*, 2006, 244-254.

Chapple, C.E. — Guigó, R. — Krol, A., «SECISaln, a Web-based Tool for the Creation of Structure-based Alignments of Eukaryotic SECIS Elements», *Bioinformatics* 29(5) (2009) 674-675.

Chavatte, L. — Brown II, B.A. — Driscoll, D.M., «Ribosomal Protein L30 is a Component of the UGA-selenocysteine Recoding Machinery in Eukaryotes», *Nature Structural & Molecular Biology* 12(5) (2005) 408-416.

Chen, J. — Berry, M.J., «Selenium and Selenoproteins in the Brain and Brain Diseases», *Journal of Neurochemistry* 86(1) (2003) 1-12.

Chen, J.T. — Fang, L. — Inouye, M., «Effect of the Relative Position of the UGA Codon to the Unique Secondary Structure in the fdhF mRNA on its Decoding by Selenocysteinyl tRNA in Escherichia coli», *J. Biol. Chem.* 268(31) (1993) 23128-23131.

Chiba, S. — Itoh, Y. — Sekine, S. — Yokoyama, S.— al., «Structural Basis for the Major Role of O-phosphoseryl-tRNA Kinase in the UGA-specific Encoding of Selenocysteine», *Molecular Cell* 39 (2010) 410-420.

Choi, I.S. — Diamond, A.M. — Crain, P.F. — Kolker, J.D.— al., «Reconstitution of the Biosynthetic Pathway of Selenocysteine tRNAs in Xenopus oocytes», *Biochemistry* 33(2) (1994) 601-605.

Clayton, P. – Davies, P., ed., *The Re-emergence of Emergence. The Emergentist Hypothesis from Science to Religion*, Oxford 2006.

Clayton, P., «Conceptual Foundations of Emergence Theory», in P. Clayton – P. Davies, ed., *The Re-emergence of Emergence. The Emergentist Hypothesis from Science to Religion*, 2006, 1-31.

Cléry, A. — Bourguignon-Igel, V. — Allmang, C. — Krol, A.— al., «An Improved Definition of the RNA-binding Specificity of SECIS-binding Protein 2, an Essential Component of the Selenocysteine Incorporation Machinery», *Nucleic Acids Research* 35(6) (2007) 1868-1884.

Cobucci-Ponzano, B. — Rossi, M. — Moracci, M., «Recoding in Archaea», *Molecular Microbiology* 55(2) (2005) 339-348.

Commans, S. — Böck, A., «Selenocysteine Inserting tRNAs: an Overview», *FEMS Microbiological Review* 23(3) (1999) 335-351.

Copeland, P.R. — Driscoll, D.M., «Purification, Redox Sensitivity, and RNA Binding Properties of SECIS-binding Protein 2, a Protein Involved in Selenoprotein Biosynthesis», *J. Biol. Chem.* 274(36) (1999) 25447-25454.

Copeland, P.R. — Fletcher, J.E. — Carlson, B.A. — Hatfield, D.L.— al., «A Novel RNA Binding Protein, SBP2, is Required for the Translation of Mammalian Selenoprotein mRNAs», *EMBO Journal* 19(2) (2000) 306-314.

Copeland, P.R. — Stepanik, V.A. — Driscoll, D.M., «Insight into Mammalian Selenocysteine Insertion: Domain Structure and Ribosome Binding Properties of Sec Insertion Sequence Binding Protein 2», *Molecular and Cellular Biology* 21(5) (2001) 1491-1498.

Copeland, P.R., «Regulation of Gene Expression by Stop Codon Recoding: Selenocysteine», *Gene* 312 (2003) 17-25.

Cummins, R., «Functional Analysis», *Journal of Philosophy* 72(20) (1975) 741-765.

Davies, P. – Gregersen, N.H., ed., *Information and the Nature of Reality. From Physics to Metaphysics*, Cambridge 2010.

Davies, P., «The Physics of Downward Causation», in P. Clayton – P. Davies, ed., *The Re-emergence of Emergence. The Emergentist Hypothesis from Science to Religion*, Oxford 2006, 35-52.

Dawkins, R., *The Selfish Gene*, Oxford 1976.

de Jesus, L.A. — Hoffmann, P.R. — Michaud, T. — Forry, E.P.— al., «Nuclear Assembly of UGA Decoding Complexes on Selenoprotein mRNAs: a Mechanism for Eluding Nonsense-mediated Decay?», *Molecular and Cellular Biology* 26(5) (2006) 1795-1805.

Deacon, T.W., «Emergence: the Hole at the Wheel's Hub», in P. Clayton – P. Davies, ed., *The Re-emergence of Emergence. The Emergentist Hypothesis from Science to Religion*, Oxford 2006, 111-150.

———, «Three Levels of Emergent Phenomena», in N. Murphy – W.R. Stoeger, ed., *Evolution and Emergence. Systems, Organisms, Persons*, Oxford 2007, 88-110.

———, «What is Missing from Theories of Information?», in P. Davies – N.H. Gregersen, ed., *Information and the Nature of Reality. From Physics to Metaphysics*, Cambridge 2010, 146-169.

Deely, J., «Umwelt», *Semiotica* 134 (2001) 125-135.

Diamond, A.M. — Choi, I.S. — Crain, P.F. — Hashizume, T.— al., «Dietary Selenium Affects Methylation of the Wobble Nucleoside in the Anticodon of Selenocysteine tRNA([Ser]Sec)», *J. Biol. Chem.* 268(19) (1993) 14215-14223.

Ding, F. — Grabowski, P.J., «Identification of a Protein Component of a Mammalian tRNA(Sec) Complex Implicated in the Decoding of UGA as Selenocysteine», *RNA* 5(12) (1999) 1561-1569.

Donovan, J. — Caban, K. — Ranaweera, R. — Gonzalez-Flores, J.N.— al., «A Novel Protein Domain Induces High Affinity Selenocysteine Insertion Sequence Binding and Elongationfactor Recruitment», *J. Biol. Chem.* 283(50) (2008) 35129-35139.

Driscoll, D.M. — Copeland, P.R., «Mechanism and Regulation of Selenoprotein Synthesis», *Annual Review of Nutrition* 23 (2003) 17-40.

Ehrenreich, A. — Forchhammer, K. — Tormay, P. — Veprek, B.— al., «Selenoprotein Synthesis in E. coli. Purification and Characterisation of the Enzyme Catalysing Selenium Activation», *European Journal of Biochemistry* 206(3) (1992) 767-773.

El-Hani, C.N. – Pereira, A.M., «Higher-level Descriptions: Why Should We Preserve Them?», in P.B. Andersen – C. Emmeche – N.O. Finnemann – P.V. Christiansen– al., ed., *Downward Causation*, Aarhus 2000, 118-142.

Emmeche, C. – Koppe, S. – Stjernfelt, F., «Levels, Emergence, and Three Versions of Downward Causation», in P.B. Andersen – C. Emmeche – N.O. Finnemann – P.V. Christiansen– al., ed., *Downward Causation*, 2000, 13-34.

Engelberg-Kulka, H. — Schoulaker-Schwarz, R., «A Flexible Genetic Code, or Why Does Selenocysteine Have No Unique Codon?», *Trends in Biochemical Sciences* 13(11) (1988) 419-421.

Etxeberria, A., «Complementarity and Closure», *Annals of the New York Academy of Sciences* 901 (2000) 198-206.

Fabbrichesi Leo, R., *Introduzione a Peirce*, Roma-Bari 20052.

Fagegaltier, D. — Hubert, N. — Yamada, K. — Mizutani, T.— al., «Characterization of mSelB, a Novel Mammalian Elongation Factor for Selenoprotein Translation», *EMBO Journal* 19 (2000) 4796-4805.

Fagegaltier, D. — Lescure, A. — Walczak, R. — Carbon, P.— al., «Structural Analysis of New Local Features in SECIS RNA Hairpins», *Nucleic Acids Research* 28(14) (2000) 2679-2689.

Favareau, D., «The Evolutionary History of Biosemiotics», in M. Barbieri, ed., *Introduction to Biosemiotics. The New Biological Synthesis*, Dordrecht 2008, 1-67.

Fletcher, J.E. — Copeland, P.R. — Driscoll, D.M. — Krol, A.— al., «The Selenocysteine Incorporation Machinery: Interactions Between the SECIS RNA and the SECIS-binding Protein SBP2», *RNA* 7(10) (2001) 1442-1453.

Fletcher, J.E. — Copeland, P.R. — Driscoll, D.M., «Polysome Distribution of Phospholipid Hydroperoxide Glutathione Peroxidase mRNA: Evidence for a Block in Elongation at the UGA/Selenocysteine Codon», *RNA* 6(11) (2000) 1573-1584.

Forchhammer, K. — Böck, A., «Selenocysteine Synthase from Escherichia coli. Analysis of the Reaction Sequence», *J. Biol. Chem.* 266(10) (1991) 6324-6328.

Forchhammer, K. — Leinfelder, W. — Böck, A., «Identification of a Novel

Translation Factor Necessary for the Incorporation of Selenocysteine into Protein», *Nature* 342(6248) (1989) 453-456.

Forchhammer, K. — Leinfelder, W. — Boesmiller, K. — Veprek, B.— *al.*, «Selenocysteine Synthase from Escherichia coli. Nucleotide Sequence of the Gene (selA) and Purification of the Protein», *J. Biol. Chem.* 266(10) (1991) 6318-6323.

Forchhammer, K. — Rücknagel, K.-P. — Böck, A., «Purification and Biochemical Characterization of SELB, a Translation Factor Involved in Selenoprotein Synthesis», *J. Biol. Chem.* 265(16) (1990) 9346-9350.

Fornerod, M. — Ohno, M. — Yoshida, M. — Mattaj, I.— *al.*, «CRM1 is an Export Receptor for Leucine-rich Nuclear Export Signals», *Cell* 90(6) (1997) 1051-1060.

Förster, C. — Ott, G. — Forchhammer, K. — Sprinzl, M.— *al.*, «Interaction of a Selenocysteine-incorporating tRNA with Elongation Factor Tu from E.coli», *Nucleic Acids Research* 18(3) (1990) 487-491.

Fourmy, D. — Guittet, E. — Yoshizawa, S., «Structure of Prokaryotic SECIS mRNA Hairpin and its Interaction with Elongation Factor SelB», *Journal of Molecular Biology* 324(1) (2002) 137-150.

Friedman, K., «Is Intertheoretic Reduction Feasible?», *British Journal of the Philosophy of Science* 33 (1982) 17-40.

Fu, L.H. — Wang, X.F. — Eyal, Y. — She, Y.M.— *al.*, «A Selenoprotein in the Plant Kingdom. Mass Spectrometry Confirms that an Opal Codon (UGA) Encodes Selenocysteine in Chlamydomonas reinhardtii Gluththione Peroxidase», *J. Biol. Chem.* 277(29) (2002) 25983-25991.

Futuyma, D.J., *Evolution*, Sunderland 2009.

Gabrielsen, O.S. — Sentenac, A., «RNA Polymerase III (C) and its Transcription Factors», *Trends in Biochemical Sciences* 16 (1991) 412-416.

Galli, G. — Hofstetter, H. — Birnstiel, M.L., «Two Conserved Sequence Blocks within Eukaryotic tRNA Genes Are Major Promoter Elements», *Nature* 294 (1981) 626-631.

Ganichkin, O.M. — Xu, X.-M. — Carlson, B.A. — Mix, H.— *al.*, «Structure and Catalytic Mechanism of Eukaryotic Selenocysteine Synthase», *J. Biol. Chem.* 283(9) (2008) 5849-5865.

Gasdaska, J.R. — Harney, J.W. — Gasdaska, P.Y. — Powis, G.— *al.*, «Regulation of Human Thioredoxin Reductase Expression and Activity by 3'-Untranslated Region Selenocysteine Insertion Sequence and mRNA Instability Elements», *J. Biol. Chem.* 274(36) (1999) 25379-25385.

Gesteland, R.F. — Atkins, J.F., «Recoding: Dynamic Reprogramming of Translation», *Annual Review of Biochemistry* 65 (1996) 741-768.

Godfrey-Smith, P., «Explanatory Symmetries, Preformation and Developmental Systems Theory», *Philosophy of Science* 67(Supp.) (2000) S322-S331.

———, «Functions: Consensus without Unity», *Pacific Philosophical Quarterly* 74 (1993) 196-208.

———, «Information in Biology», in D.L. Hull – M. Ruse, ed., *The Cambridge Companion to the Philosophy of Biology*, Cambridge 2007, 103-119.

———, «On the Theoretical Role of «Genetic Coding»», *Philosophy of Science* 67(1) (2000) 26-44.

Griffiths, P. — Gray, R., «Developmental Systems and Evolutionary Explanation», *Journal of Philosophy* 91(6) (1994) 277-304.

Griffiths, P.E., «Functional Analysis and Proper Functions», *British Journal of the Philosophy of Science* 44 (1993) 409-422.

———, «Genetic Information: a Metaphor in Search of a Theory», *Philosophy of Science* 68(3) (2001) 394-412.

Gromer, S. — Eubel, J.K. — Lee, B.L. — Jacob, J.— *al.*, «Human Selenoproteins at a Glance», *Cellular and Molecular Life Sciences* 62(21) (2005) 2414-2437.

Gromer, S. — Johansson, L. — Bauer, H. — Arscott, L.D.— *al.*, «Active Sites of Thioredoxin Reductases: Why Selenoproteins?», *Proceedings of the National Academy of Sciences* 100(22) (2003) 12618-12623.

Grundner-Culemann, E. — Martin, G.3. — Harney, J.W. — Berry, M.J.— *al.*, «Two Distinct SECIS Structures Capable of Directing Selenocysteine Incorporation in Eukaryotes», *RNA* 5(5) (1999) 625-635.

Grundner-Culemann, E. — Martin III, G.W. — Tujebajeva, R. — Harney, J.W.— *al.*, «Interplay Between Termination and Translation Machinery in Eukaryotic Selenoprotein Synthesis», *Journal of Molecular Biology* 310(4) (2001) 699-707.

Gu, Q.P. — Ream, W. — Whanger, P.D., «Selenoprotein W Gene Regulation by Selenium in L8 Cells», *Biometals* 15(4) (2002) 411-420.

Guimarães, M.J. — Peterson, D. — Vicari, A. — Cocks, B.G.— *al.*, «Identification of a Novel selD Homolog from Eukaryotes, Bacteria, and Archaea: is There an Autoregulatory Mechanism in Selenocysteine Metabolism?», *Proceedings of the National Academy of Sciences* 93(26) (1996) 15086-15091.

Gupta, M. — Copeland, P.R., «Functional Analysis of the Interplay between Translation Termination, Selenocysteine Codon Context, and Selenocysteine Insertion Sequence-binding Protein 2», *J. Biol. Chem.* 282(51) (2007) 36797-36807.

Halic, M. — Becker, T. — Frank, J. — Spahn, C.M.T.— *al.*, «Localization and Dynamic Behavior of Ribosomal Protein L30e», *Nature Structural & Molecular Biology* 12 (5) (2005) 467-468.

Hardcastle, V.G., «On the Normativity of Functions», in A. Ariew – R. Cummins – M. Perlman, ed., *Functions. New Essays in the Philosophy of Psychology and Biology*, Oxford 2002, 144-156.

Hatfield, D. — Lee, B.J. — Hampton, L. — Diamond, A.M.— *al.*, «Selenium

Induces Changes in the Selenocysteine tRNA[Ser]sec Population in Mammalian Cells», *Nucleic Acids Research* 19(4) (1991) 939-943.

Hatfield, D.L. — Gladyshev, V.N., «How Selenium Has Altered Our Understanding of the Genetic Code», *Molecular and Cellular Biology* 22(11) (2002) 3565-3576.

Hatfield, D.L. — Yoo, M.H. — Carlson, B.A. — Gladyshev, V.N.— *al.*, «Selenoproteins that Function in Cancer Prevention and Promotion», *Biochimica et Biophysica Acta* 1790 (2009) 1541-1545.

Hawkes, W.C. — Alkan, Z., «Regulation of Redox Signaling by Selenoproteins», *Biological Trace Element Research* 134 (2010) 235-251.

Heckl, M. — Busch, K. — Gross, H.J., «Minimal tRNA(Ser) and tRNA(Sec) Substrates for Human Seryl-tRNA Synthetase: Contribution of tRNA Domains to Serylation and Tertiary Structure», *FEBS Letters* 427 (1998) 315-319.

Heider, J. — Baron, C. — Böck, A., «Coding from a Distance: Dissection of the mRNA Determinants Required for the Incorporation of Selenocysteine into Protein», *EMBO Journal* 11(10) (1992) 3759-3766.

Henle, P., «The Status of Emergence», *Journal of Philosophy* 39(18) (1942) 486-493.

Hilgenfeld, R. — Böck, A. — Wilting, R., «Structural Model for the Selenocysteine-specific Elongation Factor SelB», *Biochimie* 78(11-12) (1996) 971-978.

Hill, K.E. — Lloyd, R.S. — Burk, R.F., «Conserved Nucleotide Sequences in the Open Reading Frame and 3' Untranslated Region of Selenoprotein P mRNA», *Proceedings of the National Academy of Sciences* 90(2) (1993) 537-541.

Hill, K.E. — Lyons, P.R. — Burk, R.F., «Differential Regulation of Rat Liver Selenoprotein mRNAs in Selenium Deficiency», *Biochemical and Biophysical Research Communications* 185(1) (1992) 260-263.

Hill, K.E. — Zhou, J. — McMahan, W.J. — Motley, A.K.— *al.*, «Deletion of Selenoprotein P Alters Distribution of Selenium in the Mouse», *J. Biol. Chem.* 278(16) (2003) 13640-13646.

Hirosawa-Takamori, M. — Chung, H.R. — Jäckle, H., «Conserved Selenoprotein Synthesis is Not Critical for Oxidative Stress Defence and the Lifespan of Drosophila», *EMBO Reports* 5(3) (2004) 317-322.

Hoffmann, P.R. — Berry, M.J., «Selenoprotein Synthesis: a Unique Translational Mechanism Used by a Diverse Family of Proteins», *Thyroid* 15(8) (2005) 769-775.

Houser, N. – Kloesel, C., ed., *The Essential Peirce*, vol. II Bloomington 1998.

Howard, M.T. — Aggarwal, G. — Anderson, C.B. — Khatri, S.— *al.*, «Recoding Elements Located Adjacent to a Subset of Eukaryal Selenocysteine-specifying UGA Codons», *EMBO Journal* 24 (2005) 1596-1607.

Howard, M.T. — Moyle, M.W. — Aggarwal, G. — Carlson, B.A.— *al.*, «A Recoding Element that Stimulates Decoding of UGA Codons by Sec tRNA[Ser] Sec», *RNA* 13 (2007) 912-920.

Hubert, N. — Sturchler, C. — Westhof, E. — Carbon, P.— *al.*, «The 9/4 Secondary

Structure of Eukaryotic Selenocysteine tRNA: More Pieces of Evidence», *RNA* 4(9) (1998) 1029-1033.

Hubert, N. — Walczak, R. — Sturchler, C. — Myslinski, E.— *al.*, «RNAs Mediating Cotranslational Insertion of Selenocysteine in Eukaryotic Selenoproteins», *Biochimie* 78(7) (1996) 590-596.

Hull, D., «Reduction in Genetics - Biology of Philosophy?», *Philosophy of Science* 39(4) (1972) 491-499.

Hüttenhofer, A. — Böck, A., «Selenocysteine Inserting RNA Elements Modulate GTP Hydrolysis of Elongation Factor SelB», *Biochemistry* 37(3) (1998) 885-890.

Hüttenhofer, A. — Heider, J. — Böck, A., «Interaction of the Escherichia coli fdhF mRNA Hairpin Promoting Selenocysteine Incorporation with the Ribosome», *Nucleic Acids Research* 24(20) (1996) 3903-3910.

Ibba, M. — Söll, D., «Genetic Code: Introducing Pyrrolysine», *Current Biology* 12(13) (2002) R464-R466.

Ioudovitch, A. — Steinberg, S.V., «Modeling the Tertiary Interactions in the Eukaryotic Selenocysteine tRNA», *RNA* 4 (1998) 365-373.

———, «Structural Compensation in an Archaeal Selenocysteine Transfer RNA», *Journal of Molecular Biology* 290(2) (1999) 365-371.

Ito, Y. — Mizutani, T., «Evidence of G-A/A-G Pair, not A-G/G-A Pair, in SECIS Element», *Nucleic Acids Research Supplement* 1(1) (2001) 41-42.

Itoh, T. — Chiba, S. — Sekine, S. — Yokoyama, S.— *al.*, «Crystal Structure of Human Selenocysteine tRNA», *Nucleic Acids Research* 37(18) (2009) 6259-6268.

Itoh, Y. — Sekine, S. — Matsumoto, E. — Akasaka, R.— *al.*, «Structure of Selenophosphate Synthetase Essential for Selenium Incorporation into Proteins and RNAs», *Journal of Molecular Biology* 385(5) (2009) 1456-1469.

Jablonka, E., «Information: Its Interpretation, Its Inheritance and Its Sharing», *Philosophy of Science* 69(4) (2002) 578-605.

Jakób Liszka, J., *A General Introduction to the Semeiotic of Charles Sanders Peirce*, Bloomington and Indianapolis 1996.

Jameson, R.R. — Diamond, A.M., «A Regulatory Role for Sec tRNA[Ser]Sec in Selenoprotein Synthesis», *RNA* 10 (2004) 1142-1152.

Jämsä, T., «Semiosis in Evolution», in M. Barbieri, ed., *Introduction to Biosemiotics. The New Biological Synthesis*, 2008, 69-100.

Joslyn, C., «Levels of Control and Closure in Complex Semiotic Systems», *Annals of the New York Academy of Sciences* 901 (2000) 67-74.

Juarrero, A., *Dynamics in Action. Intentional Behavior as a Complex System*, Massachusetts 1999.

Kaiser, J.T. — Gromadski, K. — Rother, M. — Engelhardt, H.— *al.*, «Structural and Functional Investigation of a Putative Archaeal Selenocysteinesynthase», *Biochemistry* 44 (2005) 13315-13327.

Kaplan, J.M. — Pigliucci, M., «Genes «for» Phenotypes: a Modern History View», *Biology and Philosophy* 16 (2001) 189-213.

Kauffman, S.A., *The Origins of Order. Self-organization and Selection in Evolution*, Oxford 1993.

Kay, L.E., *Who Wrote the Book of Life? A History of the Genetic Code*, Stanford 2000.

Keeling, P.J. — Fast, N.M. — McFadden, G.I., «Evolutionary Relationship between Translation Initiation Factor eIF-2gamma and Selenocysteine-specific Elongation Factor SELB: Change of Function in Translation Factors», *Journal of Molecular Evolution* 47(6) (1998) 649-655.

Kelly, V.P. — Suzuki, T. — Nakajima, O. — Arai, T.— *al.*, «The Distal Sequence Element of the Selenocysteine tRNA Gene is a Tissue-dependent Enhancer Essential for Mouse Embryogenesis», *Molecular and Cellular Biology* 25(9) (2005) 3658-3669.

Kemeny, J.G. — Oppenheim, P., «On Reduction», *Philosophical Studies* 7 (1956) 6-17.

Kim, H.-Y. — Gladyshev, V.N., «Methionine Sulfoxide Reductases: Selenoprotein Forms and Rolesin Antioxidant Protein Repair in Mammals», *Biochemical Journal* 407 (2007) 321-329.

———, «Methionine Sulfoxide Reduction in Mammals: Characterization of Methionine-R-sulfoxide Reductases», *Molecular Biology of the Cell* 15 (2004) 1055-1064.

Kim, H.Y. — Fomenko, D.E. — Yoon, Y.E. — Gladyshev, V.N.— *al.*, «Catalytic Advantages Provided by Selenocysteine in Methionine-S-sulfoxide Reductases», *Biochemistry* 45(46) (2006) 13697-13704.

Kim, H.Y. — Gladyshev, V.N., «Different Catalytic Mechanisms in Mammalian Selenocysteine- and Cysteine-containing Methionine-R-sulfoxide Reductases», *PLOS Biology* 3(12):e375 (2005) 2080-2089.

Kim, J., ««Downward Causation» in Emergentism and Nonreductive Physicalism», in A. Beckermann – H. Flohr – J. Kim, ed., *Emergence or Reduction? Essays on the Prospects of Nonreductive Physicalism*, Berlin 1992, 119-138.

———, «Being Realistic About Emergence», in P. Clayton – P. Davies, ed., *The Re-emergence of Emergence. The Emergentist Hypothesis from Science to Religion*, Oxford 2006, 189-202.

———, «Concepts of Supervenience», *Philosophy and Phenomenological Research* 45(2) (1984) 153-176.

———, «Psychophysical Supervenience», *Philosophical Studies* 41 (1982) 51-70.

———, «Supervenience and Nomological Incommensurables», *American Philosophical Quarterly* 15(2) (1978) 149-156.

Kim, L.K. — Matsufuji, T. — Matsufuji, S. — Carlson, B.A.— *al.*, «Methylation of the Ribosyl Moiety at Position 34 of Selenocysteine tRNA[Ser]Sec is Governed by both Primary and Tertiary Structure», *RNA* 6 (2000) 1306-1315.

Kinzy, S.A. — Caban, K. — Copeland, P.R., «Characterization of the SECIS Binding Protein 2 Complex Required for the Co-translational Insertion of Selenocysteine in Mammals», *Nucleic Acids Research* 33(16) (2005) 5172-5180.

Klein, D.J. — Schmeing, T.M. — Moore, P.B. — Steitz, T.A.— *al.*, «The Kink-turn: a New RNA Secondary Structure Motif», *EMBO Journal* 20 (2001) 4214-4221.

Knight, R.D. — Freeland, S.J. — Landweber, L.F., «Selection, History and Chemistry: the Three Faces of the Genetic Code», *Trends in Biochemical Sciences* 24(6) (1999) 241-247.

Köhrle, J. — Jakob, F. — Contempré, B. — Dumont, J.E.— *al.*, «Selenium, the Thyroid, and the Endocrine System», *Endocrine Reviews* 26(7) (2005) 944-984.

Köhrle, J., «Selenium and the Control of Thyroid Hormone Metabolism», *Thyroid* 15(8) (2005) 841-853.

Kollmus, H. — Flohé, L. — McCarthy, J.E., «Analysis of Eukaryotic mRNA Structures Directing Cotranslational Incorporation of Selenocysteine», *Nucleic Acids Research* 24(7) (1996) 1195-1201.

Korotkov, K.V. — Novoselov, S.V. — Hatfield, D.L. — Gladyshev, V.N.— *al.*, «Mammalian Selenoprotein in which Selenocysteine (Sec) Incorporation is Supported by a New Form of Sec Insertion Sequence Element», *Molecular and Cellular Biology* 22(5) (2002) 1402-1411.

Krol, A., «Evolutionarily Different RNA Motifs and RNA–protein Complexes to Achieve Selenoprotein Synthesis», *Biochimie* 84(8) (2002) 765-774.

Kromayer, M. — Neuhierl, B. — Friebel, A. — Böck, A.— *al.*, «Genetic Probing of the Interaction between the Translation Factor SelB and its mRNA Binding Element in Escherichia coli», *Mol. Gen. Genet.* 262(4-5) (1999) 800-806.

Kromayer, M. — Wilting, R. — Tormay, P. — Böck, A.— *al.*, «Domain Structure of the Prokaryotic Selenocysteine-specific Elongation Factor SelB», *Journal of Molecular Biology* 262(4) (1996) 413-420.

Kryukov, G.V. — Castellano, S. — Novoselov, S.V. — Lobanov, A.V.— *al.*, «Characterization of Mammalian Selenoproteomes», *Science* 300(5624) (2003) 1439-1443.

Kryukov, G.V. — Gladyshev, V.N., «Selenium Metabolism in Zebrafish: Multiplicity of Selenoprotein Genes and Expression of a Protein Containing 17 Selenocysteine Residues», *Genes to Cells* 5 (2000) 1049-1060.

———, «The Prokaryotic Selenoproteome», *EMBO Reports* 5(5) (2004) 538-543.

Kuiper, G.G. — Klootwijk, W. — Visser, T.J., «Substitution of Cysteine for Selenocysteine in the Catalytic Center of Type III Iodothyronine Deiodinase Reduces Catalytic Efficiency and Alters Substrate Preference», *Encocrinology* 144 (2003) 2505-2513.

Kumaraswamy, E. — Carlson, B.A. — Morgan, F. — Miyoshi, K.— *al.*, «Selective

Removal of the Selenocysteine tRNA [Ser]Sec Gene (Trsp) in Mouse Mammary Epithelium», *Molecular and Cellular Biology* 23(5) (2003) 1477-1488.

Küppers, B.-O. , *Information and the Origin of Life*, Massachusetts 1990.

———, «Understanding Complexity», *Emergence or Reduction? Essays on the Prospects of Nonreductive Physicalism*, Berlin 1992, 241-256.

Latreche, L. — Jean-Jean, O. — Driscoll, D.M. — Chavatte, L.— *al.*, «Novel Structural Determinants in Human SECIS Elements Modulate the Translational Recoding of UGA as Selenocysteine», *Nucleic Acids Research* 37(17) (2009) 5868-5880.

Lee, B.J. — Rajagopalan, M. — Kim, Y.S. — You, K.H.— *al.*, «Selenocysteine tRNA[Ser]Sec Gene is Ubiquitous within the Animal Kingdom», *Molecular and Cellular Biology* 10(5) (1990) 1940-1949.

Lee, B.J. — Worland, P.J. — Davis, J.N. — Stadtman, T.C.— *al.*, «Identification of a Selenocysteyl-tRNA(Ser) in Mammalian Cells that Recognizes the Nonsense Codon, UGA», *J. Biol. Chem.* 264(17) (1989) 9724-9727.

Lei, X.G. — Evenson, J.K. — Thompson, K.M. — Sunde, R.A.— *al.*, «Glutathione Peroxidase and Phospholipid Hydroperoxide Glutathione Peroxidase are Differentially Regulated in Rats by Dietary Selenium», *Journal of Nutrition* 125(6) (1995) 1438-1446.

Leibundgut, M. — Frick, C. — Thanbichler, M. — Böck, A.— *al.*, «Selenocysteine tRNA-specific Elongation Factor SelB is a Structural Chimaera of Elongation and Initiation Factors», *EMBO Journal* 24 (2005) 11-22.

Leinfelder, W. — Forchhammer, K. — Veprek, B. — Zehelein, E.— *al.*, «In Vitro Synthesis of Selenocysteinyl-tRNA(UCA) from Seryl-tRNA(UCA): Involvement and Characterization of the SelD Gene Product», *Proceedings of the National Academy of Sciences* 87(2) (1990) 543-547.

Leinfelder, W. — Forchhammer, K. — Zinoni, F. — Sawers, G.— *al.*, «Escherichia coli Genes whose Products are Involved in Selenium Metabolism», *Journal of Bacteriology* 170(2) (1988) 540-546.

Leinfelder, W. — Stadtman, T.C. — Böck, A., «Occurrence in Vivo of Selenocysteyl-tRNA(SERUCA) in Escherichia coli. Effect of sel Mutations», *J. Biol. Chem.* 264(17) (1989) 9720-9723.

Leinfelder, W. — Zehelein, E. — Mandrand-Berthelot, M.-A. — Böck, A.— *al.*, «Gene for a Novel tRNA Species that Accepts L-serine and Cotranslationally Inserts Selenocysteine», *Nature* 331(6158) (1988) 723-725.

Lemke, J.L., «Opening up Closure: Semiotics Across Scales», *Annals of the New York Academy of Sciences* 901 (2000) 100-111.

Leontis, N.B. — Westhof, E., «Geometric Nomenclature and Classification of RNA Base Pairs», *RNA* 7 (2001) 499-512.

Lescure, A. — Allmang, C. — Yamada, K. — Carbon, P.— *al.*, «cDNA Cloning,

Expression Pattern and RNA Binding Analysis of Human Selenocysteine Insertion Sequence (SECIS) Binding Protein 2», *Gene* 291(1-2) (2002) 279-285.

Lescure, A. — Rederstorff, M. — Krol, A. — Guicheney, P.— *al.*, «Selenoprotein Function and Muscle Disease», *Biochimica et Biophysica Acta* 1790 (2009) 1569-1574.

Lewes, G.H., *Problems of Life and Mind*, Vol. II, London 1875.

Lewin, B., *Genes IX*, Sudbury 2008.

Liu, Z. — Reches, M. — Engelberg-Kulka, H., «A Sequence in the Escherichia coli fdhF «Selenocysteine Insertion Sequence» (SECIS) Operates in the Absence of Selenium», *Journal of Molecular Biology* 294(5) (1999) 1073-1086.

Lloyd Morgan, C., *Emergent Evolution*, London 19272.

Lobanov, A.V. — Fomenko, D.E. — Zhang, Y. — Sengupta, A.— *al.*, «Evolutionary Dynamics of Eukaryotic Selenoproteomes: Large Selenoproteomes May Associate with Aquatic Life and Small with Terrestrial Life», *Genome Biology* 8:RI98 (2007) .

Lobanov, A.V. — Hatfield, D.L. — Gladyshev, V.N., «Eukaryotic Selenoproteins and Selenoproteomes», *Biochimica et Biophysica Acta. General Subjects* 1790(11) (2009) 1424-1428.

———, «Reduced Reliance on the Trace Element Selenium During Evolution of Mammals», *Genome Biology* 9:R62 (2008) .

Lobanov, A.V. — Kryukov, G.V. — Hatfield, D.L. — Gladyshev, V.N.— *al.*, «Is There a Twenty Third Amino Acid in the Genetic Code?», *Trends in Genetics* 22(7) (2006) 357-360.

Loewenstein, W.R., *The Touchstone of Life. Molecular Information, Cell Communication, and the Foundations of Life*, Oxford 1999.

Longtin, R., «A Forgotten Debate: is Selenocysteine the 21st Amino Acid?», *Journal of the National Cancer Institute* 96(7) (2004) 504-505.

Low, S.C. — Berry, M.J., «Knowing When Not to Stop: Selenocysteine Incorporation in Eukaryotes», *Trends in Biochemical Sciences* 21(6) (1996) 203-208.

Low, S.C. — Grundner-Culemann, E. — Harney, J.W. — Berry, M.J.— *al.*, «SECIS-SBP2 Interactions Dictate Selenocysteine Incorporation Efficiency and Selenoprotein Hierarchy», *EMBO Journal* 19(24) (2000) 6882-6890.

Low, S.C. — Harney, J.W. — Berry, M.J., «Cloning and Functional Characterization of Human Selenophosphate Synthetase, an Essential Component of Selenoprotein Synthesis», *J. Biol. Chem.* 270(37) (1995) 21659-21664.

Lu, J. — Holmgren, A., «Selenoproteins», *J. Biol. Chem.* 284(2) (2009) 723-727.

Mackenzie, W.L., «The Notion of Emergence», *Proceedings of the Aristotelian Society, Suppl.* 6 (1926) 56-68.

Maiti, B. — Arbogast, S. — Allamand, V. — Moyle, M.W.— *al.*, «A Mutation in the SEPN1 Selenocysteine Redefinition Element (SRE) Reduces Selenocysteine

Incorporation and Leads to SEPN1-related Myopathy», *Human Mutation* 30(3) (2009) 411-416.

Mansell, J.B. — Guévremont, D. — Poole, E.S. — Tate, W.P.— *al.*, «A Dynamic Competition Between Release Tactor 2 and the tRNA(Sec) Decoding UGA at the Recoding Site of Escherichia coli Formate Dehydrogenase H», *EMBO Journal* 20(24) (2001) 7284-7293.

Martin, G.W.3. — Harney, J.W. — Berry, M.J., «Functionality of Mutations at Conserved Nucleotides in Eukaryotic SECIS Elements is Determined by the Identity of a Single Nonconserved Nucleotide», *RNA* 4(1) (1998) 65-73.

———, «Selenocysteine Incorporation in Eukaryotes: Insights into Mechanism and Efficiency from Sequence, Structure, and Spacing Proximity Studies of the Type 1 Deiodinase SECIS Element», *RNA* 2(2) (1996) 171-182.

Martin III, G.W. — Berry, M.J., «Selenocysteine Codons Decrease Polysome Association on Endogenous Selenoprotein mRNAs», *Genes to Cells* 6 (2001) 121-129.

Martin-Romero, F.J. — Kryukov, G.V. — Lobanov, A.V. — Carlson, B.A.— *al.*, «Selenium Metabolism in Drosophila: Selenoproteins, Selenoprotein mRNA Expression, Fertility, and Mortality», *J. Biol. Chem.* 276(32) (2001) 29798-29804.

Maynard Smith, J., «The Concept of Information in Biology», *Philosophy of Science* 67(2) (2000) 177-194.

Mazumder, B. — Seshadri, V. — Fox, P.L., «Translational Control by the 3'-UTR: the Ends Specify the Means», *Trends in Biochemical Sciences* 28(2) (2003) 91-98.

McCaughan, K.K. — Brown, C.M.D., M. E. — Berry, M.J. — Tate, W.P.— *al.*, «Translational Termination Efficiency in Mammals is Influenced by the Base Following the Stop Codon», *Proceedings of the National Academy of Sciences* 92(12) (1995) 5431-5435.

McLaughlin, B.P., «The Rise and Fall of British Emergentism», in A. Beckermann – H. Flohr – J. Kim, ed., *Emergence or Reduction? Essays on the Prospects of Nonreductive Physicalism*, London 1992, 49-93.

McLaughlin, P., *What Functions Explain*, Cambridge 2001.

Mehta, A. — Rebsch, C.M. — Kinzy, S.A. — Fletcher, J.E.— *al.*, «Efficiency of Mammalian Selenocysteine Incorporation», *J. Biol. Chem.* 279(36) (2004) 37852-37859.

Meissner, W. — Wanandi, I. — Carbon, P. — Krol, A.— *al.*, «Transcription Factors Required for the Expression of Xenopus laevis Selenocysteine tRNA in Vitro», *Nucleic Acids Research* 22(4) (1994) 553-559.

Mertz, W., «The Essential Trace Elements», *Science* 213(4514) (1981) 1332-1338.

Mignone, F. — Gissi, C. — Liuni, S. — Pesole, G.— *al.*, «Untranslated Regions of mRNAs», *Genome Biology* 3(3) (2002) .

Millikan, R.G., «Biofunctions: Two Paradigms», en A. Ariew – R. Cummins –

M. Perlman, ed., *Functions. New Essays in the Philosophy of Psychology and Biology*, Oxford 2002, 113-143.

———, «In Defense of Proper Functions», *Philosophy of Science* 56(2) (1989) 288-302.

Mizutani, T. — Goto, C., «Eukaryotic Selenocysteine tRNA Has the 9/4 Secondary Structure», *FEBS Letters* 466(2-3) (2000) 359-362.

Mizutani, T. — Kanaya, K. — Ikeda, S. — Fujiwara, T.— *al.*, «The Dual Identities of Mammalian tRNA(Sec) for SerRS and Selenocysteine Synthase», *Molecular Biology Reports* 25(4) (1998) 211-216.

Mizutani, T. — Tanabe, K. — Yamada, K., «A G.U Base Pair in the Eukaryotic Selenocysteine tRNA is Important for Interaction with SePF, the Putative Selenocysteine-specific Elongation Factor», *FEBS Letters* 429(2) (1998) 189-193.

Moghadaszadeh, B. — Beggs, A.H., «Selenoproteins and Their Impact on Human Health Through Diverse Physiological Pathways», *Physiology* 21 (2006) 307-315.

Moreno, A. — Ruiz-Mirazo, K., «Metabolism and the Problem of Its Universalization», *BioSystems* 49 (1999) 45-61.

Moreno, A. – Umerez, J., «Downward Causation at the Core of Living Organization», in P.B. Andersen – C. Emmeche – N.O. Finnemann – P.V. Christiansen– *al.*, ed., *Downward Causation*, Aarhus 2000, 99-117.

Moreno, A., «Closure, Identity, and the Emergence of Formal Causation», *Annals of the New York Academy of Sciences* 901 (2000) 112-121.

Moriarty, P.M. — Reddy, C.C. — Maquat, L.E., «Selenium Deficiency Reduces the Abundance of mRNA for Se-dependent Glutathione Peroxidase 1 by a UGA-dependent Mechanism Likely to Be Nonsense Codon-mediated Decay of Cytoplasmic mRNA», *Molecular and Cellular Biology* 18(5) (1998) 2932-2939.

Morowitz, H.J., *The emergence of Everything. How the World Became Complex*, Oxford 2002.

Moustafa, M.E. — Carlson, B.A. — El-Saadani, M.A. — Kryukov, G.V.— *al.*, «Selective Inhibition of Selenocysteine tRNA Maturation and Selenoprotein Synthesis in Transgenic Mice Expressing Isopentenyladenosine-deficient Selenocysteine tRNA», *Molecular and Cellular Biology* 21(11) (2001) 3840-3852.

Murphy, N., «Reductionism: How Did We Fall into It and Can We Emerge from It?», in N. Murphy – W.R. Stoeger, ed., *Evolution and Emergence. Systems, Organisms, Persons*, Oxford 2007, 19-39.

Myslinski, E. — Schuster, C. — Huet, J. — Sentenac, A.— *al.*, «Point Mutations 5' to the tRNA Selenocysteine TATA Box Alter RNA Polymerase III Transcription by Affecting the Binding of TBP», *Nucleic Acids Research* 21(25) (1993) 5852-5858.

Nagel, E., *The Structure of Science. Problems in the Logic of Scientific Explanation*, London 1961.

Namy, O. — Rousset, J.P. — Napthine, S. — Brierley, I.— *al.*, «Reprogrammed

Genetic Decoding in Cellular Gene Expression», *Molecular Cell* 13(2) (2004) 157-158.

Nasim, M.T. — Jaenecke, S. — Belduz, A. — Kollmus, H.— *al.*, «Eukaryotic Selenocysteine Incorporation Follows a Nonprocessive Mechanism that Competes with Translational Termination», *J. Biol. Chem.* 275(20) (2000) 14846-14852.

Nicolis, G., «Physics of Far-from-equilibrium Systems and Self-organisation», in P. Davies, ed., *The New Physics*, Cambridge 1989, 316-347.

Ohama, T. — Yang, D.C. — Hatfield, D.L., «Selenocysteine tRNA and Serine tRNA Are Aminoacylated by the Same Synthetase, but May Manifest Different Identities with Respect to the Long Extra Arm», *Archives of Biochemistry and Biophysics* 315(2) (1994) 293-301.

Oliéric, V. — Wolff, P. — Takeuchi, A. — Bec, G.— *al.*, «SECIS-binding Protein 2, a Key Player in Selenoprotein Synthesis, Is an Intrinsically Disordered Protein», *Biochimie* 91(8) (2009) 1003-1009.

Oppenheim, P. — Putnam, H., «Unity of Science as a Working Hypothesis», *Minnesota Studies in the Philosophy of Science* Vol. II (1958) 3-36.

Ose, T. — Soler, N. — Rasubala, L. — Kuroki, K.— *al.*, «Structural Basis for Dynamic Interdomain Movement and RNA Recognition of the Selenocysteine-specific Elongation Factor SelB», *Structure* 15(5) (2007) 577-586.

Oyama, S., *The Ontogeny of Information. Developmental Systems and Evolution*, Duke 20002.

Paleskava, A. — Konevega, A.L. — Rodnina, M.V., «Thermodynamic and Kinetic Framework of Selenocysteyl-tRNASec Recognition by Elongation Factor SelB», *J. Biol. Chem.* 285(5) (2010) 3014-3020.

Palioura, S. — Sherrer, R.L. — Steitz, T.A. — Söll, D.— *al.*, «The Human SepSecS-tRNASec Complex Reveals the Mechanism of Selenocysteine Formation», *Science* 325(5938) (2009) 321-325.

Pap, A., «The Concept of Absolute Emergence», *British Journal of the Philosophy of Science* 2 (1952) 302-311.

Papp, L.V. — Lu, J. — Striebel, F. — Kennedy, D.— *al.*, «The Redox State of SECIS Binding Protein 2 Controls Its Localization and Selenocysteine Incorporation Function», *Molecular and Cellular Biology* 26(13) (2006) 4895-4910.

Park, J.M. — Choi, I.S. — Kang, S.G. — Lee, J.Y.— *al.*, «Upstream Promoter Elements Are Sufficient for Selenocysteine tRNA[Ser]Sec Gene Transcription and to Determine the Transcription Start Point», *Gene* 162(1) (1995) 13-19.

Park, J.M. — Lee, J.Y. — Hatfield, D.L. — Lee, B.J.— *al.*, «Differential Mode of TBP Utilization in Transcription of the tRNA[Ser]Sec Gene and TATA-less Class III Genes», *Gene* 196(1-2) (1997) 99-103.

Pattee, H.H., «Causation, Control, and the Evolution of Complexity», in P.B. Andersen

– C. Emmeche – N.O. Finnemann – P.V. Christiansen– *al.*, ed., *Downward Causation*, Aarhus 2000, 63-77.

———, «The Necessity of Biosemiotics: Matter-symbol Complementarity», in M. Barbieri, ed., *Introduction to Biosemiotics. The New Biological Synthesis*, Dordrecht 2008, 115-132.

Peirce, Ch. S. CP «The Collected Papers», Vols. I-VI, Cambridge 1931-1935; Vols. VII-VIII, Cambridge 1958.

Polanyi, M., «Life's Irreducible Structure», *Science* 160(3834) (1968) 1308-1312.

Ringquist, S. — Schneider, D. — Gibson, T. — Baron, C.— *al.*, «Recognition of the mRNA Selenocysteine Insertion Sequence by the Specialized Translational Elongation Factor SELB», *Genes & Development* 8 (1994) 376-385.

Rosenberg, A., «Reductionism (and Antireductionism) in Biology», in D.L. Hull – M. Ruse, ed., *The Cambridge Companion to the Philosophy of Biology*, 2007, 120-138.

Rother, M. — Resch, A. — Gardner, W.L. — Whitman, W.B.— *al.*, «Heterologous Expression of Archaeal Selenoprotein Genes Directed by the SECIS Element Located in the 3' Non-translated Region», *Molecular Microbiology* 40(4) (2001) 900-908.

Rother, M. — Wilting, R. — Commans, S. — Böck, A.— *al.*, «Identification and Characterisation of the Selenocysteine-specific Translation Factor SelB from the Archaeon Methanococcus jannaschii», *Journal of Molecular Biology* 299(2) (2000) 351-358.

Roy, M. — Kiremidjian-Schumacher, L. — Wishe, H.I. — Cohen, M.W.— *al.*, «Selenium Supplementation Enhances the Expression of Interleukin 2 Receptor Subunits and Internalization of Interleukin 2», *Proceedings of the Society for Experimental Biology and Medicine* 202 (1993) 295-301.

Ruse, M., «Evolutionary Biology and Teleological Thinking», in A. Ariew – R. Cummins – M. Perlman, ed., *Functions. New Essays in the Philosophy of Psychology and Biology*, Oxford 2002, 33-59.

Salthe, S.N., *Development and Evolution. Complexity and Change in Biology*, Massachusetts 1993.

———, «What Is the Scope of Biosemiotics? Information in Living Systems», in M. Barbieri, ed., *Introduction to Biosemiotics. The New Biological Synthesis*, Dordrecht 2008, 133-148.

Sandman, K.E. — Noren, C.J., «The Efficiency of Escherichia coli Selenocysteine Insertion is Influenced by the Immediate Downstream Nucleotide», *Nucleic Acids Research* 28(3) (2000) 755-761.

Sandman, K.E. — Tardiff, D.F. — Neely, L.A. — Noren, C.J.— *al.*, «Revised Escherichia coli Selenocysteine Insertion Requirements Determined by in Vivo Screening of Combinatorial Libraries of SECIS Variants», *Nucleic Acids Research* 31(8) (2003) 2234-2241.

Santaella Braga, L., «A New Causality for the Understanding of the Living», *Semiotica* 127(1-4) (1999) 497-520.

Sarkar, S., *Genetics and Reductionism*, Cambridge 1998.

Sawers, G. — Heider, J. — Zehelein, E. — Böck, A.— *al.*, «Expression and Operon Structure of the Sel Genes of Escherichia coli and Identification of a Third Selenium-containing Formate Dehydrogenase Isoenzyme», *Journal of Bacteriology* 173(16) (1991) 4983-4993.

Scharpf, M. — Schweizer, U. — Arzberger, T. — Roggendorf, W.— *al.*, «Neuronal and Ependymal Expression of Selenoprotein P in the Human Brain», *Journal of Neural Transmission* 114(7) (2007) 877-884.

Schaub, M. — Krol, A. — Carbon, P., «Flexible Zinc Finger Requirement for Binding of the Transcriptional Activator Staf to U6 Small Nuclear RNA and tRNA(Sec) Promoters», *J. Biol. Chem.* 274(34) (1999) 24241-24249.

———, «Structural Organization of Staf-DNA Complexes», *Nucleic Acids Research* 28(10) (2000) 2114-2121.

Schomburg, L. — Schweizer, U. — Holtmann, B. — Flohé, L.— *al.*, «Gene Disruption Discloses Role of Selenoprotein P in Selenium Delivery to Target Tissues», *Biochemical Journal* 370 (2003) 397-402.

Schomburg, L. — Schweizer, U., «Hierarchical Regulation of Selenoprotein Expression and Sex-specific Effects of Selenium», *Biochimica et Biophysica Acta. General Subjects* 1790(11) (2009) 1453-1462.

Schuster, C. — Myslinski, E. — Krol, A. — Carbon, P.— *al.*, «Staf, a Novel Zinc Finger Protein that Activates the RNA Polymerase III Promoter of the Selenocysteine tRNA Gene», *EMBO Journal* 14(15) (1995) 3777-3787.

Selmer, M. — Su, X.-D. , «Crystal Structure of an mRNA-binding Fragment of Moorella thermoacetica Elongation Factor SelB», *EMBO Journal* 21(15) (2002) 4145-4153.

Sengupta, A. — Carlson, B.A. — Hoffmann, V.J. — Gladyshev, V.N.— *al.*, «Loss of Housekeeping Selenoprotein Expression in Mouse Liver Modulates Lipoprotein Metabolism», *Biochem Biophys Res Commun.* 365(3) (2008) 446-452.

Sengupta, A. — Carlson, B.A. — Weaver, J.A. — Novoselov, S.V.— *al.*, «A Functional Link Between Housekeeping Selenoproteins and Phase II Enzymes», *Biochemical Journal* 413(1) (2008) 151-161.

Shannon, C.E., «Communication Theory of Secrecy Systems», *Bell System Technical Journal* 28(4) (1949) 656-715.

———, «The Mathematical Theory of Communication», *Bell System Technical Journal* 27 (3 and 4) (1948) 379-423;-623-656.

Shea, N., «Representation in the Genome, and Other Inheritance Systems», *Biology and Philosophy* 22 (2007) 313-331.

Shen, Q. — Chu, F.F. — Newburger, P.E., «Sequences in the 3'-untranslated Region of the Human Cellular Glutathione Peroxidase Gene Are Necessary and Sufficient for Selenocysteine Incorporation at the UGA Codon», *J. Biol. Chem.* 268(15) (1993) 11463-11469.

SHEN, Q. — LEONARD, J.L. — NEWBURGER, P.E., «Structure and Function of the Selenium Translation Element in the 3'-untranslated Region of Human Cellular Glutathione Peroxidase mRNA», *RNA* 1(5) (1995) 519-525.

SHEPPARD, K. — YUAN, J. — HOHN, M.J. — JESTER, B.— AL., «From One Amino Acid to Another: tRNA-dependent Amino Acid Biosynthesis», *Nucleic Acids Research* 36(6) (2008) 1813-1825.

SHERRER, R.L. — O'DONOGHUE, P. — SÖLL, D., «Characterization and Evolutionary History of an Archaeal Kinase Involved in Selenocysteinyl-tRNA Formation», *Nucleic Acids Research* 36(4) (2008) 1247-1259.

SHRIMALI, R.K. — IRONS, R.D. — CARLSON, B.A. — SANO, Y.— AL., «Selenoproteins Mediate T Cell Immunity Through an Antioxidant Mechanism», *J. Biol. Chem.* 283(29) (2008) 20181-20185.

SHRIMALI, R.K. — WEAVER, J.A. — MILLER, G.F. — STAROST, M.F.— AL., «Selenoprotein Expression Is Essential in Endothelial Cell Development and Cardiac Muscle Function», *Neuromuscul. Disord.* 17(2) (2007) 135-142.

SILBERSTEIN, M. — MCGEEVER, J., «The Search for Ontological Emergence», *Philosophical Quarterly* 49(195) (1999) 182-200.

SILBERSTEIN, M., «Reduction, Emergence and Explanation», in P. MACHAMER – M. SILBERSTEIN, ed., *The Blackwell Guide to the Philosophy of Science*, 2002, 80-107.

SKLAR, L., «The Reduction (?) of Thermodynamics to Statistical Mechanics», *Philosophical Studies* 95(1-2) (1999) 187-200.

———, «Types of Inter-theoretic Reduction», *British Journal of the Philosophy of Science* 18 (1967) 109-124.

SMALL-HOWARD, A. — MOROZOVA, N. — STOYTCHEVA, Z. — FORRY, E.P.— AL., «Supramolecular Complexes Mediate Selenocysteine Incorporation in Vivo», *Molecular and Cellular Biology* 26(6) (2006) 2337-2346.

SMALL-HOWARD, A.L. — BERRY, M.J., «Unique Features of Selenocysteine Incorporation Function Within the Context of General Eukaryotic Translational Processes», *Biochemical Society Transactions* 33 (2005) 1493-1497.

SMART, J.J.C., «Physicalism and Emergence», *Neuroscience* 6(2) (1981) 109-113.

SOLER, N. — FOURMY, D. — YOSHIZAWA, S., «Molecular Switch in Tandem Winged-Helix Motifs of Elongation Factor SelB», *Nucleic Acids Symposium Series* 51 (2007) 377-378.

———, «Structural Insight into a Molecular Switch in Tandem Winged-helix Motifs from Elongation Factor SelB», *Journal of Molecular Biology* 370(4) (2007) 728-741.

SÖLL, D., «Genetic Code: Enter a New Amino Acid», *Nature* 331(6158) (1988) 662-663.

Sperry, R.W., «Mind-brain Interaction: Mentalism, Yes; Dualism, No», *Neuroscience* 5(2) (1980) 195-206.

Squires, J.E. — Berry, M.J., «Eukaryotic Selenoprotein Synthesis: Mechanistic Insight Incorporating New Factors and New Functions for Old Factors», *IUBMB Life* 60(4) (2008) 232-235.

Squires, J.E. — Stoytchev, I. — Forry, E.P. — Berry, M.J.— *al.*, «SBP2 Binding Affinity is a Major Determinant in Differential Selenoprotein mRNA Translation and Sensitivity to Nonsense-mediated Decay», *Molecular and Cellular Biology* 27(22) (2007) 7848-7855.

Stace, W.T., «Novelty, Indeterminism, and Emergence», *Philosophical Review* 48(3) (1939) 296-310.

Stegmann, U.E., «The Arbitrariness of the Genetic Code», *Biology and Philosophy* 19 (2004) 205-222.

Steinberg, S.V. — Ioudovitch, A. — Cedergren, R., «The Secondary Structure of Eukaryotic Selenocysteine tRNA: 7/5 Versus 9/4», *RNA* 4(3) (1998) 241-245.

Steinbrenner, H. — Sies, H., «Protection Against Reactive Oxygen Species by Selenoproteins», *Biochimica et Biophysica Acta. General Subjects* 1790(11) (2009) 1478-1485.

Stephan, A., «Emergence - a Systematic View on Its Historical Facets», in A. Beckermann – H. Flohr – J. Kim, ed., *Emergence or Reduction? Essays on the Prospects of Nonreductive Physicalism*, 1992, 25-48.

Sterelny, K. — Kitcher, P., «The Return of the Gene», *Journal of Philosophy* 85(7) (1988) 339-361.

Sterelny, K. — Smith, K.C. — Dickison, M., «The Extended Replicator», *Biology and Philosophy* 11 (1996) 377-403.

Sterelny, K., «The «Genetic Program» Program: a Commentary on Maynard Smith on Information in Biology», *Philosophy of Science* 67(2) (2000) 195-201.

Stock, T. — Rother, M., «Selenoproteins in Archaea and Gram-positive Bacteria», *Biochimica et Biophysica Acta. General Subjects* 1790(11) (2009) 1520-1532.

Stoeckler, T., «A Short History of Emergence and Reductionism», in E. Agazzi, ed., *The Problem of Reductionism in Science*, 1991, 71-90.

Sturchler, C. — Lescure, A. — Keith, G. — Carbon, P.— *al.*, «Base Modification Pattern at the Wobble Position of Xenopus Selenocysteine tRNA(Sec)», *Nucleic Acids Research* 22(8) (1994) 1354-1358.

Sturchler, C. — Westhof, E. — Carbon, P. — Krol, A.— *al.*, «Unique Secondary and Tertiary Structural Features of the Eucaryotic Selenocysteine tRNA(Sec)», *Nucleic Acids Research* 21(5) (1993) 1073-1079.

Sturchler-Pierrat, C. — Hubert, N. — Totsuka, T. — Mizutani, T.— *al.*, «Selenocysteylation in Eukaryotes Necessitates the Uniquely Long Aminoacyl

Acceptor Stem of Selenocysteine tRNA(Sec)», *J. Biol. Chem.* 270(31) (1995) 18570-18574.

Sun, X. — Moriarty, P.M. — Maquat, L.E., «Nonsense-mediated Decay of Glutathione Peroxidase 1 mRNA in the Cytoplasm Depends on Intron Position», *EMBO Journal* 19(17) (2000) 4734-4744.

Suppmann, S. — Persson, B.C. — Böck, A., «Dynamics and Efficiency in Vivo of UGA-directed Selenocysteine Insertion at the Ribosome», *EMBO Journal* 18(8) (1999) 2284-2293.

Suzuki, T. — Kelly, V.P. — Motohashi, H. — Nakajima, O.— al., «Deletion of the Selenocysteine tRNA Gene in Macrophages and Liver Results in Compensatory Gene Induction of Cytoprotective Enzymes by Nrf2», *J. Biol. Chem.* 283(4) (2008) 2021-2030.

Takeuchi, A. — Schmitt, D. — Chapple, C. — Babaylova, E.— al., «A Short Motif in Drosophila SECIS Binding Protein 2 Provides Differential Binding Affinity to SECIS RNA Hairpins», *Nucleic Acids Research* 37(7) (2009) 2126-2141.

Thanbichler, M. — Böck, A. — Goody, R.S., «Kinetics of the Interaction of Translation Factor SelB from Escherichia coli with Guanosine Nucleotides and Selenocysteine Insertion Sequence RNA», *J. Biol. Chem.* 275(27) (2000) 20458-20466.

Thanbichler, M. — Böck, A., «The Function of SECIS RNA in Translational Control of Gene Expression in Escherichia coli», *EMBO Journal* 21(24) (2002) 6925-6934.

Tujebajeva, R.M. — Copeland, P.R. — Xu, X. — Carlson, B.A.— al., «Decoding Apparatus for Eukaryotic Selenocysteine Insertion», *EMBO Reports* 1(2) (2000) 158-163.

Turanov, A.A. — Lobanov, A.V. — Fomenko, D.E. — Morrison, H.G.— al., «Genetic Code Supports Targeted Insertion of Two Amino Acids by one Codon», *Science* 323(5911) (2009) 259-261.

Uy, R. — Wold, F., «Posttranslational Covalent Modification of Proteins», *Science* 198 (1977) 890-896.

Van Cleve, J., «Mind--dust or Magic? Panpsychism Versus Emergence», *Philosophical Perspectives* 4 (1990) 215-226.

Van Gulick, R., «Reduction, Emergence, and the Mind/Body Problem: a Philosophic Overview», in N. Murphy – W.R. Stoeger, ed., *Evolution and Emergence. Systems, Organisms, Persons*, 2007, 40-73.

———, «Who's in Charge Here? And Who's Doing all the Work?», in N. Murphy – W.R. Stoeger, ed., *Evolution and Emergence. Systems, Organisms, Persons*, Oxford 2007, 74-87.

Vehkavaara, T., «From the Logic of Science to the Logic of the Living. The Relevance of Charles Peirce to Biosemiotics», in M. Barbieri, ed., *Introduction to Biosemiotics. The New Biological Synthesis*, Dordrecht 2008, 257-282.

Von Sternberg, R., «Genomes and Form: the Case for Teleomorphic Recursivity», *Annals of the New York Academy of Sciences* 901 (2000) 224-236.

Walczak, R. — Carbon, P. — Krol, A., «An Essential non-Watson-Crick Base Pair Motif in 3'UTR to Mediate Selenoprotein Translation», *RNA* 4(1) (1998) 74-84.

Walczak, R. — Westhof, E. — Carbon, P. — Krol, A.—al., «A Novel RNA Structural Motif in the Selenocysteine Insertion Element of Eukaryotic Selenoprotein mRNAs», *RNA* 2(4) (1996) 367-379.

Warner, G.J. — Berry, M.J. — Moustafa, M.E. — Carlson, B.A.— al., «Inhibition of Selenoprotein Synthesis by Selenocysteine tRNA[Ser]Sec Lacking Isopentenyladenosine», *J. Biol. Chem.* 275(36) (2000) 28110-28119.

Watson, J.D. — Crick, F.H.C., «Genetical Implications of the Structure of Deoxyribonucleic Acid», *Nature* 171(4361) (1953) 964-967.

———, «Molecular Structure of Nucleic Acids. A Structure for Deoxyribose Nucleic Acid», *Nature* 171(4356) (1953) 737-738.

Weiss, S.L. — Sunde, R.A., «Cis-acting Elements Are Required for Selenium Regulation of Glutathione Peroxidase-1 mRNA Levels», *RNA* 4(7) (1998) 816-827.

Wen, W. — Weiss, S.L. — Sunde, R.A., «UGA Codon Position Affects the Efficiency of Selenocysteine Incorporation into Glutathione Peroxidase-1», *J. Biol. Chem.* 273(43) (1998) 28533-28541.

Westhof, E. — Dumas, P. — Moras, D., «Crystallographic Refinement of Yeast Aspartic Acid Transfer RNA», *Journal of Molecular Biology* 184 (1985) 119-145.

Whanger, P.D., «Selenocompounds in Plants and Animals and Their Biological Significance», *Journal of the American College of Nutrition* 21(3) (2002) 223-232.

Willis, I.M., «RNA Polymerase III. Genes, Factors and Transcriptional Specificity», *FEBS* 212 (1993) 1-11.

Wilting, R. — Schorling, S. — Persson, B.C. — Böck, A.— al., «Selenoprotein Synthesis in Archaea: Identification of an mRNA Element of Methanococcus jannaschii Probably Directing Selenocysteine Insertion», *Journal of Molecular Biology* 266(4) (1997) 637-641.

Wingler, K. — Böcher, M. — Flohé, L. — Kollmus, H.— al., «mRNA Stability and Selenocysteine Insertion Sequence Efficiency Rank Gastrointestinal Glutathione Peroxidase High in the Hierarchy of Selenoproteins», *European Journal of Biochemistry* 259 (1999) 149-157.

Wright, L., «Functions», *Philosophical Review* 82(2) (1973) 139-168.

Wu, X.Q. — Gross, H.J., «The Length and the Secondary Structure of the D-stem of Human Selenocysteine tRNA Are the Major Identity Determinants for Serine Phosphorylation», *EMBO Journal* 13(1) (1994) 241-248.

———, «The Long Extra Arms of Human tRNA((Ser)Sec) and tRNA(Ser) Function as Major Identify Elements for Serylation in an Orientation-dependent, but Not Sequence-specific Manner», *Nucleic Acids Research* 21(24) (1993) 5589-5594.

Xu, X.-M. — Carlson, B.A. — Irons, R. — Mix, H.— *al.*, «Selenophosphate Synthetase 2 Is Essential for Selenoprotein Biosynthesis», *Biochemical Journal* 404(1) (2007) 115-120.

Xu, X.-M. — Carlson, B.A. — Mix, H. — Zhang, Y.— *al.*, «Biosynthesis of Selenocysteine on Its tRNA in Eukaryotes», *PLOS Biology* 5(1):e4 (2007) 96-105.

Xu, X.-M. — Carlson, B.A. — Zhang,Y. — Mix, H.— *al.*, «New Developments in Selenium Biochemistry: Selenocysteine Biosynthesis in Eukaryotes and Archaea», *Biological Trace Element Research* 119(3) (2007) 234-241.

Xu, X.-M. — Mix, H. — Carlson, B.A. — Grabowski, P.J.— *al.*, «Evidence for Direct Roles of Two Additional Factors, SECp43 and Soluble Liver Antigen, in Theselenoprotein Synthesis Machinery», *J. Biol. Chem.* 280(50) (2005) 41568-41575.

Yamada, K., «A New Translational Elongation Factor for Selenocysteyl-tRNA in Eucaryotes», *FEBS Letters* 377(3) (1995) 313-317.

Yoshizawa, S. — Rasubala, L. — Ose, T. — Kohda, D.— *al.*, «Structural Basis for mRNA Recognition by Elongation Factor SelB», *Nature Structural & Molecular Biology* 12(2) (2005) 198-203.

Young, P.A. — Kaiser, I.I., «Aminoacylation of Escherichia coli Cysteine tRNA by Selenocysteine», *Archives of Biochemistry and Biophysics* 171 (1975) 483-489.

Yuan, J. — O'Donoghue, P. — Ambrogelly, A. — Gundllapalli, S.— *al.*, «Distinct Genetic Code Expansion Strategies for Selenocysteine and Pyrrolysine Are Reflected in Different Aminoacyl-tRNA Formation Systems», *FEBS Letters* 584 (2010) 342-349.

Yuan, J. — Palioura, S. — Salazar, J.C. — Su, D.— *al.*, «RNA-dependent Conversion of Phosphoserine Forms Selenocysteine in Eukaryotes and Archaea», *Proceedings of the National Academy of Sciences* 103(50) (2006) 18923-18927.

Zavacki, A.M. — Mansell, J.B. — Chung, M. — Klimovitsky, B.— *al.*, «Coupled tRNASec-dependent Assembly of the Selenocysteine Decoding Apparatus», *Molecular Cell* 11 (2003) 773-781.

Zhang, Y. — Baranov, P.V. — Atkins, J.F. — Gladyshev, V.N.— *al.*, «Pyrrolysine and Selenocysteine Use Dissimilar Decoding Strategies», *J. Biol. Chem.* 280(21) (2005) 20740-20751.

Zhang, Y. — Gladyshev, V.N., «General Trends in Trace Element Utilization Revealed by Comparative Genomic Analyses of Co, Cu, Mo, Ni, and Se», *J. Biol. Chem.* 285(5) (2010) 3393-3405.

Zhang, Y. — Romero, H. — Salinas, G. — Gladyshev, V.N.— *al.*, «Dynamic Evolution of Selenocysteine Utilization in Bacteria: a Balance Between Selenoprotein Loss and Evolution of Selenocysteine from Redox Active Cysteine Residues», *Genome Biology* 7(10):R94 (2006) .

Zhang,Y. — Zhou, Y. — Schweizer, U. — Savaskan, N.E.— *al.*, «Comparative

Analysis of Selenocysteine Machinery and Selenoproteome Gene Expression in Mouse Brain Identifies Neurons as Key Functional Sites of Selenium in Mammals», *J. Biol. Chem.* 283(4) (2008) 2427-2438.

Zinoni, F. — Birkmann, A. — Leinfelder, W. — Böck, A.— *al.*, «Cotranslational Insertion of Selenocysteine into Formate Dehydrogenase from Escherichia coli Directed by a UGA Codon», *Proceedings of the National Academy of Sciences* 84 (1987) 3156-3160.

Zinoni, F. — Heider, J. — Böck, A., «Features of the Formate Dehydrogenase mRNA Necessary for Decoding of the UGA Codon as Selenocysteine», *Proceedings of the National Academy of Sciences* 87 (1990) 4660-4664.

AUTHOR INDEX

Index

PART III: CASE STUDY

Finito di stampare nel mese di settembre 2016
presso Mediagraf spa - Noventa Padovana (PD)